P9-DDO-532

Our Children's Toxic Legacy

Our Children's Toxic Legacy

How Science and Law Fail to Protect Us from Pesticides

John Wargo

Yale University Press New Haven and London

Designed by Sonia Scanlon
Set in Times Roman type by The Marathon Group, Durham, North Carolina.
Printed in the United States of America by BookCrafters, Inc., Chelsea, Michigan.

Library of Congress Cataloging-in-Publication Data
Wargo, John, 1950–
Our children's toxic legacy : how science and law fail to protect us from pesticides / John Wargo.
p. cm.
Includes bibliographical references and index.
ISBN 0-300-06686-4 (cloth : alk. paper)
1. Pesticides—Toxicology. 2. Pesticides—Law and legislation—United States. 3. Children—Health risk assessment. 4. Pesticides—Environmental aspects. I. Title.
RA1270.P4W34 1996
363.17'92—dc20 96-24990
CIP

A catalogue record for this book is available from the British Library.

The paper in this book meets the guidelines for permanence and durability of the Committee on Production Guidelines for Book Longevity of the Council on Library Resources.

10 9 8 7 6 5 4 3 2 1

To Linda, Adam, and Kate

Contents

Preface ix
Acknowledgments xv

Part 1 Defining the Problem
1 The Global Experiment 3
2 The Urgency of Malaria 15
3 Resistance: A Race Against Time 43

Part 2 Evolving Law
4 Beyond Control: Pesticide Law Before 1972 67
5 EPA as the Gatekeeper of Risk 86
6 Risk Assessment and Tolerance Setting: The Delaney Paradox 104

Part 3 Evolving Knowledge
7 The Human Ecology of Pesticide Residues 131
8 The Susceptibility of Children 172
9 The Diet of a Child 201
10 Averaging Games: Simplification of Exposure and Risk 219
11 The Complex Mixture Problem 235

Part 4 Managing the Coevolution of Knowledge and Law
12 Fractured Law, Fractured Science: Restating the Pesticide Problem 251
13 Toward Reform 270

Epilogue 289
Abbreviations 301
Notes 303
Index 371

Preface

During the first half of the twentieth century, most of the world's rural population faced enormous health risks from infectious insect-borne diseases. Malaria alone claimed more than 50 million lives during that period, and most of those who died were children under age five. The discovery of the insecticidal effects of dicholorodiphenyl-trichloroethane (DDT) by Paul Müller in 1939 marks the beginning of the modern chemical industrial revolution and a turning point in both public health and agricultural history. The immediate success of DDT in controlling typhus, malaria, and yellow fever during World War II was a stunning example of the new control humans could exercise over disease.

By mid-century, pesticide manufacturers and government officials also had demonstrated the potential of the new biocides to control an extremely diverse collection of crop-damaging insects, diseases, and predators. Herbicides introduced in the 1940s allowed farmers to simply spray their fields to kill nutrient-robbing weeds rather than tilling weeds under the soil. The dust storms of the 1930s in the southern plains states had swept away millions of tons of nutrient-rich topsoil from tilled farmlands. Herbicides often made tilling unnecessary, leaving the soil undisturbed and reducing erosion by wind and water. Soon after their introduction, herbicides conveyed an image of responsible and scientific land stewardship.

Synthetic pesticides became symbols of progress during the postwar years and provided a level of control over environmental risks—both natural and manmade—never before experienced. Their benefits were easily recognized by lives saved, increased crop yields, decreased soil loss, and economic growth. Once released to the environment, their invisibility supported the impression that risks to health or the environment quickly vanished through chemical disintegration or simple dilution.

As the second half of the century evolved, experts slowly pieced together a patchwork quilt of evidence suggesting that pesticides and their effects were far more difficult to control than had been anticipated. The first serious surprise was the rapid evolution of insects' resistance to insecticides, which created a continual demand for new products. Second, scientists discovered that residues often persisted longer and traveled farther in the environment than initially expected. Third, residues accumulated in plant, animal, and human tissues. Fourth, experts underestimated the age-related variance among humans in their susceptibility to adverse health effects from pesticides. Fifth, scientists underestimated the diversity of ways that humans could encounter and accumulate pesticide residues from exposures to contaminated food, water, air, and soil. Finally, the fact that some people face far higher risks than others was normally disregarded by public policymakers. By the close of the century, the image of progress was tarnished and fading and was increasingly replaced by a competing view that pesticides threaten human health, biological diversity, and basic ecological processes.

Since 1947, the most important law governing the licensing of pesticides has demanded that products be effective and that their hazards be accurately labeled. It required that benefits exceed risks before government licensed new pesticides or re-registered old ones. The U.S. Department of Agriculture before 1970, and EPA after that time, each interpreted this law as giving them permission to "balance" positive and negative effects—in effect, to use a utilitarian standard. As pesticide risks became more clearly understood, however, the dilemma became more pronounced. How should society manage the mixture of risks and benefits that result from the release of pesticides into the environment? Responding to this question is the ultimate purpose of this book.

I have approached the question historically, in the belief that future policy will be most effective if it incorporates lessons from the past. Part 1 explores expert and public perception of pesticide benefits near mid-century—perceptions shaped by pesticides' short-term effectiveness in reducing insect-borne disease and protecting crops, and by experts' limited comprehesion of their risks. The evolution of key laws and regulations governing pesticides are traced in part 2. In this section, I became especially interested in how the absence of evidence of risk justified a sea of pesticide registrations and use entitlements. Part 3 explores the evolution of our understanding of pesticide-related risks: a series of unexpected findings eventually led to challenges of earlier registrations and tolerances. Part 4 considers how fractured pesticide law and fractured science have managed to distract government agencies and the public from identifying and managing significant risks. Suggestions for reform follow, which include principles for producing knowledge of risk, distributing that knowledge in a way that informs and empowers the public, and enacting laws and regulations to manage these dangers.

The Human Ecology of Food

With every meal, we are intimately affected by the quality of the food and water we consume. The complex mixture of chemicals we eat and drink may include both nutrients essential for life and other chemicals that may jeopardize our health. Just as individual choices about foods influence one's personal health, aggregated choices about what we eat influence agricultural practices, the use of technology, and environmental quality wherever food is produced. Although the decisions to use biocides may appear perfectly rational to the farmer, these choices convey risks—unrecognized by human senses—into the local environment and the global marketplace.

Among all hazardous chemicals, pesticides are especially interesting. First, they include an extremely diverse set of synthetic and natural toxins or "biocides," including insecticides, fungicides, rodenticides, herbicides, and algicides. Second, they must normally be released into the environment to be effective, which makes contamination of food, water, air, soil, and wildlife difficult to control. Pesticides are commonly applied outdoors to crops, forests, wetlands, lawns, gardens, parks, golf courses, lakes, ponds, swimming pools, and along rail, highway, and powerline corridors. They are also often sprayed inside buildings including offices, hospitals, schools, and day-care centers, as well as within transport vehicles such as aircraft cabins, ships, railcars, and trucks, especially those crossing international borders. They are deliberate components of many building materials, paints, shampoos, drugs, wallpapers, rugs, shower curtains, veterinary products, clothing, dry cleaning fluids, and even mattresses. Anyone growing up in the United States during the 1950s probably spent more than a few nights sleeping between a mattress and blanket, each impregnated with DDT. These diverse uses for pesticides mean that we may be exposed to them in numerous and unexpected ways.

Since 1947, the U.S. government has licensed or registered over 600 different pesticide "active ingredients," and 325 of these are allowed to remain as residues in the food supply. In addition, the government has set maximum contamination limits for individual pesticides on individual foods, and this list now includes 9,300 pesticide-food "tolerances."Any one pesticide may be a component of hundreds or thousands of separately registered products, and some are licensed for use on as many as one hundred different crops. Apples and milk, for example, are each allowed by law to contain residues of nearly one hundred different pesticides.

The diversity of statutory and administrative standards used by the Environmental Protection Agency to set pesticide tolerances in food and water became so confusing by 1985 that EPA asked the National Academy of Sciences to explore the level of protection that law and regulation provided against cancer risks from pesticide residues in food. The Delaney Committee found that nearly all registrations for pesticide use on food crops were set using a risk-benefit balancing standard contained within the Federal Insecticide, Fungicide, and Rodenticide Act (FIFRA), and therefore were not required to protect the public from significant health risk. Only if pesticides were suspected of "inducing cancer" and they "concentrated during food processing" were limits set to protect against significant cancer risks. This limitation was required by the Delaney clause contained within the Federal Food, Drug, and Cosmetic Act (FFDCA). However, the Delaney Committee found that the more protective cancer risk standard was applied in fewer than 3 percent of tolerance-setting decisions. Even in these few cases, risks were calculated using methods that assumed their average distribution across the population—and without regard for the probability that minorities such as children may face higher than average levels of risk.

Thus neither the risk-benefit balancing standard contained in FIFRA nor the Delaney "zero-risk" standard contained within FFDCA have been implemented in a way that has ensured protection for all against significant cancer risk. Further, it has become clear that other human health risks—including neurological disease, immune dysfunction, birth defects, and reproductive failure—were all balanced against economic benefits of pesticide use when EPA and FDA had issued registrations and set tolerances.

Pesticides and Children

I had the good fortune to work with the Delaney Committee between 1985 and 1987, and near the end of the project I began to explore the ways that diet could influence pesticide exposure. I was surprised to find that children normally consume more of fewer foods than adults and suspected that these dietary patterns could increase children's exposure to mixtures of pesticides. In 1988, a second National Academy of Sciences committee was formed to examine childhood exposure to pesticides and their possible heightened vulnerability to toxic effects.

I spent the following three years working with this "Kids Committee," evaluating data and designing methods to estimate childhood exposure to complex mixtures of pesticides that collectively posed cancer or neurological risks. The committee concluded that children are more vulnerable than adults to several important adverse effects that pesticides may cause. Also, the committee found that children may often be more exposed to certain pesticides than adults. These two findings—elevated susceptibility and higher exposure—led to the unambiguous conclusion that children face higher risks from pesticides than experts had anticipated.

When the Kids Committee report was released in 1993, the administrative heads of EPA, FDA, and USDA immediately endorsed its findings and conclusions. Still, substantive change in the standards used for making decisions about pesticide use has not followed. Deregulatory fervor and slashed budgets stalled the planned statutory and regulatory reforms, which had promised to change the massive complex of pesticide regulations into an umbrella of health protection. Given the history of debate over benefits and risks described in the chapters that follow, the stalling of reforms that would threaten use rights is hardly surprising.

This book is an attempt to place the academy's findings within a broader historical and legal context—one that might clarify the human potential to know and control environmental health risks. Many questions need to be asked. How did the fractured body of law governing pesticides and the "crazy quilt pattern of regulation" evolve? How did experts' understanding of different types of health and ecological risks develop? Can a pattern of influence—of science on law, or law on science—be discerned? If we want to create responsible and responsive public policies on pesticides, this history must be explored.

The history told here is unsettling because it demonstrates the enormous amount of research needed to know the risks we have created. Thousands of toxins and hazardous products are regularly released to the environment. Moreover, funding for the development of new technologies that carry risk has overwhelmed that for research to understand the magnitude, distribution, and effects of contamination. As the power, authority, and reach of international markets expand, our understanding of the environmental effects of hazardous technologies appears to be diminishing. We now have more specialized information, but understand it less. This paradox may pose the greatest challenge for managers of environmental health and quality in the next century.

Experimentation

Enormous uncertainty will always surround pesticide licensing decisions. If twentieth-century pesticide science has taught us anything, it is that we have very limited knowledge of the fate of residues, patterns of human exposure, and their adverse health effects. Pesticide licensing under conditions of such uncertainty has been an act of uncontrolled human and ecological experimentation. My intent is to question the underlying expectation that risks can be controlled by knowing and limiting exposure. Past regulatory behavior assumed that experts could know when exposure to invisible toxins would occur and when the level of exposure was dangerous; it also took for granted that we had the organizational capacity to control exposure. Legislators and regulators gave very little consideration to the social, cultural, and ecological conditions that surround the release of pesticides to the environment—conditions that generally make human exposure extraordinarily difficult to predict and manage.

I do not pretend to provide definitive estimates of pesticide risks facing children or adults. Although understanding risks is important, the debate over their magnitude and distribution will never be fully resolved because of the prohibitively high costs of obtaining complete information. While accepting the premise that human exposure and risk estimates will remain uncertain, it is still possible to place defensible bounds around current forecasts and to suggest precautionary public policy.

This book is my attempt to trace the origin and evolution of the pesticide experiment, and my interpretations lead to one final question: Does everyone deserve protection from contamination that poses significant health risks? Given the scale of pesticide use in the world, the diversity of ways most people encounter residues, and the many types of toxic outcomes they may cause, a positive response would demand dramatic changes in the behavior of states and markets. Perhaps the same question should be asked for any technology that distributes significant risk.

Acknowledgments

I am indebted to many who have helped to make this book possible. Linda Evenson Wargo, my wife, provided enormous support, commenting thoughtfully on risk assessment and management, reviewing manuscript drafts, and constantly encouraging me in my work. All these efforts on my behalf were a distraction from her career in hazardous waste risk management. Adam and Kate, our two children, deserve recognition for their patience and discipline while their dad spent thousands of hours at his computer instead of playing catch with them or reading them stories.

My understanding of the special vulnerability of children to pesticides results from my associations with numerous pediatricians and obstetricians. Chief among them are Richard Jackson, Philip Landrigan, Donald Mattison, and William Weil. Michael Gallo offered important advice on toxicology, especially in the area of organophosphate pesticides. Daniel Krewski of Health and Welfare Canada provided critical and cautious guidance in the design of statistical analyses of childhood exposures and risks. His associate Sheryl Bartlett reviewed and clarified material on pesticide mixtures. James Seiber and Richard Schmidt assisted in interpreting residue data.

I am deeply indebted to William Burch, who encouraged my examination of pesticides from the different perspectives offered by the social, policy, ecological, and health sciences. A theme that spans his career—to consider the effects of policy on the individual—has had an obvious influence on this work. F. Herbert Bormann has shaped my understanding of ecology, and our long conversations on systems thinking and science policy became important to the structure of the book. Kristiina Vogt provided important advice about how the methods and principles of ecological risk assessment are transferrable to environmental health analysis. James Scott helped me to understand that pesticide regulation is a type of agrarian reform with many historical precedents and that EPA's pattern of risk averaging is but one of many efforts by the state to simplify its image of the society it attempts to regulate. This book would not have been completed without the help of Jared Cohon, who provided policy insight into congressional behavior, gave technical advice in the area of risk assessment, and offered careful reflection on suggestions for reform.

William Cronon has provided critical guidance over the past decade and has always managed to ask provocative questions about purpose, structure, and reform. Garry Brewer introduced me to the policy sciences and to ways of understanding and managing technically complex problems. I thank William Smith for his cautious interpretations of claims that contamination has induced damage. John Gordon provided important support for the project, especially in its early stages. Graeme Berlyn was especially helpful in interpreting ecological risks from pesticides. Brian Leaderer provided essential critical reviews of methods of exposure assessment. Numerous conversations with Stephen Kellert helped me to understand better how discernable values and ideologies shape scientific inquiry. Jason Shogren provided insight and criticism of methods of risk assessment and proposals for management reform. Tim Clark, Joseph Miller, and Leonard Doob provided substantive reactions to early analyses.

Charles Benbrook attracted me to this area of research in 1985 while serving as the director of the National Academy of Sciences Board on Agriculture. Few people have his understanding of the politics and economics of both domestic and international agriculture. Richard Wiles has been a source of expert advice on agriculture and pesticides for nearly a decade. James Aidala influenced the theme of the project while he worked for the Congressional Research Service. Mark Childress became an important legal guide to the intricacies of both the Delaney Clause and the Federal Insecticide, Fungicide and Rodenticide Act.

Brian Young and David Bruce provided invaluable help in navigating Yale University's diverse computing facilities. I also wish to thank my students. Those who performed important research roles include Jennifer Allen, Ross Brennan, Stewart Dary, Deborah Guber, Jill Humphrey, Jonathan Kaplan, Allan Kaufman, Jonathan Labaree, Philip Liu, Kyle Lonergan, Katherine Reilly, and Wesley Taylor.

I am especially grateful for the enthusiasm, encouragement, and insight of Jean Thomson Black, science editor at Yale University Press. Laura Jones Dooley and Julie Carlson provided excellent suggestions to clarify technical concepts and language.

PART I

Defining the Problem

CHAPTER 1

The Global Experiment

In this century, several hundred billion pounds of pesticides have been produced and released into the global environment.[1] Nearly 5 billion pounds of the insecticide DDT alone have been applied both indoors and out since it was introduced in 1939, and DDT is only one of nearly six hundred pesticides currently registered for use in the world.[2] By 1969, almost sixty thousand different products were sold containing some combination of pesticides along with their inert ingredients.[3] As we approach the twenty-first century, an additional 5 to 6 billion pounds of insecticides, herbicides, fungicides, rodenticides, and other biocides are added to the world's environment each year, with roughly one-quarter of this amount released or sold in the United States.[4]

The structure of the pesticide industry makes it extremely difficult to regulate. Pesticides have been traded internationally during most of the twentieth century. As of today approximately 80 percent of the world's pesticide production is controlled by only twenty companies, whose annual sales total nearly $25 billion.[5] Although this statistic may suggest a concentrated industry, corporate ownership and responsibility for single chemicals is often highly fractured. For example, a compound such as DDT—prohibited from use within the United States—may be synthesized by as many as a dozen different manufacturers and be a component in hundreds of different foreign pest-control products. Production facilities are now dispersed worldwide as multinational corporations purchase or build subsidiary plants close to growing markets. These plants often manufacture in low-income nations where labor is inexpensive and environmental health regulations are weak. Ownership of companies, facilities, and specific pesticides is now often divided among many investors.[6]

Moreover, the legal relation between a parent company and a subsidiary is often unclear, making responsibility for risks difficult to assign. In 1984, for example, the release of a toxic chemical at a Union Carbide pesticide plant in Bhopal, India, killed more than two thousand and injured over one hundred thousand. A highly public and international debate followed, and an attempt was made to assign blame among Union Carbide U.S., Union Carbide India, Ltd. (the U.S. firm's subsidiary), and the Indian government. The tragedy led to a $470 million settlement by the U.S. parent company, but many Indians now believe that lax regulatory policies in India were partially responsible.[7] The fragmentation of property rights between parent companies and their subsidiaries—which are often located in different parts of the world—diffuses and confuses accountability and responsibility.

A cobweb of conflicting national regulations is hardly in the interest of multinational corporations, so they press for international uniformity at the least possible level of restriction. This "lowest common denominator" phenomenon also occurs within the United States. Here corporate support for federal regulation that preempts the right of states to set more restrictive controls grows from the fear that strict regulation by a single state may force costly production changes, especially if adopted by other states. The gradual adoption of California's air quality standards by other populous states is a good example of how the most restrictive standard may become the industry and national standard. "Free trade" advocates point to health-protective regulations in industrialized nations as "trade barriers," and press instead for less restrictive but uniform standards.

Surprisingly, some U.S. companies manufacture pesticides that the U.S. Environmental Protection Agency (EPA) has prohibited from domestic use. If EPA decides to ban a product already licensed, manufacturers will normally ship remaining stocks abroad and continue domestic production, as allowed by U.S. law.[8] These banned pesticides, however, may reappear as residues in imported foods, a phenomenon termed the "circle of poison" by U.S. environmental and consumer interest groups.[9] U.S. Customs records, for example, demonstrate that DDT, banned by EPA in 1972 from domestic use, has been shipped from U.S. ports at an average rate of one ton per day in 1996.[10] This export policy has haunted us, because the proportion of the average U.S. diet made up of imported foods—particularly those from tropical or semitropical countries where pesticides are heavily used—has increased significantly during the past several decades, and imported foods have a higher chance of containing residues of more toxic, unregistered pesticides than those produced in the United States.[11]

Although chemical and food markets are global, legal authority to control pesticides is held by individual nations. Unfortunately, each system of rules is unique. Within the United States, pesticide use and residues in food, water, and the workplace are now regulated by the EPA. Before EPA was created in 1970, however, pesticides were licensed or "registered" by the U.S. Department of Agriculture (USDA), commonly with little understanding or questioning of their health or ecological effects.[12] The absence of evidence of risk became a rationale for issuing licenses or "registrations" to pesticide manufacturers, and by 1970, USDA had registered nearly sixty thousand separate products containing pesticides. By 1990, nearly forty-five thousand separate products were registered for sale in the United States.[13] This number has dropped to about twenty thousand to thirty thousand separate registrations as companies have voluntarily withdrawn discontinued products or have decided that continued registration is not worth the expense of producing the updated environmental health data required by EPA.[14] In 1995 there were nearly six hundred active and 1,600 inert pesticide ingredients registered in the United States, and these were often mixed in various combinations along with inactive ingredients to produce the desired effect.

The legal criterion for pesticide registration was changed in 1972 and requires EPA to consider protection of public health and environmental quality when balancing the risks and benefits of product use.[15] This change prompted EPA to demand new information on environmental health effects as a basis for deciding whether to reregister pesticides formerly licensed by USDA. Nearly nineteen thousand studies have been submitted to EPA under this "Data-Call-In" effort; but 7,500 of these had not yet been reviewed by the agency by the end of 1994.[16]

Current Toxins

Today, nearly 325 active pesticide ingredients are permitted for use on 675 different basic forms of food, and residues of these compounds are allowed by law to persist at the dinner table. Nearly one-third of these "food-use" pesticides are suspected of playing some role in causing cancer in laboratory animals, another one-third may disrupt the human nervous system, and still others are suspected of interfering with the endocrine system. Because nearly 9,300 separate regulations now limit the maximum residues permitted to exist on raw and processed foods,[17] during the early 1970s the legal responsibility to judge the health and ecological effects of so many compounds overwhelmed the newly formed EPA. Since that time, the agency has reviewed each active ingredient one at a time, in a process that commonly drags on for a decade or more while scientific evidence of damage is accumulated. By the end of 1994, EPA estimated that 4,500 food tolerances—more than 50 percent—remained to be evaluated in light of more recent environmental health data.[18]

The scale, diversity, and intensity of pesticide use since 1945 raise many important questions regarding how pesticides have been and should be managed. Even the simplest of questions are often difficult to answer with confidence: Where are the pesticides produced, shipped, and used? Once released to the environment, where do pesticides go? What effects do they have on other species and on human health? What benefits do they provide? How are the risks and benefits distributed? Are the risks worth the benefits?

How should we control a technology that both benefits and harms our health and environment?

Although EPA has tried to answer these questions for individual pesticides, the agency has never attempted to control public exposure to the "complex mixture" of pesticides we normally encounter from residues in our food, drinking water, clothing, homes, workplaces, and the outdoors. My gradual recognition that government understands little about human exposure to pesticides or their health effects while continuing to license their use captured my curiosity. Given this ignorance, why would policymakers presume that pesticides—intentionally toxic substances—pose no significant threat to human health?

Pesticides and Food

The twentieth-century agricultural revolution was built in large part on the use of chemical technologies to prevent crop failure. Fertilizers were designed and applied to assure adequacy of nutrient availability. By 1850, it was common to apply chemical pesticides to diminish insect, viral, and fungal damage to crops; and by 1950, "miracle" herbicides reduced or eliminated the need to till fields for weed control. EPA attempts to cancel registrations of pesticides—due to evidence of health risks—have been met by strident protests from farmers. Historically, farmers have fought to protect their access to chemical crop protection technologies because they view the use of these chemicals as essential to their economic security.

Today, pests—including insects, plant pathogens, and weeds—destroy annually approximately 37 percent of all food and fiber crops in the world.[19] There are more than one hundred thousand distinct diseases caused by viruses, bacteria, mycoplasma, fungi, algae, and parasitic higher plants. Nearly thirty thousand different species of plants are classified as "weeds," and approximately 1,800 of these cause significant economic damage. Almost ten thousand different species of insects cause damage by directly eating crops. Losses are higher in tropical countries, largely due to the increased diversity and abundance of pests. If no pesticides were used, various experts have estimated that losses would increase between 10 and 100 percent depending on crop and location.

Yet a 1989 National Academy of Sciences study found that farms that do not use synthetic insecticides, herbicides, or fungicides may be more profitable than farms that do.[20] Insecticide use may actually increase crop losses over time, because insects may rapidly develop resistance to specific pesticides and require either more concentrated applications or a switch to a different compound. Between 1945 and 1989, for example, insecticide use in the

United States has increased tenfold, while crop losses from insect damage have almost doubled, from 7 percent to 13 percent.[21] It is revealing that this increase in use has been concentrated on relatively few crops. By 1964, two-thirds of all insecticides applied in the United States were used on only three crops: cotton, corn, and apples.[22] Despite a thousandfold increase in the use of insecticides on corn, losses to insects have increased 400 percent.[23]

The Delaney Paradox

In 1984, the EPA asked the National Academy of Sciences (NAS) to convene a panel of experts to judge the effectiveness of federal law in controlling cancer risks from pesticides in food.[24] I began working with this group early in 1985, trying to understand how the public is exposed to pesticide residues in food. We also questioned whether existing regulations prevented public exposure to "significant" levels of cancer risk.

The entire panel was at first confused by the government's inability to answer very basic questions, such as: Which pesticides "induce cancer"? Which of those that induce cancer also concentrate during food processing—such as when apples are processed to make apple juice? How many pesticides are allowed to persist as residues in the diet? How is cancer risk related to dietary habits and patterns of food contamination? We began by examining a large computer data set provided by EPA that contained a list of all federal regulations, or "tolerances," that set maximum limits for pesticide residues in food.[25] Our first surprise was that EPA, and USDA before it, had set nearly ten thousand separate tolerances for different pesticide residues in different foods. Some foods, such as apples and milk, are still allowed to contain residues of nearly one hundred separate pesticides. Whereas the active ingredients in pesticides are regulated by federal law, few of their metabolites and virtually none of their roughly 1,600 inert ingredients have been examined for their health effects and still remain unregulated as possible food contaminants.

Understanding which pesticide residues were allowed to persist on individual foods led us naturally to question what people eat. We turned to a national dietary survey of thirty thousand people conducted in 1977 and 1978 by USDA, which was undertaken primarily to estimate nutrient intake. To forecast pesticide exposure, we needed to merge food intake data and allowable residue data. What should have been a simple process quickly turned into a research nightmare.[26] The food intake data had been collected by USDA, the tolerance data by EPA, and the residue data by the U.S. Food and Drug Administration (FDA). Each agency used a different computer language, and no codes existed to join the data sets. Even more surprising, none of the agencies had much interest in collaborating to estimate pesticide exposure. The different computer languages served well to define territorial boundaries among the bureaucracies and were enormously successful in preventing outsiders from understanding agency behavior related to pesticide regulation. Inadvertently, we had stumbled on the realization that EPA had very limited knowledge of the risks posed by pesticides, because they had very poor data on residues, chemical toxicity, and human exposure.

Although the academy's Delaney Committee was intrigued by these findings, it was most interested in evaluating how EPA had regulated carcinogenic pesticides.[27] Cancer risk at that time was calculated assuming a lifetime average exposure to a carcinogen. Because food was the dominant source of exposure, estimating cancer risk required assumptions about food intake and about food contamination by residues. To estimate food intake, EPA developed average food consumption estimates for different foods and assumed that these patterns persisted over a seventy-year period. They then developed estimates of average pesticide-residue levels believed to contaminate various foods, based upon FDA's residue data and studies, which estimated the effects of food processing on residues. These residue data were prepared almost exclusively by manufacturers seeking pesticide registrations.[28]

A quick analysis of USDA's national dietary survey demonstrated that no one consumed an average level of all of the 5,000 foods sold in the marketplace. Instead, food intake varied by ethnic group, region, season, and age for some foods. The Delaney Committee had neither the time nor the expertise to fully evaluate these findings, other than to note them in an appendix of their final report.[29] Instead, the group concentrated its attention on only 28 of 53 pesticides that EPA believed to be carcinogenic and chose those compounds with the highest quality toxicity studies for further analysis.[30] Inconsistencies in FDA sampling practices, analytical methods, and recordkeeping led the committee to conclude that FDA's data were not suitable for estimating lifetime patterns of pesticide exposure and associated cancer risks. Instead, the committee forecast risk by combining food intake data with tolerance data—legally allowable residue levels—and cancer potency estimates for individual pesticides that EPA had classified as carcinogens.[31] Using these methods, the cancer risks allowed by legal tolerances appeared to be higher than expected for some pesticides and foods.

The primary conclusion of the committee was that the statutory framework for regulating pesticides has been confused by conflicting standards. The first standard is contained within the Federal Insecticide, Fungicide, and Rodenticide Act (FIFRA), and it requires that risks be weighed against benefits when deciding to register specific pesticide uses. Another statute, the Federal Food, Drug, and Cosmetic Act (FFDCA), governs the setting of food tolerance levels and demands that two separate standards for setting tolerances be applied. The first standard requires that tolerances be established for raw agricultural crops at a level necessary to protect the public health, while considering the need for an "adequate, wholesome and economical food supply."[32] The second guideline contains the Delaney prohibition against any pesticide residues shown to be cancer-inducing or concentrated during food processing.[33] The Delaney clause is the most famous environmental law because of its clarity and stringency in demanding no cancer risk. Only 150 tolerances had been set to control residue concentration in processed food, however, and the Delaney clause has rarely been invoked to prohibit pesticide use.[34] Instead, EPA had followed FDA and allowed carcinogenic pesticides on crops, even if the residues were concentrated, provided the risk did not exceed some unstated agency threshold of significance. Thus all pesticide *registrations* are set based upon FIFRA's risk-benefit balancing test, and nearly all food *tolerances* have been set pursuant to a FFDCA's risk-benefit balancing test for raw agricultural commodities. Only those pesticides that are both carcinogenic and are concentrated during food processing are prohibited by the Delaney clause. As chapter 6 explains, however, EPA has generally permitted even these pesticides to remain as food residues because the agency considers their risks to be trivial.

In its concluding report, the Delaney Committee published a table listing the twenty-eight compounds classified as probable or possible human carcinogens and estimated the additional cancer risk associated with average dietary exposure over a lifetime. Given the limited focus of the study, the greatest cancer risk appeared to come from fungicides, followed by herbicides and then insecticides. In fact, the universe of probable or possible carcinogens appeared to be expanding as new toxicity studies were submitted to the agency by manufacturers seeking product reregistration. Also, new data on processing effects demonstrated an increasing number of cases in which residues concentrated as raw crops were processed.

These findings led the panel to conclude that EPA's inconsistent application of FIFRA's risk-benefit balancing standard was responsible for the vast majority of the pesticide-related cancer risk in the American diet. By contrast, the severity of the Delaney clause, and the general absence of high-quality data demonstrating how pesticides concentrate during food processing, had prevented the agency from using it to ban specific pesticide uses.[35] Reflecting on this history of regulation, the committee concluded that a single non-zero (but trivial) risk threshold should replace the zero-risk Delaney standard. Further, these analyses demonstrated that the application of a single "negligible risk" standard across the twenty-eight pesticides studied would reduce cancer risk by 98 percent from that allowed in 1987.

The committee's final report, entitled *Regulating Pesticides in Food: The Delaney Paradox,* presented cancer-risk estimates not only for specific pesticides, but also for specific crops such as apples and lettuce. It was met by howls of protest from nearly every interest group. Pesticide manufacturers, food processors, grocery corporations, and farmers considered the report a scare tactic by environmentalists. Their complaints were most strenuous over the committee's decision to estimate cancer risks based on *legally allowable residue levels* rather than *actual detected residue levels,* which they argued were far lower in the marketplace. Environmentalists, by contrast, did not trust the panel's call for a negligible risk standard, believing that it granted the agency too much discretion to calculate risk. If EPA, through its management of the tolerance system, had permitted or condoned the cancer risks reported by the panel, why should environmentalists yield them further discretion?

The tone and volume of the debate spilled into the scholarly and popular media. Editorials and articles in *Science* claimed that natural carcinogens in food were a greater hazard than pesticides. Environmentalists captured the public's attention when they presented counterclaims on *60 Minutes* that EPA allowed children to accumulate cancer risks at unacceptable rates.[36] Amid the confusion and competing arguments over what we should worry about, the real contributions of the Delaney Committee were obscured. Their research provided the first systematic review of a highly confused environmental health issue that touches every individual at every meal. By pulling together data on food intake, residues, and pesticide toxicity, the committee exposed the double standard that EPA had applied to raw and processed foods, as well as the agency's consistent avoidance of the Delaney clause.[37] Perhaps most importantly, the research effort demonstrated that no one in the federal government had a clear understanding of the magnitude or distribution of pesticide residues in the food supply or the public health threat they posed.

Pesticides and Children

While the debate continued over appropriate public-policy responses to the Delaney Committee recommendations, in 1987 Congress quietly funded a new NAS study.[38] The academy was asked to explore whether pesticides posed any special risks to children, by focusing on three questions: Are children more heavily exposed to pesticides than adults? Are children more susceptible to toxic effects of pesticides than adults? And third, do current laws and decision practices sufficiently protect children? Although these questions appear simple, the responses took twenty people five years to prepare.

The earlier Delaney Committee analyses pointed to significant differences in food intake between children and adults but did not explore the implications this variance had for childhood exposure to pesticide residues. Wondering first if there were any differences among the dietary patterns of children and adults, we asked the computer to list the twenty-five most consumed foods for infants, children ages one to five, and adults in the entire population sampled (nearly thirty thousand people). The results were striking and unmistakable. The dietary diversity of very young children is extremely low, meaning that they consume more of fewer foods than adults. Dietary diversity increases during the first several years of life, approaching adult levels by age five. Infants initially have a diet comprised primarily of breast milk, infant formula, and cow's milk, but they move on to fruit juices, pureed fruits and vegetables, then quickly to the table foods of their parents. Adult diets, by contrast, are dominated by beef, poultry, potatoes, and wheat products.[39]

Did these age-related differences in dietary patterns tell us anything about pesticide exposure?[40] Which pesticides were likely to persist as residues on the foods that children eat most often? We knew which pesticides were allowed to persist on specific foods, and these data suggested that fungicides and some insecticides were used more often on fruits and vegetables than on other foods such as grain crops. We then wondered if pesticide levels in food exceeded legal limits. Answering this question slowed the committee's conclusions by at least two years while several of us sifted through enormous volumes of government and industry residue data sets. Cases where multiple pesticide residues existed on a single food were not systematically recorded by FDA, and there was no relationship between the frequency of sampling for residues and the foods consumed by either adults or children. For example, over a two-year period FDA tested only seventy-two samples of bananas for the presence of benomyl, a suspected carcinogen. During this same period nearly 25 billion bananas were imported into the United States.[41]

The committee finally agreed with many U.S. General Accounting Office (GAO) studies that the nation's pesticide residue-monitoring program, managed by FDA, was inadequate to estimate patterns of exposure.[42] We can still be confident, however, that children's dietary patterns result in predictable patterns of pesticide exposure that are very different from those experienced by adults or some fictitious U.S. average individual.

Children are especially vulnerable to health damage from pesticides. Following conception, a child's susceptibility changes as organ systems grow and certain functions mature, such as the detoxification potential of the liver or the filtration potential of the kidneys.[43] Children may be especially vulnerable to carcinogens during periods when their cells are normally reproducing most rapidly, generally between conception and age five.[44] They may be more susceptible to loss of brain function if exposed to neurotoxins during critical periods of development. This is suggested by irradiation, drug, fetal alcohol, and lead studies.[45] And their reproductive systems appear to have special periods of vulnerability, both very early in life and later near puberty or menarche.

Two classes of pesticides, fungicides and insecticides, are of significant concern due both to their common use and their toxic effects. Many fungicides are classified as "probable human carcinogens," and many insecticides inhibit functions of the nervous system by depressing an enzyme known as acetyl-cholinesterase. Herbicides may also pose a significant threat to children in areas where drinking water has been contaminated. A 1995 study of national drinking-water quality estimated that as many as 14 million U.S. residents are currently drinking water contaminated by herbicides that have migrated from fields to surface and subsurface water supplies.[46] Because children drink more water a day than do adults—often mixed with concentrated fruit juices—contaminated water translates into a greater risk for children than for adults. Several herbicides have also been classified as possible or probable human carcinogens, including the most commonly used triazine compounds. Finally, some chlorinated compounds—including many pesticides—are now suspected of somehow disrupting the human endocrine system and possibly encouraging the development of tumors in reproductive organs.[47] Whereas the precise mechanism of toxicity is still unknown, these compounds appear to act by either mimicking or possibly blocking human estrogen.[48] Reflecting on the toxicity and exposure data, the NAS panel concluded that our knowledge of the risks that children face from pesticides is highly inadequate. Most studies conducted by pesticide manufacturers to support their product registrations have been performed on sexually mature animals; very little research has been done on the effects of pesticides on the neurologic, immunologic, and endocrine systems of infants and children.

When the final analyses were released in the 1993 book *Pesticides in the Diets of Infants and Children,* the academy held a press conference in Washington, D.C. The group had deliberated for five years and had concluded that the poor quality of information about foods consumed, as well as the inconclusive data on pesticide residues and toxicity, prevented them from knowing with confidence the level of risk faced by children. Still, they were able to conclude that children are exposed to certain chemicals more than are adults. They also concluded that some organs and biological functions, during well-defined periods of development, are more susceptible to damage.

The president of the NAS Institute of Medicine concluded the press conference by assuring Americans that the U.S. food supply was safe and that the committee's recommendations, if followed, would make it safer. But this assurance rang hollow to those who studied the committee's findings. If the nation's experts do not fully understand the level of contamination by pesticides in our food and drinking water, and if they have not conducted or interpreted toxicity tests that can accurately predict the toxic effects of these compounds on children, how can we conclude that our government is adequately protecting the health of children?

The Challenge: Controlling the Experiment

> The field of pesticide toxicology exemplifies the absurdity of a situation in which 200 million Americans are undergoing life-long exposure, yet our knowledge of what is happening to them is at best fragmentary and for the most part indirect and inferential. **—Mrak Commission, 1969**

When the insecticidal effects of DDT were discovered in 1939, public health experts hoped that diseases such as malaria, yellow fever, and typhus—responsible for killing millions of people each year—might be completely eradicated. The new pesticides, many introduced during or just after World War II, were enormously effective in reducing crop damage and disease incidence. Innovative technologies, including specially fitted aircraft, led to broad application of the modern insecticides to vast areas of the world's landscape.

Experts and government officials, eager to protect crops and reduce disease, were quick to prescribe pesticides without a clear understanding of their ecological or human health consequences. Where biocides were most heavily used, pests—including insects, parasites, weeds, and fungi—often developed resistance to the chemical assault and required new generations of pesticides and drugs, which drew us deeply into an addictive cycle of chemical dependence.

Pesticide law, described in part 2, was built on early enthusiasm for their immediate benefits and required the government to weigh benefits against risks when licensing new products. Pesticides were registered one at a time, each with the presumption that somehow we could manage human exposure to prevent health damage after the chemicals were released into the environment. Effective control of exposure, however, requires detailed information concerning where pesticides are used; where they move and come to rest; their concentration in air, food, water, or soil; and their toxicity to humans and species not considered to be pests. Understanding these effects for a single pesticide may easily cost millions of dollars. Understanding them for tens of thousands of separate licensed products along with their combinations is a virtual impossibility.

The absence of pesticide contamination data was commonly used to justify decisions to avoid toxicity testing. Inadequate toxicity information reinforced the perception among scientists and regulators that risks were minimal or easily managed. When more sensitive detection technology or careful sampling designs demonstrated that residues did not simply disappear, regulators demanded toxicity studies. These studies often suggested the need to more carefully manage human exposure, especially for vulnerable populations such as children.

Pesticide law has evolved in a way that compounds the problem of knowing exposure and risks. Pesticides have been licensed individually, and contamination standards were set for separate environmental media such as water, food, or air. Law has therefore neglected the risks posed by chemical mixtures commonly dispersed into the environment, and instead has directed scientific attention toward very narrow questions—such as the potential for a single pesticide to contaminate a single medium and to induce a single type of effect such as cancer. This history of incremental decisionmaking leaves little optimism that public officials know which pesticides pose the greatest risks, how we accumulate risks, how risks are distributed, or even if use prohibition will result in a pattern of pesticide substitution that poses greater or lesser peril.

The complexity and scale of these problems demand an entirely new legal architecture for producing knowledge of risk and for managing its distribution, as suggested in the final chapter. Although we will never have a perfect understanding of the risks created by pesticide use, our knowledge could and should be much clearer. Government's traditional response to this uncertainty has been to license pesticide use anyway and to assume that exposure may be accurately predicted and carefully managed. The story that follows challenges the wisdom of this assumption and suggests that fundamental reforms in environmental science and law are needed to bring the pesticide experiment under control.

CHAPTER 2

The Urgency of Malaria

In 1948, the Nobel Prize in Medicine was awarded to Paul Müller, who in 1939 had discovered the effectiveness of DDT as an insecticide.[1] Pesticides—especially DDT—enjoyed enormous prestige in the 1940s because of their success in combating the many epidemic and endemic infectious diseases transmitted by insects. Malaria, for example, is caused by a microscopic parasite hosted by female *Anopheles* mosquitoes. Yellow fever is carried by *Aedes* mosquitoes. Sleeping sickness is conveyed carried by Tsetse flies; bubonic plague by a rat flea; Chagas's disease by assassin bugs; epidemic typhus by the human louse; onchocerciasis by black flies; hemorrhagic fever by mites and ticks; and dysentery by flies.[2] Our gradual understanding of the parasitic causes of these diseases—how they are acquired by humans and how they can be managed with drugs and pesticides—raised expectations among public-health experts that the parasites and their insect messengers could be eradicated.

Among all insect-borne diseases, malaria has been the most deadly in modern history. During this century alone, it has killed between 100 and 300 million people, mostly infants and small children, and it infects and debilitates hundreds of millions of others each year.[3] To place this loss in some perspective, combat-related deaths totaled 4 million during World War I, 15 million during World War II, and 2 million during the Korean War.[4] The enormous death toll from malaria has resulted not only from our misunderstanding its causes, but also from our inability to interrupt the parasite's life cycle. The disease is caused by four species of the microscopic parasite *Plasmodium*. Although the relative roles of parasites and mosquitoes were recognized by 1900, this understanding did little to reduce the incidence of malaria during the first half of the century. This failure is not easily explained, but it is clear that both the pesticides and anti-parasitical drugs of the day were unable to break the cycle of transmission. During the 1930s, as many as 50 million people worldwide died of malaria, one of the worst rates of incidence ever recorded. Before the introduction of DDT and other chlorinated insecticides during the 1940s, control of epidemics required almost superhuman efforts.

In 1995, the World Health Organization (WHO) estimated that 40 percent of the world's population, nearly 2 billion people within one hundred different nations, are still exposed to malaria. Moreover, nearly 10 percent of these persons live in areas where malaria is endemic and where no national malaria program has ever been implemented.[5] In 1995, nearly 300 million people carried one of the four species of the parasite *Plasmodium* in their blood, and approximately 120 million clinical cases of malaria were reported. Many epidemiologists, however, believe the actual number of cases to be at least three times that large.[6] Further, the incidence of malaria is not declining; in fact, it may be growing for complex reasons related to the use of biocides. Some have suggested as well that global warming may be expanding the habitats conducive to malaria transmission.

The primary problem in malaria management continues to be that human and the *Anopheles* mosquito live in the same areas. Wherever the *Plasmodium* parasites reside in human reservoirs there is a risk of malaria, and this risk is amplified when the *Anopheles* habitat, which includes slow-moving or still bodies of water, overlaps this parasite's territory. Many cities are located on or near seacoasts, riverbanks, lake shores, or wetland areas that are breeding grounds for mosquitoes. Rural development projects, irrigated fields, and ponds used for aquaculture also provide ideal places for *Anopheles* to thrive.[7]

The Ecology of Malaria

Throughout most of human history, swamps, bogs, and marshes were considered at best wastelands and at worst sources of disease and suffering. Our late-twentieth-century affection for swamps as "wetlands" that serve valuable ecological functions such as storing floodwaters and providing habitat for a diverse array of wildlife and plant species counters a long tradition of belief that somehow these landscapes harbored disease. Hippocrates, for example, in the fifth century B.C., associated stagnant water with the occurrence of fevers and enlarged spleens. Empedocles, a Sicilian physician and philosopher in the mid-fifth century B.C., is credited with draining marshes and running two rivers through the town of Selinunte, and thereby preventing water from stagnating and becoming a breeding ground for mosquitoes. He also broke a gap in the rock wall behind his native town of Agrigentum so that the "healthy north wind could blow the fever-bearing vapours of the plain far out to sea."[8]

Throughout the next two millennia, doctors speculated that malaria was somehow associated with marshland air. This gave rise to the Italian name for the disease— "mal'aria" or "bad air," known to pay deadly visits to Rome every summer. In 1717, Giovanni Lancisi suggested that microscopic "bugs" or "worms" entered the bloodstream and caused malaria. Louis Pasteur's discovery that microorganisms were required for fermentation in the mid-nineteenth century led him to believe that they could also effect disease. The malaria "germ," however, remained undetected in the air, water, or soil throughout the nineteenth century.[9]

In 1880, Charles Lavaran, a British physician, was examining a fresh blood smear from a malaria patient when he noticed something moving. Switching to a more powerful lens on his microscope, he saw that spindle-shaped bodies, recognized earlier by Heinrich Meckel to be associated with malaria, were moving so rapidly that they caused the entire cell to jump about. He also saw that one of the dark pigmented bodies was expelling several moving filaments. Although unknown to Lavaran, these filaments were male spermatozoa, which fertilize female gametes in the mosquito's stomach. Still, he believed he was looking at the live parasite that causes malaria.

Lavaran's claim met with great skepticism. Other scientists' beliefs that soil microorganisms were to blame had not yet been discredited, and many of the followers of Louis Pasteur, who held enormous respect at the time, had clearly demonstrated that many infectious diseases were transmitted by bacteria. Lavaran's findings were widely disbelieved, but soon similar-looking parasites were found in the blood of malaria patients around the world, a discovery that eventually discredited earlier claims that bacteria caused the infection. Camillo Golgi working in northern Italy demonstrated that the parasite invades red blood cells; grows rapidly within the cell, collecting pigment as the cell is attacked; and divides into numerous segments. When these segments separate, the cell disintegrates and bursts, with individual parasites moving to other red blood cells to repeat the process. In 1886 Golgi correlated clinical symptoms of malaria with stages of parasitic development. Chills coinciding with the process of mass segmentation, as well as fevers, occurred in response to the mass destruction of red blood cells. Golgi also believed that periodic recurrence of fever was caused by different species of the parasite. Even Lavaran, who was eventually awarded the Nobel Prize in 1907 for his discoveries, disputed Golgi's multi-species source theory, which proved to be true and to have devastating consequences for malaria-control efforts during the twentieth century.[10]

Finding the parasite in the blood of humans infected with malaria did little to further scientists' understanding of how the disease is acquired. Patrick Manson, a well-known Scottish physician, pursued his research on filariasis, a disfiguring tropical disease caused by a parasitic worm that can live in the human lymphatic system. Occasionally a prematurely voided egg blocks a lymphatic duct and causes a gross enlargement of a limb. Manson wondered whether the worms developed within or outside the body. He saw that the young larvae were more successful in escaping their surrounding sheaths if he chilled them first with ice. This suggested to him that somehow their development might require temperatures cooler than those found inside the human body. Manson then reasoned that because flagella of the parasite appeared only outside humans, this feature was part of an extraordinary reproductive strategy.[11]

Manson heard of Lavaran's discovery by 1890 and assumed that the filaments somehow prepared the parasite for the next stage in its life cycle. By 1894, Manson suggested that mosquitoes transmitted the parasite from one person to another. He allowed them to feed on his infected gardener and successfully demonstrated the parasite's vitality in the mosquito's stomach. Manson then made a wrong turn. He supposed that as the mosquitoes died they contaminated water, which then became the vector for infecting humans.

Manson encouraged Ronald Ross, a young physician in the Indian Medical Service, to explore further the mosquito transmission hypothesis.[12] Ross first discovered that the sexual form of reproduction witnessed by Lavaran occurred far more readily in the stomach of a mosquito than in human blood. Then he found rounded masses on the outside of the mosquito's stomach, another stage in the parasite's life cycle. Next, he discovered that the round bodies on the outer wall of the mosquito's stomach were linked physically to its salivary glands. This suggested that the parasite was transmitted through the proboscis of the mosquito during their blood meals, which Ross demonstrated in birds in 1898. Manson meanwhile had collected infected mosquitoes outside of Rome during the malarial season. They were sent to London and allowed to bite Manson's son, Patrick Thorburn, who soon developed malaria—and thereby demonstrated the possibility of controlling the disease by managing mosquitoes.

Although understanding the role that mosquitoes play in parasite transmission was an enormous scientific achievement, it introduced a complex set of new questions, all relevant to managing malaria. Which species of mosquitoes are involved? Are some species more efficient transmitters than others? What are the common territories, lifestyles, and feeding habits of each species? Does the incubation period for the parasite in the mosquito vary by species? What are the different species of *Plasmodia* parasite? How are they geographically distributed, and do their habitats overlap? Is there any significance to genetic variance within a single species? What will determine the rate at which any *Anopheles* species develops resistance to specific insecticides, or the rate at which different species of the parasite will develop resistance to therapeutic drugs? How probable is it that malaria will move from one habitat or region of the world to another, especially given modern transportation?

Manson, for example, believed that among the hundreds of species of *Anopheles* mosquitoes only several were especially important and efficient transmitters. This proved to be a gross underestimate. By 1990, sixty different *Anopheles* species were known to be capable of transmitting malaria.[13] Other factors also were important, including the presence of domesticated animals (which also served as food for the insects), and the distinctive feeding habits of the species: some preferred day and others night, some indoors, others outdoors.[14] The findings of Lavaran, Manson, and Ross at first seemed to simplify the problem of malaria management. Paradoxically, the opposite occurred as scientists slowly came to understand the complexity of the parasite's ecology. The diversity and variability of factors influencing the risk of malaria transmission seemed to escalate with each new research finding.

The Malaria Life Cycle

Knowledge of the malaria life cycle is important to understand why pesticides were used during eradication and management campaigns. Types of malaria are distinguished by the four species of parasite that cause it, because each has its own distinctive life cycle.[15] The most common is *Plasmodium vivax,* which causes "tertian" fever—fever that recurs every third day—and this type rarely causes death.[16] *Plasmodium falciparum* is the most deadly and is more commonly responsible for cerebral infection, or blackwater fever.[17] The highest fever, often 106 to 107 degrees Fahrenheit, is caused by *Plasmodium malariae,* which is rarely fatal but has a tendency to recur every fourth day, and later in life as well.[18] The last species, *Plasmodium ovale,* produces a tertian infection that tends to occur at night at regular intervals.[19] A daily fever suggests infection by more than one strain of one species or by more than one species.

The life cycle of all species of the malaria parasite is believed to include four phases, each of which results in the appearance of new invasive parasites (fig. 2–1). The first phase, known as fertilization, takes place in the mosquito's stomach. The second phase occurs in the outer stomach wall and body cavity of the mosquito. The third takes place in the human liver, and the fourth in human red blood cells. In each of the four phases, the parasite feeds, grows rapidly, and divides into numerous invasive forms that burst from the host cell and invade new hosts to initiate the next phase.

As a mosquito bites, it injects saliva into the skin of its host. The saliva contains a chemical that both damages capillaries and acts as an anticoagulant. If the victim has malaria, the blood-borne parasites are ingested along with the blood meal. It is here in the mosquito's stomach that the female and male gametes fuse to form a zygote that soon bores through the stomach wall and attaches to its outer surface. The rounded, semi-transparent body is now known as a oocyst, and as it grows as many as a thousand young parasites

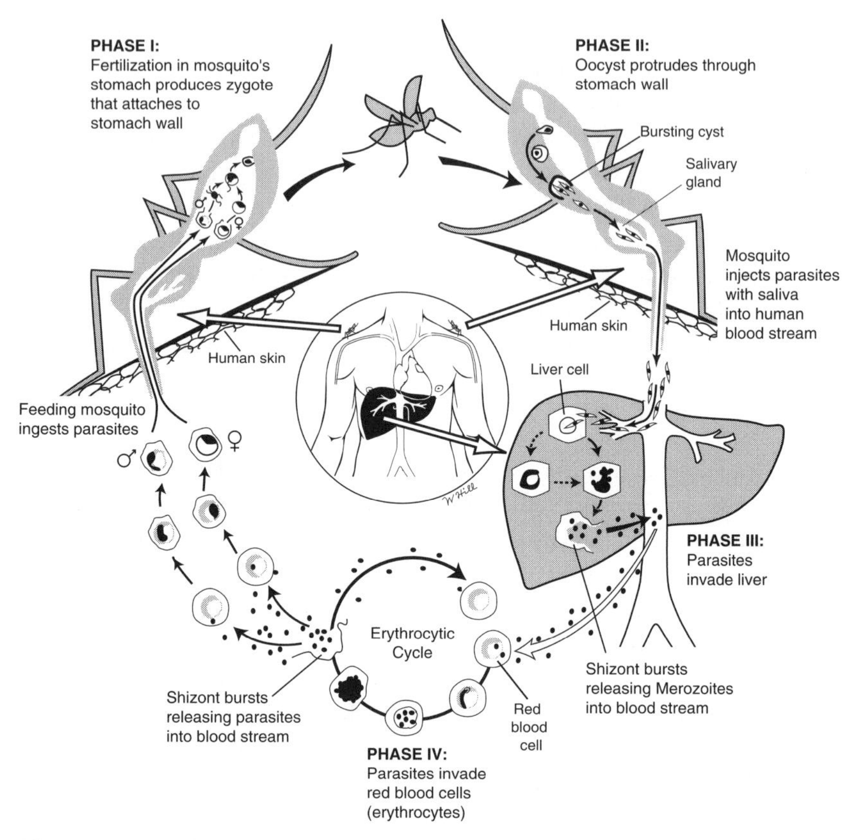

Figure 2–1. The life cycle of malaria. The life cycle of all four species of the malaria parasite occurs within mosquitoes and humans. After a mosquito ingests parasites from the blood of an infected host, the male and female forms fuse to produce an egg mass that grows and penetrates the stomach wall. This sac eventually bursts, releasing more than ten thousand sporozoites that migrate to the mosquitoes' salivary glands, where they may be injected into the next host during a meal. Inside humans, the parasites quickly invade liver cells. Over the next five to fifteen days, each parasite produces more than ten thousand "daughter" parasites that move from the liver to invade red blood cells. Once inside the cell, each parasite produces between eight and thirty-two offspring that eventually rupture the cell, releasing the young parasites into the bloodstream, where they invade additional red blood cells, continuing the cycle.

form within it. These young parasites eventually burst into the body cavity of the mosquito and reach the salivary glands of female *Anopheles*. From here they are injected into the human bloodstream as mosquitoes feed, and many enter liver cells where they further develop, multiply, and eventually break back into the bloodstream. The parasites then break into red blood cells, consuming hemoglobin and eventually bursting through cell walls, producing characteristic fevers. If the individual is bitten by another female *Anopheles,* the cycle of transmission continues. *Falciparum* parasites commonly invade the brain through small blood vessels and induce swelling, coma, seizures, and if untreated, death.

Controlling malaria requires interrupting this cycle. One strategy is to kill the parasites within infected humans using drugs. A second is to kill all of the mosquitoes capable of transmitting the parasite to humans. A third method is to prevent uninfected humans from being bitten by infected mosquitoes through the use of clothing, repellents, bednets, or window screens. A fourth strategy is to manage areas where people live in a way that discourages *Anopheles* mosquitoes from breeding. A final method is to treat people living or traveling in endemic areas with prophylactic doses of drugs so that if bitten, the parasite is unlikely to survive.

This gradually expanding knowledge about the transmission of malaria did lead to successful eradication efforts in temperate and some semitropical regions of the world.[20] Further, the new understanding of the ecology of malaria raised the hope that the disease might be controlled through the use of biocides: pesticides to kill the mosquito vector, and antibiotic drugs to kill the parasite within infected humans or to prevent their infection. But within many tropical regions, malaria was intertwined with deeply rooted social problems such as poverty, famine, warfare, migration, and the absence of an institutional infrastructure that could deliver aid.

Gorgas and Malaria Control Before DDT

There is perhaps no better example of malaria's potential to impede economic development than the struggle to complete the Panama Canal. Malaria and yellow fever together had killed thousands during the French attempt to construct the canal through dense tropical forests and swamplands. Before the French finally abandoned the project, it was not uncommon to see ships quietly anchored in the Panamanian harbor of Colón, their entire crews having succumbed to the disease. The U.S. decision to pick up where the French had left off was made with the expectation that these diseases could somehow be managed.[21]

The effort to "sanitize" the Canal Zone was lead by Dr. William Gorgas, an exceptional leader and health scientist known for his successful campaign against similar insect-borne diseases in Havana.[22] But the crossing of Panama posed a problem of much greater scale than any he had experienced. By 1904, the discoveries of Manson and Ross were not yet well assimilated into the thinking of U.S. military leaders, and Gorgas found himself laboring to convince the head of the Panama Canal Commission of even the fundamental premise that malaria and yellow fever were carried by mosquitoes.[23] During the winter of 1904, the mortality rate rose so quickly that the workers panicked and construction stopped. Commission members realized that they were on the brink of duplicating the French failure and desperately turned to Gorgas.

Gorgas formed his anti-mosquito brigade and was helped by the passage of law that made anyone harboring mosquito larvae punishable by a fine. He gave one man the responsibility for eradicating mosquitoes in each clearly delineated zone of Colón. Rubbish collection and burial were instituted, and other surfaces capable of retaining water were drained, removed, or covered. Emergency response teams (literally "swat" teams) were organized. Anyone who found a mosquito but failed to catch it was required to call an emergency telephone number. Within minutes a sanitary engineer with six or seven laborers would arrive to search the entire house with flashlights and trap, kill, and catalogue all that were found. It became a misdemeanor, punishable by a five-dollar fine, to harbor mosquito larvae on any premise. Yet the mosquitoes found other places to breed, particularly in the concave recesses of plants. Gorgas responded to this obstacle by baiting the insects. He placed hundreds of water barrels outside as breeding grounds and then killed the larvae with petroleum oils as they were ready to hatch (fig. 2–2).

It was easy for people to believe that filthy swamp water was the source of the disease, but it was more difficult to convince them that clear, newly fallen rainwater posed the same threat. Gorgas knew that North American cities such as New York and Philadelphia did not effectively control yellow fever until public water was transported by underground piping. When he went to Panama, the two main cities, Panama City and Colón, both derived their water from rain that drained off of rooftops and into tin troughs and rain barrels. Additionally, inside Panamanian houses water stood in jars, pitchers, dishes, and flower vases.[24] Attacking yellow fever and malaria was not just a matter of managing standing water in a huge tropical development project; it required fundamental changes in the residents' lifestyle. This effort was hardly helped by heightened Latin immunity to the disease, which led residents to believe the Americans had some other motive for their seeming fanaticism. When American inspectors arrived to order that water must never be left exposed, their reception was anything but warm.[25]

Malaria and yellow fever raged on, demonstrating the enormity of the problem, and Gorgas came dangerously close to becoming a political scapegoat for the operation's failure. Privately he despaired to his wife. Secretary of War William Howard Taft and the new governor of the Canal Zone both urged President Roosevelt to replace Gorgas, to at least give the impression that something was being done. Roosevelt decided to visit Panama and to

Figure 2–2. Installing underground drainage in Colón. Because it is located at sea level, Colón faces enormous drainage problems. Standing water provided excellent habitat for mosquitoes that carried both malaria and yellow fever. As part of his war on mosquitoes, Dr. William Gorgas drained and paved the streets and built an enclosed water supply system. (From *Panama and the canal* [New York: Syndicate Publishing, 1913], 255)

meet with Gorgas personally. Roosevelt immediately recognized Gorgas's leadership and organizational skills. Instead of removing him, Gorgas was promoted to a position on the Canal Commission.[26]

Finally in 1908, yellow fever began to subside. This disease is transmitted by *Aedes* mosquitoes, and Gorgas had learned in Havana that *Aedes* tended to breed in relatively clean water, near dwellings. Receptacles suitable for *Aedes,* such as cans, barrels, bottles, and other refuse, were easily identified and removed. Malaria, by contrast, was far more difficult to control, partly due to the variety of *Anopheles* suspected of carrying the parasite, and in part because of the wide variation in their breeding and feeding habits. Gorgas and his team, for example, found that the dominant carrier species was *Anopheles albimanus,* which unfortunately was also the most abundant and could squeeze its body through tight places such as ill-fitting screens or doors left ajar.

When Gorgas arrived in Panama, he found that the hospital facilities themselves contributed to an "average constant sick rate" of 33 percent among the French workforce.[27] Few buildings were screened, and the Ancon hospital had a policy of grouping patients by nationality rather than by disease, which contributed to the spread of any infectious disease. Moreover, the hospital was located on a hill and was well known for its landscaping. To combat the ants attracted by the gardens' flowers, staff placed cups of water beneath the legs of tables, beds, and chairs. These cups, along with several thousand pottery rings filled with water and tropical plants, were ideal breeding sites for mosquitoes.

Again seeming overwhelmed, Gorgas and his crew made another discovery. By painting the feet of mosquitoes with aniline dye and setting out traps, they were able to track the insects' flight patterns and range. This method was later improved with the use of fluorescent paint, which allowed the insects to be tracked like fireflies. Unfortunately, however, they found the infected mosquitoes capable of flying more than a mile, even into a stiff wind. This discovery meant that the boundary of their control area had to be extended into the jungle swamplands. Gorgas turned his crews' attention to draining the marshes, where they eventually dug more than 8 million feet of ditches. They also attacked the brush and shrubs that provided protection for adult mosquitoes and designed a flame thrower of sorts to burn the gardens that were commonly planted next to doorways and that often harbored sleeping mosquitoes. The many bodies of water that could not be drained were treated with oils, normally mixtures of kerosene and crude oil. Elaborate mechanical devices were designed to drip oil slowly into streams and pools to create a lasting residue (fig. 2–3). Even some of the hoofprints of cattle and horses created depressions that retained water long enough during the rainy season to permit the laying and hatching of larvae. Hundreds of workers were sent out to oil roadside ditches and even standing pools of water in fields.

Meanwhile Gorgas continued to study the ecology of the region and began breeding spiders and ants that he found consumed mosquitoes. Lizards were commonly distributed to each room, where they would climb walls to dine on one of their favorite meals. Bats were also encouraged because they consumed large numbers of flying mosquitoes. Slowly malaria yielded to Gorgas, and yellow fever was eradicated. Work proceeded on the canal, which was completed in 1914.[28] Gorgas's efforts had obviously made the difference between success or failure for the Panama Canal, and he taught several lessons important for understanding later attempts to manage insect-borne diseases with pesticides.[29] One such lesson was that successful control depended upon knowledge of the insect ecology of the area, because the parasite was transmitted by different species with different feeding, resting, flight, and reproductive habits in different parts of the world. Gorgas also verified the intuition of Hippocrates and Empedocles when he discovered that

Figure 2–3. Canisters of fuel oil drip into Panamanian streams and marshes. Fuel oil creates a thin film on water surfaces that suffocates mosquito larvae. (U.S. Army Signal Corps)

patterns of human settlement affected the incidence of disease. Agriculture, livestock grazing, water supply and irrigation techniques, solid waste disposal practices, and even the planting of gardens all influence the virulence of *Anopheles* mosquito populations. Understanding these relations required extensive research into the ecology and behavior of humans, insects, and parasites. When combined with extraordinary planning, organization, vigilance, and leadership, Gorgas's intervention was highly effective.[30]

The Brazilian Campaign

By 1930, *Anopheles gambiae* was well known as a dominant malaria vector in Africa, so its discovery in a mosquito trap in Natal, Brazil, was a deeply disturbing surprise. The invader was believed to have hitchhiked on one of the destroyers employed for rapid mail transport between Dakar and Natal in 1930. *Gambiae* was first detected in March among hayfields created by diking the river's tidewaters. By May an epidemic was raging in the workers' suburb of Alecrim, where over 30 percent of the *gambiae* tested were infected. Nearly every household had at least one afflicted family member, and public-health officials organized to provide food and medicine for the community. By the following year, the infected mosquitoes extended their range to six square kilometers, and soon ten thousand of the twelve thousand people in Alecrim reported infections.[31]

The incidence of malaria declined in 1932, primarily because many districts received less than 20 percent of their normal rainfall that year. But drought replaced one misery with another; the crop failure that followed caused a famine, which in turn induced mass migration. The government responded by creating refugee camps, one of which clustered over half a million people without clean water and adequate sewage disposal. Given these problems, official malaria control efforts were abandoned. Meanwhile, *gambiae* not only survived the drought close to Natal and along the northern coast, but had extended its territory into the Assú, Apodí, and Jaguaribe river valleys, again hitchhiking on sailboats that moved goods along the rivers, or possibly by truck or rail.

The full effect of the *gambiae* invasion was not felt until 1938. Those living in the newly infiltrated river valleys had no immunity to the disease. Often entire families fell ill at once, leaving no one to provide meals or go for medicine. As one observer noted, "In a population always suffering from undernourishment, the inability to work for even a few days resulted in further reduction of food supplies, entailing in may cases complete absence of food. Illness, poverty, hunger, starvation, and death were all close associates in this fulminant epidemic of Northeast Brazil."[32]

Stocks of quinine and atabrine were quickly exhausted, and starving families blocked the roads begging for medicine. Of 1,060 inhabitants of one rural village, 1,012 carried the infection. During 1938 in the state of Rio Grande do Norte, over one hundred thousand people were ill and twenty thousand died. By 1939, official figures registered 185,000 cases, which made this outbreak similar in magnitude to the horrific epidemics that swept Mauritius in 1867, Punjab in 1908, and Ceylon in 1935.[33]

In 1939 the Rockefeller Foundation was asked by the Brazilian government to control eradication efforts. Brazil made the personnel and equipment of the Brazilian Yellow Fever Service available to Fred Soper of the foundation, who organized a military-style campaign named the Malaria Service of the Northeast.[34] The effort was highly dependent on pesticide application. In contrast to Brazilian species of *Anopheles,* the *gambiae* fed principally on

Figure 2–4. Dispersing Paris green (arsenic) by hand. Paris green—copper acetoarsenite—was the insecticide most commonly used outdoors to control *A. gambiae*. (From F. L. Soper and D. B. Wilson, *Anopheles gambiae in Brazil* [New York: Rockefeller Foundation, 1943], 122)

humans, most commonly between two and four o'clock in the morning, and then retired to the darkest and most remote corners of buildings. The propensity of *A. gambiae* to breed and feed close to dwellings also made it vulnerable to control efforts, which primarily involved the use of two pesticides, copper arsenate (Paris green) and pyrethrum (fig. 2–4).[35]

Soper divided the entire region into zones and clearly defined the responsibilities for application and recordkeeping. Crews sampled adult mosquitoes and larvae to ensure that the insects' life cycle had been broken. A cartographic staff mapped the incidence of disease and charted the progress of spraying efforts, which were coordinated using a concentric zone theory (fig. 2–5). Outlying areas were first sprayed in the hope that further migration of the insects could be prevented. Then efforts turned toward the central zones, where malaria was most prevalent. Once a week, inspectors moved from house to house, spraying the interiors with a cloud of pyrethrum. Umbrella makers were commissioned to produce square, white umbrellas without handles that could be hung upside down in ceiling corners to collect mosquitoes killed by the insecticide; these mosquitoes were then catalogued (fig. 2–6). Medical teams, by contrast, started their work at the village centers, dispensing quinine and atabrine as well as food and general medical care to the weakened population.

This simultaneous attack on mosquitoes using pesticides, and on parasites using drugs, immediately lowered mortality rates. By the end of 1940, eradication of *A. gambiae* was complete. (Later occasional outbreaks of the disease were caused by indigenous species of *Anopheles*.) Yet the incident

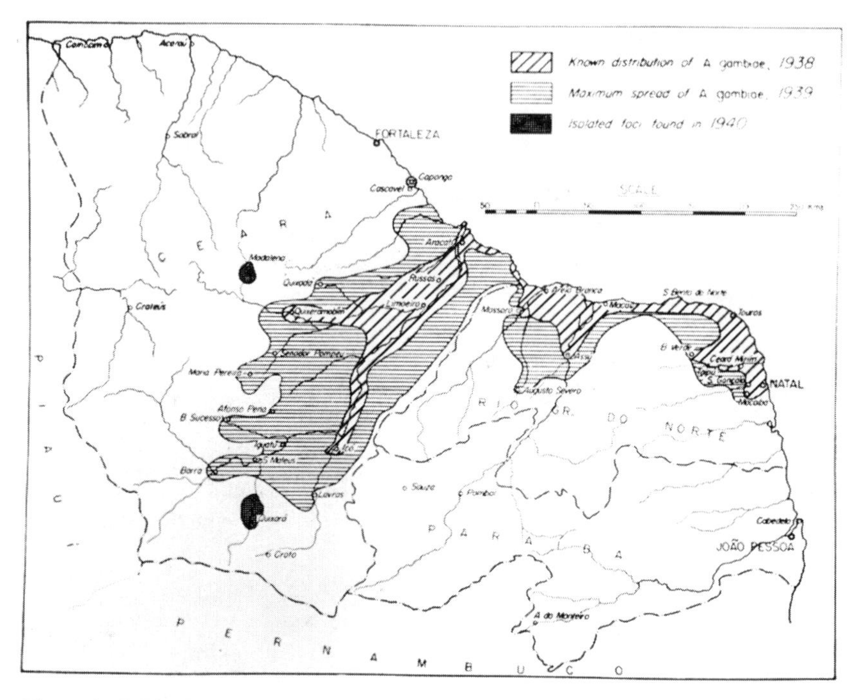

Figure 2–5. Distribution of *A. gambiae* in northeastern Brazil. *A. gambiae* spread up river valleys from Natal during the 1930s, distributed by winds and by hitchhiking on river boats. Intensive eradication efforts by the Rockefeller Foundation and the Brazilian government between 1938 and 1940 reduced its known habitat to the few well-defined foci outlined above. (From *A. gambiae eradication in Brazil* [New York: Rockefeller Foundation, 1943], 155)

demonstrated the extreme vulnerability of those living in any climate hospitable to species of *Anopheles* capable of transmitting malaria. Population migration, modern transportation, expanding markets, and warfare all increased the chances that somehow the malaria parasite would move beyond its historical boundaries.

Following the epidemic, Brazilian health authorities required that all ships and planes from Africa be fumigated. Pilots were required by law to keep all doors and windows closed until the interior of the plane had been sprayed (fig. 2–7). Although this might seem excessive, many airports—both in Africa and Brazil—are located near coastlines, marshlands, bays, or rivers that provide ideal breeding areas for *Anopheles*. African authorities relied only on personal protection measures such as screening, insect repellents, and prophylactic drugs, rather than the application of larvicides to marshlands or waterbodies. Further, aircraft were commonly parked next to swamps with their windows or cargo bays left open. By 1939, for example, over 225 dif-

Figure 2–6. Tracking the epidemic: mosquito collection techniques. A square inverted umbrella was designed to capture mosquitoes falling from ceiling corners after they were sprayed with pyrethrum. Mosquitoes were then tested for the malaria parasite, and the results were mapped to track the spread or contraction of the disease. (From F. L. Soper and D. B. Wilson, *Anopheles gambiae in Brazil* [New York: Rockefeller Foundation, 1943], 113)

ferent species of mosquitoes were found in commercial aircraft providing service between Africa, Latin America, and Miami.[36] In Natal alone during a three-month period in 1942, 1,493 insects representing forty-four different species of arthropods were found on aircraft arriving from Africa.[37] By 1944, American, British, and French forces had joined to create malaria-free zones surrounding most major airports in northern and western Africa, particularly those along the Gold Coast, Nigeria, and Senegal.

Malaria and Warfare

Malaria has also been an important influence on the outcome of war.[38] Military leaders, including Alexander the Great, Julius Caesar, Henry II, Napoleon, Hitler, Hirohito, Eisenhower, and MacArthur, all learned that malaria was likely to claim more troop casualties than combat. On the eve of World War II, U.S. military leaders were anxiously aware of their inability to control the disease. The bombing of Pearl Harbor drew millions of U.S.

Figure 2–7. Fumigating airliners with pyrethrum. Recognizing that the Brazilian epidemic was caused by malaria imported from Africa, the government adopted policies requiring the fumigation of airliners. Insecticides were soon routinely mixed with planes' fresh air supplies. (U.S. Public Health Service)

troops into combat within malarious parts of the world, especially the South Pacific and the Mediterranean Basin.

Conditions of war favor malaria transmission in several ways. Troops live outdoors and are therefore easy targets for mosquitoes; once they move into malarious regions, disease incidence and severity among soldiers is normally high due to lack of immunity; troops are concentrated, which makes transmission more likely; and combatants are generally weakened by fatigue, poor nutrition, and poor sanitation. In addition, military objectives normally include control over ports, rivers, and water supplies, which are the primary habitats of mosquitoes. Troop movement itself can exacerbate the spread of the disease by inadvertently carrying the parasite into regions previously malaria-free. Military tactics such as widespread shelling, bombing, and more recently, herbicide-induced forest defoliation all increase standing water, the preferred breeding site for mosquitoes.

In 1915, during World War I, the British and French decided to attack the German and Austrian soldiers at the Macedonian front. Despite warnings that troops would be entering a highly malarious region, little attention was paid to prevention. Few even understood what malaria was. Hospitals were unscreened, and hardly any used quinine. At least 50 percent of the troops contracted the disease during the summers of 1916 and 1917, with one hundred thousand cases among British soldiers and sixty thousand among the French. When the final statistics on British casualties were determined, 28,000 had been wounded or were missing, killed, or captured, whereas 162,000 had contracted malaria. Thousands of sick soldiers were evacuated to British and French hospitals at enormous cost and further, carried the risk of malaria home. When the French General Sarrail was ordered to attack, he responded in desperation, "Mon armée est immobilisée dans les hôpitaux." By 1917, refusal to take quinine prophylactically was punishable as "disobeying orders in the face of the enemy." Urine tests were used to judge compliance, which soon approached 100 percent.[39]

The devastation of Macedonia was on the minds of the chief U.S. military officers on the eve of World War II. These were risks the government knew it was unprepared to manage. The U.S. command estimated that if nothing was done, 50 percent of their troops would succumb to the disease, leaving them incapacitated for between two and three weeks, with likely relapses and reinfections. Controlling malaria meant attacking the mosquito directly and attacking the parasite once it had infected humans. But the painstaking methods used by Gorgas in Panama were believed to be nearly impossible under combat conditions; large-scale drainage of swampland was infeasible, and sleeping quarters were tents under the best of conditions, and under the worst either a bedroll or wide-meshed hammock. Mosquito nets and insect repellents, often experimental, were initially the only technologies used to control insect-borne disease.

The American landing in the New Hebrides islands became one of the first tests of Allied malaria-control efforts in the Pacific. The New Hebrides were strategically important for resisting Japanese advances from the south, and the primary objective of the occupation was to construct an airfield. The site of the airfield was relatively flat and surrounded by marshlands and shrub-filled streams. The indigenous population carried chronic malaria, providing a reservoir of parasites for the swarms of mosquitoes that greeted the troops. Unfortunately, mosquito netting and screening for buildings were packed deeply within the hold of transport ships, which took several weeks to unload. Although atabrine and quinine were available, drug prophylaxis was not required, and attempts to spray adjacent waterbodies with oil and Paris green to control larvae were poorly organized (fig. 2–8).

Within two months, nearly two-thirds of the men contracted malaria.[40] Someone noticed that mosquitoes were breeding rapidly in broken coconut shells that covered the ground of the island. Men and equipment were quickly assigned to collect and burn all the shells. Before much had been accomplished, however, an entomologist arrived and discovered that the mosquitoes breeding in the shells were not *Anopheles*; instead, *Anopheles*

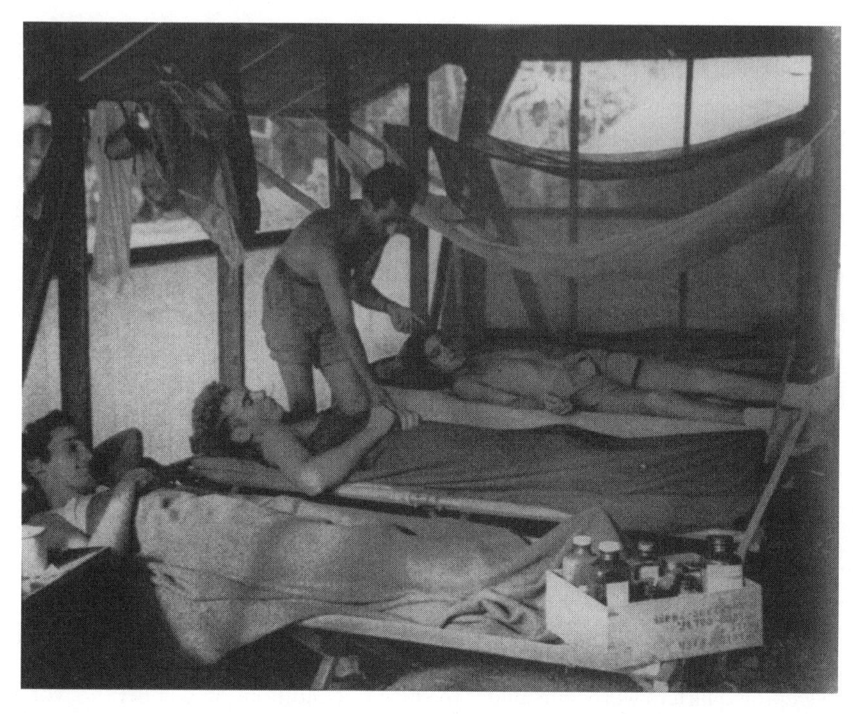

Figure 2–8. Marines in sick bay with malaria in the southwest Pacific. (U.S. Marine Corps)

were breeding within a limited area of nearby streams. Once vegetation was cleared from the stream banks and larvicides were applied, the incidence of malaria quickly declined.

During the same year, the U.S. command learned the extreme difficulty of managing malaria under intense combat conditions when American troops landed on Guadalcanal in the Solomon Islands. Japanese greeted the Americans with a rain of mortar shells, which left the beach front pockmarked. Mosquitoes quickly found the new breeding sites, along with the reservoir of parasites among Japanese troops. Within three months, malaria among Allied troops reached epidemic proportions with an incidence five times higher than combat casualties. Over one hundred thousand men contracted the disease, and each experienced, on average, two attacks. Moreover, in the case of Guadalcanal, the epidemic escalated because of the attitudes of commanding officers, one of whom said, "We are here to fight the Japs, and to hell with mosquitoes"—a comment strikingly similar to those expressed by early military leaders of the Panama Canal project. Yet the Japanese were suffering perhaps more than the Americans; nearly every captured soldier carried a chronic infection. The Japanese efforts to control malaria were reported to be even less sophisticated than the Americans', which perhaps helps account for their eventual loss of the island.

Gradually, the U.S. command pulled together an anti-malaria force that at its peak included five thousand men, among them doctors, entomologists, parasitologists, sanitary engineers, explosives experts, heavy equipment operators, welders, and carpenters. Their obstacles were enormous. They often had to fight within their own institutions to divert equipment from other road, airfield, or sanitary construction projects. Shellholes, foxholes, wheel ruts, and ditches gave mosquitoes new habitat in addition to the marshlands, estuaries, and streams that already supported a healthy population of insects. Whenever possible, bivouacs for troops were erected far from mosquito habitat, and by the end of the war these were supplied with screens that were treated with insecticides. Reliance on indigenous populations for labor occasionally proved costly because some were chronic carriers of the parasite. This situation led to policies that segregated local populations from Allied troops.

Insecticide stocks at the beginning of the war were low, as were supplies of repellents. Before the introduction of DDT, diesel fuel was the dominant larvicide, and nearly 1,800 fifty-five-gallon drums were emptied on water bodies during one wet season on Guadalcanal alone. After streams were cleared of adjacent vegetation by bulldozers that dragged heavy chains across the banks, crews carrying the diesel fuel or Paris green in backpack tanks were sent out periodically to spray water surfaces.

The need for an effective repellent was enormous. Over eight thousand combinations of insecticides and solvents were tested by the army, navy, and Department of Agriculture in an effort to find a smell that would repel mosquitoes but not nauseate the soldiers (fig. 2–9). Volunteers were tested by having a well-defined area of their arm covered by the experimental compound. The arm was then extended into a chamber filled with mosquitoes, and the number of mosquitoes landing on the patch was compared with the number landing in a similar untreated area.

Fabrics impregnated with insecticides were tested in a different way (fig. 2–10). Volunteers were dressed similarly but had different compounds on their shirts. They would dash out of a tent and run to a single line of chairs next to a swamp known to be infested with mosquitoes. Once seated, each would stare at the back of the soldier in front of him, counting the mosquitoes landing on a well-marked patch. At the sound of a whistle, they would then run back to the tent. Several compounds were found to be effective, but many contained solvents, one of which was strong enough to dissolve both watch crystals and metal cigarette cases. The burning and stinging of solvents was worsened by the profuse sweating common in the humid tropics,

Figure 2–9. Experimental repellents tested on GI's. Finding an effective repellent that did not burn, smell too wretched, or attract the Japanese was a challenge for military chemists. (From *National Geographic* 86 [Feb. 1944], 173; photo by Terris Moore; courtesy National Geographic Society)

Figure 2–10. What color shirt do mosquitoes like best? The army tested the affinity of mosquitoes for fabrics of different colors. The results? In thirty seconds, four mosquitoes landed on the white shirt patch, fifteen on the black, thirty on khaki, and ninety on a bare back. (From *National Geographic* 86 [Feb. 1944], 172; photo by David Tutrone; courtesy National Geographic Society)

which also reduced their effectiveness. These limitations were compounded by the fear among South Pacific combat troops that the repellents would attract the attention of another serious threat, the Japanese.[41]

Over fourteen thousand separate drugs were tested for their effectiveness, normally on volunteer prisoners who submitted to malaria infection and subsequent clinical trials. Although atabrine was the most commonly used prophylactic drug, quinine derived from cinchona bark was still most commonly prescribed to treat infection. Warfare for U.S. troops in the South Pacific was made even more dismal by the fall of Java to the Japanese, because Java was the site of enormous Dutch-owned cinchona plantations that at the beginning of the war produced 20 million tons of bark per year (figs. 2–11 and 2–12).[42]

The Promise of DDT

The experiences of Gorgas in Panama, the Brazilian epidemic of 1938, and the difficulty of managing malaria during World War II together contributed to a desperate, urgent search for new chemical technologies to manage the disease. Both pesticides and antibiotic drugs such as quinine were central components of control strategies, but each had limited effectiveness.

As late as 1940, there were very few insecticides that were both effective in controlling insects and "safe" for humans. Those in common use included nicotine, pyrethrum, rotenone, sulfur, hydrogen cyanide gas, cryolite, petroleum oils, and several metals, primarily the arsenicals.[43] In the 1930s Brazilian state-of-the-art technology in the fight against malaria included the use of Paris green (copper arsenate)—known for its extraordinary acute and chronic toxicity—to kill mosquito larvae, and the spraying of pyrethrum indoors to kill adult mosquitoes. Hundreds of tons of Paris green were used as a larvicide. It was spread, often by hand, on water surfaces, marshlands, and fields.[44]

As late as 1939, the United States imported nearly 4 million pounds of rotenone from Japan, a supply that was soon cut off. Japan also controlled large quantities of pyrethrum, and during the 1930s provided most of the U.S.-imported supply.[45] The worldwide demand for pyrethrum grew rapidly as warfare penetrated parts of the world with endemic malaria. Following Pearl Harbor, the Japanese supply of pyrethrum was stopped, and by 1942 the insecticide was declared to be a strategic war material for malaria control. Poor weather in Kenya reduced the harvest of chrysanthemums (from which pyrethrum was derived), and war-related shipping interruptions made supplies even more difficult to obtain.[46]

Hundreds of compounds were examined for their insecticidal potential, with few promising much improvement over those that had dominated the

Figure 2–11. Quinine and the fever tree. Powdered bark from the South American cinchona tree was recognized as a cure for malarial fevers in the early 1600s. One classification system for fevers is depicted by the "fever tree," designed by Francesco Torti, an Italian physician, in 1712. Chinchona bark was believed to be therapeutic for the fevers listed among the branches with bark on the left of the tree. It was not believed to cure those fevers listed among the denuded branches. The active ingredient in the bark is quinine, which directly attacks the parasite in red blood cells. (From Institutum Ad Propagandum Usum Chinini, *Malaria et chininum* [Amsterdam, 1927], 18)

Figure 2–12. Civilian donations of quinine stocks being categorized for use in the South Pacific Theater, 1943. A massive national campaign to collect civilian stocks of quinine followed Japanese control over Java, which had formerly provided supplies of the drug to the United States. (From *National Geographic* 84 [Feb. 1943], 614; photo by Alfred Palmer; courtesy National Geographic Society)

marketplace for decades. No more efficient method was available for combating typhus, for example (which is transmitted by lice), than steaming clothing, a treatment introduced during World War I. Within this context of enormous risk to national security and public health, the insecticidal effects of DDT were recognized in 1939. Although DDT was first synthesized in 1874 by Othmar Zeidler, a young doctoral student at the University of Strasbourg, its effectiveness as an insecticide was not known until Paul Müller discovered its toxic effects on potato beetles and moths while searching for new textile mothproofing agents.[47] The enthusiastic claims of effectiveness as an insecticide made by Müller and others at Geigy inspired disbelief, so in 1942 he arranged for several hundred pounds of spray and powder to be sent to the Department of Agriculture for independent testing.

DDT is a contact poison that acts on the nervous system of insects, causing overstimulation of neurons and rapid death. Insects were said to get "the DDT's," a combination of twitching and convulsions followed quickly by coma and death. The compound is especially toxic to mosquitoes, but it is equally effective at killing thousands of species of insects, including disease-carrying flies and lice. In one informal experiment, DDT was placed in a jar of water at a concentration of 1 part per million. The jar was emptied and replaced with fresh water. Mosquito larvae were placed on the surface and all had died by the next day. The jar was then emptied again and refilled with fresh water, and residues were still potent enough to kill the larvae.

In a joint effort to test the health effects of the compound, the U.S. National Institutes of Health, FDA, the Public Health Service, and the National Research Council immediately began toxicity testing with an emphasis on the acute effects of high exposures, especially to applicators and others directly involved in production. The persistence of DDT, its accumulation in the body, and its chronic toxicity were also studied, but given the demand for the compound and the short time available to conduct chronic toxicity tests, these studies were given lower priority. In general these concerns were believed insignificant compared to the benefits that the chemical seemed to promise.[48]

Not only did the claims of toxicity to insects prove to be true, but for some unknown reason the acute or immediate toxicity of the compound to mammals, particularly humans, seemed to be exceptionally low. When the choice to deploy DDT was made by military, health, and agricultural officials, they could not have known that their fundamental logic of choosing near-term, certain benefits over uncertain long-term risks was to be adopted by USDA as it licensed tens of thousands of products containing hundreds of newly discovered synthetic pesticides. This logic would support if not propel the agricultural chemical industry through its most rapid period of growth and innovation between 1945 and 1970.

The United States began a massive DDT production effort in 1943. The synthesis of the compound proved to be relatively simple, which contributed to its low cost, and by 1944 American firms were producing nearly 2 million pounds of DDT per month. The chemical was allowed only for military use until the end of the war in 1945. One of its first military uses was as a louse powder that helped to bring the 1943–44 typhus epidemic in Naples under control (figs. 2–13 and 2–14).

A new type of fumigating gun forced DDT into the sleeves, waistbands, pant legs, and hair of more than 3 million troops and citizens of Naples in less than a year. DDT received credit and enormous publicity for ending the epidemic, although extensive and earlier use of pyrethrum probably already had reduced the incidence of typhus before substantial shipments of DDT had arrived. News of the epidemic's decline stimulated high hopes for the chemical's potential and even speculation that the major insect-borne diseases in the world could be eradicated.

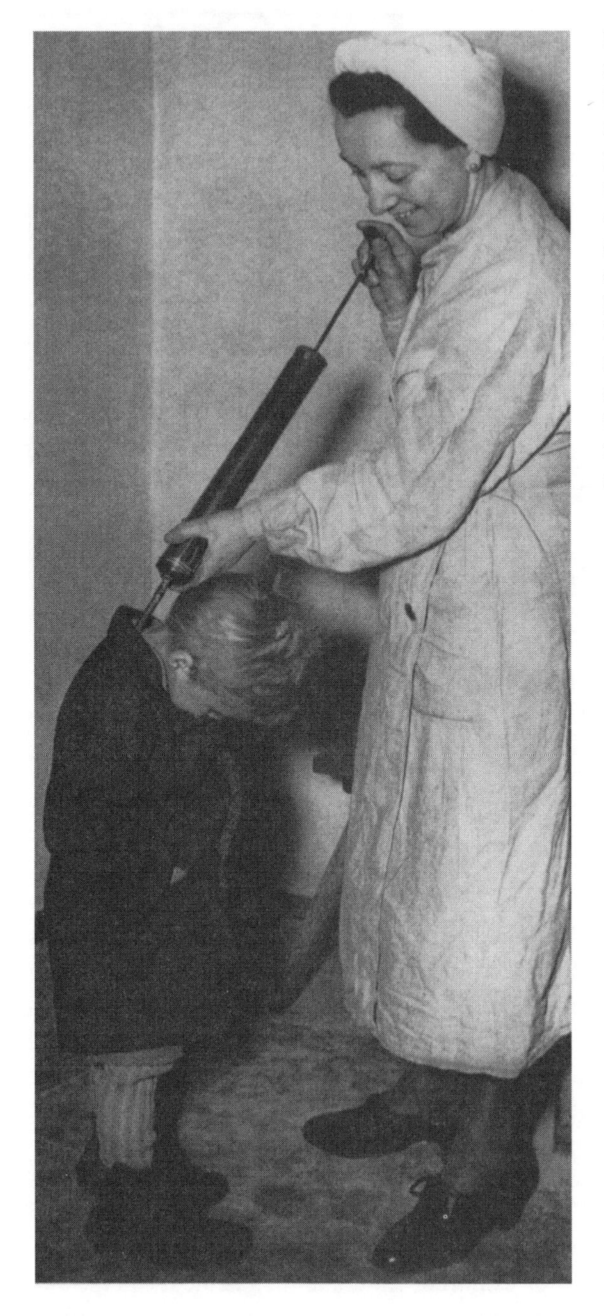

Figure 2–13. Fumigating children in Berlin with DDT. DDT was enormously effective in stopping the Naples typhus epidemic. Public health officials routinely fumigated millions of refugees as they migrated through Europe after World War II. Fumigating guns forced the compound into collars, sleeves, and trouser legs in an effort to control lice, the vector for typhus. (U.S. Army Signal Corps)

Figure 2–14. DDT fumigation in Japan, 1945. The American army sprayed DDT through the streets of U.S.-occupied Japanese cities following Japan's surrender. The sign on the Jeep, "Sanitation," suggests the health-promoting image of DDT that followed its well-publicized wartime effectiveness in controlling a variety of life-threatening insect-borne diseases. (John Baylor Roberts; © National Geographic Society Image Collection)

As soon as the potential effectiveness of DDT was recognized, government stimulated production and designed new technologies to facilitate its broadscale application. Bombers were retrofitted with tanks and spraying devices invented to quickly blanket the landscape with insecticides. Engineers fashioned new nozzles for these aircraft so that a mixture of aerosol particles would be produced—some large enough to ensure that the mixture would fall to the earth and other droplets small enough to increase dispersion. DDT could easily be mixed with oils, fuels, or kerosene. One ton of the insecticide in kerosene could kill mosquito larvae over an area two hundred yards wide by eighteen miles long. Under many conditions, only one quarter to one-half pound per acre was needed to eliminate adult mosquitoes (fig. 2–15).[49] By 1944, the year of its introduction into the Pacific theater, entire islands were sprayed prior to the landing of U.S. troops, which otherwise would have been greeted by clouds of mosquitoes carrying malaria, yellow fever, and dengue fever.[50]

Figure 2–15. DDT being applied to control flies. After World War II, DDT was quickly adopted for nuisance insect control. In 1945, children frolic in mists of DDT sprayed to control biting flies at Jones Beach, New York. (From *National Geographic* 88 [Sept. 1945], 410; photo by Acme Photo; UPI/CORBIS-BETTMAN)

Risk Reduction or Risk Imposition?

DDT entered a world wracked by insect-borne disease and torn by global warfare. The insecticide appeared to be miraculous in its ability to "sanitize" huge areas of the landscape. All of the mystery confronting Lavaran, Manson, and Ross regarding the complexity of the malarial life cycle suddenly seemed irrelevant. If broadcast widely and regularly, DDT was deadly not only to the mosquitoes carrying the disease but also to many other nuisance insects.

Gorgas in Panama, and Soper in Brazil, each had learned through painstaking efforts that understanding the vector's ecology was essential to controlling the spread of both malaria and yellow fever. Knowledge of the local breeding, feeding, and territorial habits of insects was once the key to pest management, both in agriculture and in malaria control. DDT changed that. Its ease of application by air and low cost made knowledge of local ecol-

Figure 2–16. The caption that accompanied this photo of 1945 read, "Mrs. Gee Goldstein of Brooklyn, shown using the Army's DDT 'bomb.' . . . Her son, Robert, 5, watches from bed and illustrates the fact that the spray is not harmful except to winged pests." (UPI/Bettman)

ogy almost irrelevant. DDT quickly became the atomic bomb of insecticides, but more importantly, it would become a profoundly effective teacher of ecology as we gradually understood its unanticipated fate and effects—a history that unfolds in part 3. The rapid evolution of insect and parasite resistance; the chemical's persistence, mobility and "bioaccumulation" (accumulation in the fats of plants, animals, and humans); and human exposure through the food supply all were unpleasant surprises that resulted from rapidly deploying a new technology with little understanding of the environmental changes it would cause (fig. 2–16).

DDT provided clear and immediate benefits. It gave insecticides a health-promoting image, which in turn shaped the statutes and decision standards chosen to control pesticides since 1947. It has probably saved millions of children's lives, especially in tropical areas where insect-borne diseases still take their worst toll. In many ways, the discussion that follows traces the changing image of pesticides from a technology that reduces risk to one that imposes it. The next chapters explore how and why this transformation evolved, as well as the management problems that result when knowledge of risk emerges long after government has licensed broad use of a hazardous technology.

CHAPTER 3

Resistance A Race Against Time

The surrender of Germany and Japan to Allied forces in 1945 marked the beginning of a new global campaign to fight insects with synthetic chemicals. Military expertise, equipment, and DDT stockpiles were quickly redirected toward insect-borne disease control and crop protection. The urgency of war had led to numerous breakthroughs in biocide development. DDT, discovered in 1939, was quickly followed by benzene hexachloride (BHC) in 1941; the herbicide 2,4-D in 1942; the fungicide zineb in 1943; the herbicide 2,4,5-T in 1944; and the insecticides chlordane in 1945, parathion in 1946, toxaphene in 1947, and dieldrin in 1948.[1] Global markets for these products developed quickly following their invention, and new uses for each active ingredient were continually found. Most of these compounds are now well known to environmental historians because after EPA was formed in 1970 major legal battles were fought to prohibit their use in the United States.

When insecticides were introduced, there was little appreciation of how the variable susceptibility within a species or the rapid evolution of genetic resistance could affect efforts to control insect-borne disease and protect crops from pests. This chapter presents a brief history of how we came to better understand resistance to two forms of biocides—insecticides used to control malaria-carrying *Anopheles* mosquitoes, and drugs used to control parasitic infections in humans. The combination of insect resistance to insecticides and parasitical resistance to drugs now pose grave danger to public health in several parts of the world where malaria incidence is resurging.

The speed of innovation and technology diffusion resulted from an extraordinary research partnership between the federal government and industry. The U.S. government's awkwardness in regulating the industry it was promoting, however, had a profound and lasting influence on the institutions that evolved to control pesticides.

The Role of the World Health Organization

Following the war, the United Nations, newly formed in 1945 with the primary aim of preventing or resolving international conflict, played an important role in distributing pesticide technology to areas of the world where insects posed serious public health risks. Within the United Nations two important technical divisions, the World Health Organization (WHO) and the Food and Agriculture Organization, focused on the potential of pesticides to help them achieve their respective central missions—disease reduction and increased crop yields.[2]

Even before WHO was formally created, a committee of scientific experts on malaria was formed to judge the severity of the insect-borne disease problem and to consider alternative control strategies. By 1947, their enthusiasm for DDT was enormous: "The new synthetic drugs and insecticides have not only reduced the costs of malaria control, they have made it possible to combat the scourge in villages and hamlets, and they raise hopes of the complete eradication of the disease—even of *Anopheles*—from entire countries. . . . It is believed that a decrease in morbidity and mortality from diseases other than malaria may be obtained by the wide use of DDT, just as after the introduction of better water supplies and chlorination a reduction was observed in diseases other than typhoid fever, diarrhea and dysentery."[3]

The United Nations' global war on malaria quickly grew to rely on chlorinated pesticides, especially DDT, and to a lesser extent dieldrin.[4] One of the earliest recognized benefits of DDT was its "residual action"—the long duration of its toxic effect. This feature distinguished it from pesticides formerly used to control malaria. Pyrethrum—which is derived from the crushed heads of chrysanthemum flowers—has a potent "knock-down" effect, but it degrades quickly and therefore requires frequent applications to be effective. Other chlorinated pesticides such as benzene hexachloride (BHC) were also available, but BHC has a shorter duration of effectiveness than DDT, as well as an offensive odor, which made it less desirable for indoor use. From the perspective of those attempting to manage insect-borne disease, the extended residual life of DDT was attractive. This trait, however, led industrial nations to restrict or prohibit its use during the 1970s and 1980s.[5]

WHO termed the period 1946–54 one of "*malaria control*." Control implied disease management rather than eradication, however, and acceptable mortality and morbidity rates were never clearly articulated.[6] The postwar control strategy avoided large-scale aerial spraying of DDT such as had occurred on many South Pacific islands during the war. Surprisingly, WHO experts also argued against broadscale application of DDT to water bodies and standing pools. Instead, they chose to concentrate on spraying building interiors to kill adult mosquitoes, because they believed this strategy would disrupt the cycle of parasite transmission without incurring the much higher costs of broad, aerial spraying.[7]

These choices had important ecological consequences that were then little known to the experts. In particular, primary reliance on indoor spraying slowed the evolution of insect resistance to the chlorinated insecticides.

This tactic, however, also could expose people to high levels of insecticide residues, which could be absorbed through inhalation or contaminated food, water, clothing, or bedding. Evidence of pesticides' persistence, bioaccumulation in food chains, accumulation in human body fat, excretion in human milk, and their estrogenic properties emerged slowly in the late 1940s and early 1950s; meanwhile DDT remained in the U.S. and international marketplaces.[8]

WHO estimated that in 1955, 2 million people died of malaria, while 200 million others contracted it. The disease was declared by WHO to be the "world's greatest single cause of disablement." In response, WHO's expert committee directed increasing attention to the heightened susceptibility of infants and children to the disease. They recognized that in Africa *10–15 percent of the children under age four died* from its direct effects and many more perished from pneumonia, dysentery, and typhus—all of which opportunistically follow malaria in weakened children.[9]

WHO also began calculating the indirect economic effects of the disease, estimating that in 1955 India had lost 130 million person-days of labor, Thailand 9.5 million, and Mexico 4 million. The experts also calculated that huge tracts of rich agricultural land lay uncultivated in Asia, Central America, and equatorial Africa due to endemic malaria. Regional success in reducing disease incidence in Ceylon, Java, and Mexico permitted planting in large areas of previously unused but highly fertile lands.

Meticulous indoor spraying combined with vigilant outdoor larvicidal control resulted in near eradication of the disease in Greece by 1946. A shortage of DDT supplies interrupted the program, but unexpectedly no resurgence was experienced. This encouraged malariologists to discontinue spraying once transmission of the disease had been stopped and raised hopes that eradication could be accomplished during a short, fixed time period.

WHO formally adopted the objective of worldwide malaria eradication in 1955 at its Eighth World Health Assembly in Mexico City. The campaign became known as the "vertical approach to malaria control" and included four phases. First came the planning phase, during which detailed surveys were made, personnel were trained, and strategies formed. Second came the attack itself, during which every home was sprayed with insecticides, usually twice a year, with necessary attention devoted literally to the nooks and crannies of buildings. Third, antimalarial drugs were to be used to eliminate all human reservoirs of the disease through rapid diagnosis and treatment. Fourth and finally, once the parasites and vectors were destroyed vigilance and rapid intervention were prescribed.[10]

The goal of malaria eradication was endorsed by all nation-members of WHO in 1955, and "control programs" were transformed into "eradication campaigns" in the Americas, Europe, and most nations in Asia and Oceania. Africa continued to pose enormous and unique problems to malaria experts, in part due to high levels of endemicity and the absence of basic infrastructure to carry out and monitor the effectiveness of any campaign. Throughout the entire second half of the twentieth century, Africa alone could easily have consumed all international resources devoted to eradication.

The results of the WHO efforts were extraordinary. By 1970, malaria had been eliminated from Europe; the Asian portion of the Soviet Union; some Middle Eastern states; much of North America (including all of the United States); major portions of Mexico and most of the Caribbean nations; the far northern and southern portions of South America; Japan; Singapore; Korea; Taiwan; and Australia. Worldwide, an area inhabited by nearly 700 million people that was malarious in 1950 was, by 1970, virtually free from the disease and was under a condition of "surveillance" by public-health officials. Many lives were likely saved by the effort. In addition, agricultural productivity rose and general public health improved as other insect-borne diseases such as leishmaniasis and plague also were reduced.[11]

Eradication efforts were most successful where the problem was least significant and in countries with thriving economies. During the 1970s malaria rebounded in several South Asian nations, in Latin America, and in Asian portions of Turkey, while eradication was maintained in the United States and throughout Europe. Worse, the global incidence increased 2.5 times between 1973 and 1977, partly due to the deterioration of surveillance programs and partly because of rising energy prices and increasing costs of insecticides.[12]

Other concerns emerged over the effects and costs of a potentially endless control effort. The cost of continual spraying was enormous. F. L. Soper and D. B. Wilson had recognized during the campaign to eradicate malaria from northeastern Brazil in 1939 that it is far easier to secure funds to combat the disease in an area formerly malaria-free than to continually request additional funds to manage chronic incidence.[13] Gradual understanding of the evolution of mosquito resistance to pesticides made WHO planners anxious to achieve eradication quickly.

The Evolution of Insect Resistance

By 1948, the arsenal of insecticides used to combat *Anopheles* mosquitoes included DDT, BHC, and dieldrin: cost, duration of toxic effect, and evidence of pest resistance determined which was used. Mosquitoes initially appeared to develop resistance to one of the two groups of chlorinated pesticides: DDT and methoxychlor, or chlordane, dieldrin, aldrin, and BHC. In the early

1950s the possibility that any one species of *Anopheles* might become resistant to compounds in both groups seemed remote.

DDT became well known as a broad-spectrum insecticide, meaning that it is toxic to many different genera and species of insects. When pesticides of this type are repeatedly sprayed over large areas, there is always the potential that genetically resistant strains will survive and flourish in the absence of competition and predators. As more generations are treated with the same pesticide, the likelihood that resistance will develop increases. Exposing insects to less-than-lethal doses of an insecticide also promotes resistance.

Resistance to DDT was largely unsuspected. It was first formally acknowledged by WHO in 1953 based upon evidence collected from Italy and Greece as early as 1950.[14] The WHO expert committee reviewed the data and was initially highly critical of the quality of analyses, particularly the absence of "a specific measurement of the resistance by laboratory techniques." The experts did, however, acknowledge the disturbing change in behavior among *Anopheles gambiae* in Africa. Instead of remaining on pesticide-treated surfaces long enough to absorb a lethal dose of DDT, *gambiae* would quickly retreat, escaping from open windows or doorways. Because this species had been highly susceptible to DDT, the committee assumed it was finding behavioral evidence of reduced susceptibility.[15]

Under these circumstances, dieldrin was often substituted, because it is more volatile than DDT. This attribute represents both an advantage and a disadvantage. As the compound is released into the air, it can have a lethal fumigating effect on insects that do not land on the treated surface. (BHC is another possible substitute that is initially more potent than DDT or dieldrin and is used for this reason.) However, both dieldrin and BHC have shorter effective life spans due partly to their higher volatility, and therefore they require frequent applications. Both substitutes were also more expensive to purchase than DDT.[16]

The highest cost of insect resistance to pesticides is the increase in human disease that can occur before resistance is recognized and managed. Given the normal inadequacy of sampling efforts and insect susceptibility testing, important increases in disease incidence often occurred before resistance was suspected. Other significant costs were also incurred because resistance is normally followed by a change to compounds that are more expensive to purchase, to apply, and to maintain at effective concentrations. In 1984, for example, compared to DDT, bendiocarb was 7.5 times more expensive to purchase, fenitrothion 5 times, malathion 3 times, permethrin 12 times, pirimiphos-methyl 11 times, and propoxur 19 times the initial cost. Further, each of these alternatives was estimated to be twice as costly as DDT to transport and apply, and each required more frequent applications.

The case of Greece illustrates well the difficulties posed by resistance. There a malaria control program was initiated in 1945 and was based exclusively upon the use of DDT to spray the interior of every house.[17] Within three years, the disease was almost eradicated; reports of *Anopheles sacharovi* resistance to DDT, however, emerged by 1951. Switching to dieldrin or HCH caused DDT resistance to evolve more quickly. This same mosquito was found to have cross-resistance to chlordane and BHC in laboratories, raising additional fears of uncontrollable epidemics.[18]

By 1956, WHO experts were clearly concerned by overwhelming evidence of the diminishing susceptibility of *Anopheles* to insecticides.[19] Anopheline resistance within a malarious area posed a grave danger because at that time there were only two general classes of residual insecticides. Switching from one class to the other constituted the "attack of last resort" and therefore demanded aggressive and intensive eradication efforts. In these cases, the 1956 committee first suggested that organophosphorus compounds such as malathion or diazinon might be tried. Combinations of pesticides across groups were also recommended based on the theory that insects surviving exposure to one class of compound would be killed by the other. The most intensive attack combined the use of drugs and larvicides with the spraying of interior spaces.

Understanding the evolution of insect resistance to pesticides raised an ominous possibility. DDT was so inexpensive, easily produced and transported, and safely applied that alternatives could be too costly or risky to continue the progress made with DDT. The result might be a resurgence of pesticide-resistant strains of *Anopheles* that could reverse steps made toward controlling the malarial cycle. Eradication became a race against time.

In 1960 WHO called insecticide resistance the greatest threat to the future of malaria eradication. By this time entomologists had learned that a species may be resistant only within a limited geographic area while remaining susceptible elsewhere. They also recognized that resistance among insects other than mosquitoes, such as bedbugs, flies, cockroaches, and fleas, could grow into a serious nuisance or even pose other health risks, which in turn could cause a community to resist participation in eradication efforts.

By 1966, a standardized test of insect susceptibility demonstrated that twenty-four important anopheline species had become resistant to one or both of the main groups of chlorinated hydrocarbon insecticides. This finding appeared in several cases to be related to agricultural uses of the insecticides, and resistance to both of the major groups was found in eleven species distributed in sixteen countries. Double resistance—to DDT and the dieldrin/HCH group—was found in El Salvador, Guatemala, Honduras, Iran, Mexico, and Nicaragua, and it was clearly impeding those nations' eradication campaigns.[20]

Throughout the 1960s scientists gradually came to understand that resistance was often regionally specific. Further, the duration of the problem varied with the success or failure of the resistant strain. This tension between the short-term need to control malaria and the long-term struggle to eradicate the *Anopheles* was demonstrated in Iran during the late 1950s. Malaria was controlled in Iran with DDT between 1949 and 1957. DDT resistance was detected shortly thereafter, however, and the resulting explosion in the mosquito population led quickly to a malaria epidemic. Dieldrin was substituted for DDT, but by 1960 the mosquitoes demonstrated resistance to this compound as well. When DDT was reintroduced, its effectiveness rose from 0 percent to 57 percent. Because this rate of control was still not enough to prevent malaria transmission, however, malathion, an organophosphorus compound, was sprayed inside homes, larvae were attacked using both chemicals and biological controls, and antimalarial drugs were broadly distributed. These efforts resulted in a 30–90 percent reduction in the incidence of malaria, but they drastically slowed eradication efforts.

Six countries in Central America—Costa Rica, El Salvador, Guatemala, Honduras, Nicaragua, and Panama—initiated malaria eradication campaigns between 1957 and 1959. These efforts were plagued not only by financial and organizational problems, but by resistance to dieldrin and DDT. Mosquitoes in the cotton- and rice-growing coastal regions extending from southern Mexico to Costa Rica, which were heavily sprayed by pesticides for crop protection, were especially resistant. Because many of the dwellings were not fully enclosed, and because DDT had an irritating rather than fatal effect on these mosquitoes, they simply escaped out of doors, except to feed.[21]

By 1973, 375 species of *Anopheles* mosquitoes had been recognized, but only twelve years later five hundred species were identified.[22] This new information raised the enormous management problem of identifying the most important vectors and focusing eradication and control efforts accordingly. By 1985, fifty species of *Anopheles* were reported to be resistant to one or more insecticides, nearly double the number reported in 1966. It is even possible that some species were resistant before they had been identified by entomologists.

Cross-resistance to different classes of pesticides is a potentially devastating problem for malaria control. Cases of resistance to multiple chemicals normally appeared in or adjacent to agricultural areas where significant quantities of pesticides were applied from the air. This method of application creates tremendous evolutionary pressure because larvae are exposed in wet breeding areas while adults are exposed on plant surfaces where they rest.[23] As a result, resistance had spread to all major classes of insecticide by 1986: organochlorines, organophosphorus compounds, carbamates, and pyrethroids.

Fortunately, this lack of susceptibility appears to have occurred only within these well-defined areas.[24]

By 1992, WHO restated its concern over the seemingly intractable problems in Africa, where the three main mosquito vectors—*Anopheles arabiensis, A. gambiae, and A. funestus*—had all developed broadscale resistance to dieldrin and HCH. Additionally, *A. gambiae* had developed a narrower resistance to DDT. The absence of broader resistance to DDT is most likely due to its limited use; malaria control in Africa relied far more heavily on the therapeutic or prophylactic use of drugs than on the more expensive strategy of repeatedly applying pesticides indoors.

Meanwhile, in the Americas, four important malaria-carrying species were found to have developed the ability to endure exposure to one or more pesticides, again in areas where insecticides were heavily used for agriculture. In the eastern Mediterranean region, malaria is still the most important insect-borne disease, and of eighteen malaria-carrying species in the region, fifteen have developed resistance to one or more insecticides.

In Southeast Asia as well, many countries used DDT extensively during the 1950s and 1960s, resulting in widespread resistance. There are nineteen anopheline species that are either primary or secondary vectors of malaria in southeast Asia. Organophosphorus insecticides were often substituted, but they also induced resistance; further, their unpleasant odor discouraged people from using them. The costs of alternatives such as malathion were six to seven times that of DDT, which meant that effective doses were seldom applied. Thus insect resistance, cost, and the unpopularity of specific insecticides all appear to have contributed to an accelerated incidence of disease.

Increasing resistance quickly translated into greater numbers of malaria cases within endemic areas. In Southeast Asia, extensive spraying with organochlorine compounds was followed by an increase in incidence from one hundred thousand cases in 1969 to 7.2 million cases in 1976. In the Azerbaijan lowlands, malaria had nearly been eradicated in the 1960s, but early in the 1970s it began to reappear—coinciding with new evidence of DDT resistance among the key vectors, particularly *A. sacharovi*. In India, *A. culicifacies,* an important malaria vector, is now resistant to DDT in an area inhabited by 262 million people; is no longer susceptible to DDT and HCH in an area with 125 million people; and is unaffected by DDT, HCH, and malathion in an area with 7 million people.[25]

The control and eradication campaigns of WHO revealed that the causes of malaria were complex, with highly significant variation among regions, and even within endemic areas over time. The clearing of land for settlement, agriculture, and forestry was found capable of promoting mosquito reproduction because it created stagnant pools of surface water, especially during rainy seasons. The conversion of tropical forests to pasture in the Amazonian state of Rondonia, Brazil, precipitated malaria outbreaks between 1970 and 1995. And development projects designed to increase agricultural productivity often have involved the building of roads, irrigation and drainage systems, and dams—each of which may provide standing pools of water perfect for incubating mosquito larvae. Thus, poorly conceived economic development projects have paradoxically hindered economic progress and settlement.

By 1983, almost 5 billion pounds of DDT had been sprayed worldwide. Nearly 80 percent of this amount had been used in agriculture, and the remaining 20 percent had been applied to control insect-borne diseases.[26] Despite the efforts of WHO to constrain application to the interior of buildings, nearly 4 billion pounds had been sprayed outdoors, an act that had unquestionably accelerated the pace of evolving pest resistance. By 1959, seven *Anopheles* species were resistant to DDT and eight to dieldrin; by 1962, nine species were resistant to DDT and twenty-six to dieldrin; and in 1969, the numbers increased to fifteen and thirty-seven, respectively. As of 1994, fifty species of *Anopheles* had demonstrated some resistance to DDT.[27]

Of all insecticides produced in the world, 90 percent are used to protect crops. Among all crops, cotton and high-yield rice demand the largest proportion of pesticides used in world agriculture. Cotton deserves special attention, because it accounts for 90 percent of insecticide use in some countries. By 1964 in the United States, cotton accounted for more than half of total insecticide use, and many of these pesticides are either highly toxic or highly persistent. This one crop in 1968 accounted for 80 percent of total methyl parathion use, 86 percent of endrin use, 70 percent of DDT use, and 69 percent of toxaphene use.[28] All of these compounds have now been banned or severely restricted for agricultural uses in the United States but are commonly used in the tropics.

Rice cultivation poses unique problems for managing mosquito-borne disease because flooded fields provide an excellent breeding habitat. For this reason, fields are often treated with larvicides in addition to compounds that protect the rice. Combining applications in this way, however, increases the probability that resistant insects will evolve. This trend may also be exacerbated when breeding occurs in water contaminated by pesticide-laden runoff from the fields—or when insects land on plants that have been treated with pesticides. In either case, those mosquitoes that happen to be slightly more resistant to the pesticides have greater opportunities than others to breed and flourish.

In Japan extensive use of insecticides to protect rice crops began in 1966. This strategy controlled *Culex tritaeniorhynchus* until the late 1970s, when this species developed resistance to all agricultural insecticides. This effort may also have translated to greater resistance of malaria-ridden pests. In East Asia, *Anopheles subpictus* and *A. nigerrimus,* which both normally live outdoors, developed resistance to numerous insecticides, indicating that the cause was most likely agricultural use.[29]

The case of Sri Lanka adds a new twist to the story of resistance to insecticides. Sri Lankan rice paddies provide breeding sites for *A. subpictus* and *A. nigerrimus,* both of which frequently developed broad-spectrum cross-resistance. *A. subpictus,* for example, has become highly resistant to DDT, malathion, and most other pesticides used extensively in agriculture. This means that they commonly survive exposure to numerous types of insecticides applied to protect rice crops. The most curious aspect of this development is that some pesticides such as malathion and fenitrothion were for years reserved for use as weapons of last resort to fight epidemics. Heavy application of other pesticides surprisingly induced mosquito resistance to malathion and fenitrothion. All are toxic via the same biochemical mechanism: cholinesterase inhibition—the disruption of normal signal transfer across nerve cells.[30]

Insect Resistance Lessons

The history of evolving insect resistance to pesticides has taught a us a number of extremely important lessons.[31] The primary concept that underlies most current recommendations is that pesticides should be thought of as a scarce resource. Their use tends to induce vector resistance, decreasing management options. It is now clear that pesticides that induce narrow-spectrum resistance should be used before those such as DDT that are capable of causing broad-spectrum resistance. Further, applying a pesticide repeatedly against the same species in the same area can result in more rapid evolution of resistance.

Before a pesticide is chosen, it is also important to consider what other agricultural chemicals are being used in the target area. If similar pesticides are used for controlling insect-borne diseases and agricultural pests, the likelihood of insect resistance increases. Residual pesticides that degrade slowly are more likely to induce insect resistance than pesticides that break down quickly. The residual effect makes these pesticides desirable for combating chronic problems such as continual invasion by mosquitoes. Broad landscape application should be avoided. Instead, spraying should be limited to areas known to be the feeding, resting, or reproductive habitats of the offensive species.

Rotating the use of different pesticides can also help delay or avoid the evolution of insect resistance. The theory underlying this strategy is that if two pesticides are used sequentially, resistance to the first will fall while the second is being used. Biological control agents such as *Bacillus thuringiensis* (*B.t.*) are toxic to a wide variety of agricultural pests, including many species of mosquito and blackfly larvae. Mosquitoes and blackflies that are resistant to synthetic insecticides tend not to be resistant to *B.t.*, which makes it an attractive compound for rotation among synthetic pesticides.[32]

Finally, manipulation of the environment to make it less suitable for insect breeding and disease transmission can reduce populations of insects and the incidence of disease transmission. This tactic in turn reduces the need for insecticides and slows the development of resistance.[33]

Although these suggestions seem obvious now, concerns over resistance to insecticides played little role in malaria management prior to 1960. Continuing to assume that new pesticides will be developed to replace those that are no longer effective is clearly a deadly gamble.

Parasite Resistance to Antimalarial Drugs

By 1995, at least one of the four species of the *Plasmodium* parasite lived in the blood of nearly 300 million people. Drugs that kill the parasite, or prevent its reproduction, have played an important role in controlling rates of malaria transmission. In addition to their more obvious effects of relieving suffering and preventing deaths, these drugs may also play a more subtle but important role in reducing the reservoir for the disease in the human community.

Before it was widely understood that malaria is transmitted by mosquitoes, medical attention was directed toward treating the illness and finding a cure. A drug to control the fevers was known to indigenous South Americans in the early 1600s. Made from the powdered bark of the cinchona tree, its active ingredient was unknown until 1820, when Joseph Pelletier, a Parisian apothecary, and Joseph Caventou, a twenty-two-year-old professor of toxicology at the Ecole de Pharmacie, collaborated to test extracts from gray-barked and yellow-barked cinchona and isolated the alkaloids quinine and cinchonine. Quinine directly attacks *Plasmodium* in red blood cells; recurrence of fever, however, commonly resulted from an inadequate dose, which in turn was caused either by a poor understanding of the relative abundance of quinine in various species of cinchona or by deceptive marketing practices.[34] In 1850, the Society of Pharmacy of Paris offered four thousand francs to the first chemist to synthesize quinine, yet its artificial production eluded researchers until 1944.[35]

The first reports of resistance to quinine were reported in 1910, but these were rather isolated. Pressure to develop new synthetic antimalarial drugs evolved more from a fear of losing access to supplies of quinine than from any concern over resistance.[36] The initial absence of a synthetic source of quinine created a great demand for cinchona bark, which, as mentioned, prompted the Dutch to establish extensive plantations for its cultivation in the East Indies, especially in Java. Later, atabrine was developed by the Germans during World War I, and chloroquine was synthesized as a result of Allied research efforts during World War II. Chloroquine was inexpensive and effective both in treating infections and as a prophylactic agent against all four species of *Plasmodium*.[37]

In 1960, WHO announced that some malaria parasites, namely *Plasmodium falciparum,* had become resistant to chloroquine, which had become the most commonly prescribed antimalarial drug of the time. Resistance was soon detected in Brazil, Venezuela, and Colombia.[38] A diminished susceptibility to chloroquine was first suspected in Thailand during 1957 near the Cambodian border and in West Malaysia in 1962, when military personnel stationed at the Thai-Malayan border relapsed following normal treatment with chloroquine. As U.S. involvement in Vietnam increased during the 1960s, military leaders became increasingly interested in the drug resistance problem. During this period, nearly 50 percent of the U.S. troops inflicted with *P. falciparum* experienced a recurrence of the disease, despite being treated with chloroquine. Just as insect resistance to chlorinated hydrocarbon insecticides motivated rapid innovation in the pesticide industry, *Plasmodia* resistance to chloroquine motivated the search for alternative antimalarial drugs.

By 1967, the WHO Expert Committee on Malaria seemed cautious but still optimistic about control efforts. They suggested that although *P. falciparum* may be resistant to chloroquine, increasing the dosage or the frequency of its administration could solve the problem. Also, they explained, resistance might be geographically specific as it is with pesticides, and they optimistically recognized that the other three species of *Plasmodium* appeared to remain vulnerable to the drug. The committee criticized questionable evidence of parasitic resistance to drugs in Africa and cited results of WHO-sponsored trials in Upper Volta, Liberia, and Uganda that found normal response to chloroquine. Finally, they noted that following twelve years of prophylactic administration to more than a million children in Madagascar no evidence of resistance had emerged. Still, the experts alluded to the worst possible nightmare: a lethal combination of pesticide-resistant mosquitoes and drug-resistant parasites.[39]

Chloroquine resistance spread through Latin America during the 1960s and 1970s, but Africa continued to pose the most difficult and intractable problems.[40] Chloroquine resistance in Africa was not identified until 1979, when a tourist contracted a resistant strain in Kenya. Other visitors to East Africa, where resistance was unanticipated, have contracted a resistant strain of the parasite. Resistance has spread rapidly since 1979 due to the common prophylactic use of antimalarial drugs and the general absence of insect-control programs.[41]

When a parasite can survive concentrations of the drug at the maximum dose tolerated by the host, it is considered "drug resistant." Thus effective treatment requires that the drug must reach the parasite at a lethal concentration, an outcome that depends on a variety of "pharmacokinetic" factors such as human drug absorption, distribution, metabolism, and elimination—as well as the ability of the drug to permeate the parasite's external membrane. Complete cure relies upon the delivery of an effective concentration maintained for a specific time period. Yet even at effective doses, drugs are rarely lethal to all of the parasites, and the immune system's response completes the cure. Moreover, an infected individual may eliminate even drug-resistant strains of the parasite as a result of a highly efficient immunological response. By contrast, the spread of drug-resistant strains of the parasite appears to be overwhelming the immune response even among the most healthy, resulting in a resurgence of infections regionally.[42]

Just as *Anopheles* mosquitoes have become "cross-resistant" to structurally similar insecticides so too have *Plasmodia* parasites become less sensitive to structurally similar drugs.[43] Drug prescriptions for parasitic infections normally include two components: the concentration of the drug, which is normally based upon the weight or body-surface area of the patient, and the duration of treatment. A mistake in either part can result in the drug's failure and provide an environment that selects for resistant parasite strains. These conditions can result from errors in prescribed dosages, patient failure to take medicine as prescribed, and biochemical variance among patients, all of which influence the concentration of the drug that reaches the parasite.

Mass prophylactic drug administration exerts enormous selection pressure favoring resistant strains of the parasite. Historic examples include the mass administration of medicated salts—especially pyrimethaminized salt, which rapidly produced resistant strains of *P. falciparum*. And chloroquinized salts, used over longer periods of time, contributed to the first cases of chloroquine-resistant *P. falciparum*.[44] In Tanzania, where intense malaria transmission is especially serious for children, mass drug administration to children under age ten produced resistance within three years—a situation that required increases in dosage and a lengthening of administration interval, both of which were limited by human tolerance to toxic effects. Self-treatment and compliance failure, especially in rural areas with little health-care infrastructure or basic understanding of principles of modern medicine, result in lower than effective concentrations and rapidly evolving resistant strains of the parasite.[45]

Resistance is also encouraged in areas of severe epidemic malaria after drug therapy has been discontinued. For example, following drug treatment an individual may be reinfected by a mosquito while drugs with long half-lives remain in the blood stream in diminishing concentrations. If these concentrations are less than levels necessary to fight off the new invasion, selection of resistant parasites is favored. This effect might explain the rapid evolution of *P. falciparum* resistance to sulphadoxine-pyrimethamine in Kenya. It also has been offered as a reason for rapid outbreak of mefloquine resistance along the Thailand-Cambodian and Thailand-Myanmar borders.[46] This situation demonstrates the need for rapidly acting drugs that also have short half-lives.

Once resistance evolves, how it spreads becomes a fundamental concern. In Brazil, the distribution of resistance appears to be related to the natural and human-created habitats of the primary vector *Anopheles darlingi,* especially within the Amazon River basin. Malaria incidence in Brazil grew dramatically during the last decades of the twentieth century—from fifty thousand cases in 1970 to six hundred thousand in 1990—with the latter number representing 10 percent of the world's cases outside of Africa.[47] The reasons for this outbreak are complex, but they include the rapid colonization and deforestation of the Amazon region and inappropriate housing for workers engaged in agriculture, forestry, or mineral extraction within malarious regions.

The migration of chloroquine-resistant malaria was more rapid in tropical Africa than in South America. In Cameroon, for example, *P. falciparum* resistance was first reported in 1985, and only eight months later this strain was responsible for 86 percent of childhood cases of malaria.[48] By 1991, resistant strains of the parasite were reported in every African nation, and many experts believe that the mass administration of drugs common in many of these countries during the 1970s and 1980s played at least a contributing role. For example, a reduction in the price of chloroquine in Madagascar in 1987 led to its widespread and uncontrolled use, which might explain the rapid rise in chloroquine-resistant *P. falciparum* from 10 percent to 30 percent of the parasites. In 1988, one of the most severe epidemics of the decade occurred in that country, claiming nearly twenty-five thousand lives. By 1990 between 25 percent and 60 percent of children in Africa may have become resistant to chloroquine. Nearly 90 percent of recently recorded cases of malaria in Madagascar were caused by *P. falciparum,* yet this strain remains quite sensitive to chloroquine, suggesting that the island's biogeography might have a protective effect. It has often been unclear whether resistance evolves locally or is imported by infected mosquitoes or humans. And this confusion has important implications for disease management.[49]

Immunity is crucial to the incidence of malaria, and it varies by age. On average, children under age five experience from four to nine attacks per year, as compared with one to two attacks per adult.[50] Further, children usually have malaria for about five days, whereas adults on average recover after only 3.5 days. Luckily, maternal antibodies protect infants during roughly the first three months of life; afterward, however, immunity diminishes dramatically. Infections in infants older than three months are often severe and fatal. Those who survive enjoy some immunity, but usually specific to that strain of *Plasmodium* parasite.[51] In humid sub-Saharan Africa, those older than age five have generally developed enough immunity to make their experience of malaria much less severe than that which afflicts native infants and toddlers, or migrants and visitors.

Drugs can also play an important role in the development of immunity to malaria. Some drugs such as chloroquine suppress the immune system, and early studies suggested that prophylactic use of antimalarial drugs in children caused later infections to be more severe. Later studies, however, found that if children did not receive drugs in a preventive way they would have a more severe bout of malaria if infected. These effects have been seen in adults as well.[52]

In the drier savanna regions to the north and south of humid tropical areas, malaria tends to be more prevalent during the wet season, a trend that may produce corresponding increases in immunity (although this relationship is not fully understood). The seasonality of the disease, for example, may explain why generalized immunity is seldom acquired until age ten. Further, drought tends to diminish immunity, because there are fewer mosquitoes—and therefore fewer insect bites—during these times. (The intense epidemics that followed the extended droughts in the Sudan and Mali during 1986 are cases where decreased immunity combined with a falloff in prophylactic drug use to increase the population's susceptibility.) The number of bites per day may range between less than one during the dry season to an average of 9.5 during the wet season. In rural African areas where homes are unscreened and bed nets are not used, a person can be bitten hundreds of times each year from infected mosquitoes.[53]

Cross-Resistance and Drug Choice

A parasite's resistance to numerous drugs poses a grave danger to malaria control. Within many endemic areas, parasites have developed cross-resistance to several compounds that have been given prophylactically. The combination of sulfadoxine and pyrimethamine, for example, proved initially to be a cure for malaria and to help prevent infections by *P. falciparum*-resistant malaria. Resistance to this combination, however, was first reported in Thailand in 1978, with a decline in cure rates to 32–42 percent. Resistance to this particular combination has also been reported in Brazil, Kenya, and Tanzania.

The rapid evolution of resistant strains of *Plasmodium,* especially those no longer susceptible to pyrimethamine, proguanil, and chloroquine, has complicated the process of choosing drugs for prophylactic treatment. The primary problem is that doses now needed to kill the parasite are toxic to humans. Chloroquine, for example, accumulates in the body and is implicated in damage to the retina, so it is not recommended for use beyond three years without an ophthalmological examination.

Maloprim and mefloquine are among the more effective alternatives to chloroquine, but they are also among the most toxic. Four groups appear to be at special risk from these drugs: (1) pregnant women living in malarious regions; (2) children younger than age five living in these areas; (3) those moving into malarious regions from non-malarious regions—for example, migrants or military personnel; and (4) travelers who visit or move to malarious parts of the world. The choice of prophylactic drug will also depend upon the region of interest. For example, chloroquine is still effective in Mexico, while mefloquine is recommended in Brazil.[54] "Standby treatment"—using drugs only at the onset of fever—may be preferable to prophylaxis for people especially susceptible to adverse drug effects.

WHO originally intended that mefloquine be reserved for treating multiresistant *falciparum* malaria. But it is now used prophylactically in East Africa, the Amazon, and Southeast Asia because parasites in these areas have become less vulnerable to other drugs. Mefloquine is controversial; it may cause neuropsychiatric complications such as dizziness, and it may have contributed to birth defects in Thailand. Moreover, because mefloquine, quinine, and quinidine are pharmacologically similar, their interaction can potentially cause cumulative toxicity.

Despite these concerns, especially susceptible populations—such as very young children living in areas that have high rates of transmission and those without access to modern health care—are still thought to benefit from receiving preventive antimalarial drugs. In one study, for example, prophylactic drug treatment saved more Gambian children than did malaria treatment. In this study, the children who died had an average duration of illness of only three days, demonstrating the need for rapid diagnosis and proper treatment. Yet concerns remain over the side effects of these drugs and the likelihood that under-dosages will result in resistant parasites. Only an average of 30–60 percent of patients take the drugs as prescribed. Reasons for this noncompliance include the bitter taste of the drugs, side effects such as nausea or dizziness, and general concerns about the effects of long-term treatment.[55]

A number of other alternatives to chloroquine pose similar problems. For example, pregnant women are commonly discouraged from using mefloquine, which is commonly prescribed for *P. vivax* and some *P. falciparum* infections that are resistant to chloroquine and sulfadoxine-pyrimethamine. Doxycycline is especially effective against strains of the parasite that have evolved in Thailand (where sulfadoxine-pyrimethamine is no longer effective), but it is contraindicated for pregnant women and children under age eight. Sulfadoxine-pyrimethamine has been used with chloroquine where *P. vivax* and *P. falciparum* exist together. Concerns over its safety have discouraged its use by pregnant women, newborns, and breastfeeding mothers.[56]

In 1985, resistance of *P. falciparum* to most antimalarial drugs had become the most significant technical problem in controlling the disease. Chloroquine had been both effective and inexpensive, but the cost of its alternatives ranged from amodiaquine, which is 2.3 times more expensive; sulfadoxine/pyrimethamine (4 times higher in cost), quinine alone (15.5 times higher), quinine and tetracycline (19.8 times as expensive); and mefloquine plus sulfadoxine/pyrimethamine, which costs an astonishing 33.3 times that of chloroquine.[57]

Vaccines

Optimism about the potential of vaccines to protect populations against malaria was high even during the 1940s.[58] Yet the obstacles to a safe and effective vaccine have proven enormous. Most often these stumbling blocks are related to the complexity of the parasite's life cycle. In the past decade, scientists have demonstrated that successful vaccines complement the human immunological response to the parasite at each stage of its life.[59] For example, once injected by the mosquito into the bloodstream, the parasites, or sporozoites, are vulnerable outside the liver for less than sixty minutes. Once in the liver, they remain for at least five days, where they are vulnerable to attack by the immune system's T cells. Following their release from the liver, the merozoites are free for only a short time before they attach to or invade erythrocytes. Thus effective vaccines focus on the stages in which the parasite is most vulnerable and help arm appropriate immune mechanisms for attack.[60]

Research during the early and mid-1990s revealed that the most deadly parasite species, *P. falciparum,* has an extraordinary ability to dodge and deceive the human immune system. Researchers trying to understand how *P. falciparum* is able to resist chloroquine discovered how human antibodies—or T cells—are tricked by the parasites' ability to vary proteins often enough and quickly enough to stay ahead of human immune defenses. The parasite may vary a large set of genes—known as "var genes" for their variability—to produce millions of antigenic proteins. Red blood cells, damaged by hemoglobin-consuming parasites, would normally be transported to the spleen for destruction as the body attempts to restore normalcy. Any parasites feasting on the inside of the red blood cells would also be destroyed by the spleen. The parasite impedes this process by growing protein knobs that allow the infected cell to attach to the interior surface of blood vessels, preventing their transport to the spleen. Each parasite has between 50 and 150 different var genes that are rearranged in various combinations. As the parasite reproduces in a petri dish, for example, with each division it changes the combination of var genes by roughly 2 percent. By the time the victim's immune system recognizes and attacks one form of protein anchor, the parasite has varied its genes to create a new form of adhesive that is again temporarily unrecognized by the immune system. Researchers have recently concentrated on understanding how the parasite recombines its genes, in order to discover a way of preventing it from doing so.[61]

This resilience may explain variance in the success of a recently synthesized vaccine, SPf66. Used in trials in Tanzania, it reduced the incidence of malaria by 31 percent in vaccinated children.[62] The same vaccine was used in an area of Ecuador where malaria is endemic; there it was found to be safe, to induce anti-SPf66 antibodies, and to produce a protective effect in 66.8 percent of those vaccinated.[63] Similarly, in a malaria-endemic area of Venezuela, SPf66 was found to be 55 percent effective against *P. falciparum* and 41 percent against *P. vivax*.[64] When tested in Colombia, among those vaccinated it reduced the incidence of *P. falciparum* malaria by 10 percent, compared with controls.[65] SPf66 offered almost no additional protection to Gambian children ages six to eleven months.[66] Although these effects are less than hoped for, in areas where inoculation rates are high even minor levels of protection may save many lives and significantly reduce suffering.

Late-Twentieth-Century Incidence

In 1991, WHO estimated that 10 percent of the world's population—primarily those living in tropical Africa, where national antimalarial programs had never been attempted—lived in conditions where malaria continued to rage. An additional 1.6 billion people lived in areas where control or eradication was attempted but transmission resurged, especially during the 1970s

—and where the conditions by 1991 were either unstable or deteriorating. Approximately 1.65 billion others inhabited parts of the world where malaria was once endemic but had disappeared following control efforts. These areas remain malaria-free. Finally, approximately 1.4 billion live where malaria has either disappeared without intervention or never existed.[67]

The early success of WHO eradication efforts, especially in the temperate part of the world, was striking. By 1970 nearly 727 million people—predominantly those living in Europe, North America, the Asian portion of the former Soviet Union, Japan, Australia, portions of China, Korea, Singapore, and Taiwan—had been freed from the risk of malaria. These efforts probably saved tens of millions of lives that would have been lost to malaria and other insect-borne diseases, especially plague and leishmaniasis. Elsewhere the intertwined problems of insect resistance, parasite resistance, and the lack of continuous funding led to the transformation of eradication campaigns into campaigns to simply "control" or manage the disease.

Although the temperate world has been relieved of malaria, by the middle of the 1990s problems appeared to be unstable or worsening in three areas of the world: tropical Africa, Latin America (especially Brazil), and Southeast Asia. Sub-Saharan Africa suffers enormously, experiencing the majority of the world's morbidity and mortality. WHO avoids providing statistics on malaria incidence in Africa by citing the insufficiency and irregularity of reporting.[68]

Tropical Africa has always posed the most intractable problems for those trying to manage the human suffering associated with malaria. Nearly a half billion people now live within a region with the highest level of endemic malaria in the world. Over 250 million cases are estimated to occur each year, and 90 percent of these are caused by *P. falciparum,* the most deadly of the parasite species. More than 1 million children under age fourteen died of the disease in tropical Africa as late as 1995. Contributing factors have included the chronic inadequacy of health care and the unwillingness of wealthier nations and donor organizations to assist with distributing and managing pesticides and drugs.

In a review of recent literature on malaria epidemiology in Africa, case estimates were found ranging between 35 million and 189 million per year. Actual cases are known to exceed reported figures. For example, only 8 percent of people with malaria in Guinea, 21 percent in Togo, and 41 percent in Rwanda reported to a health-care center. Mortality statistics are confusing as well, because most nations do not require death certificates and many patients never make it to the hospital. Further, malaria commonly induces other illnesses—such as pneumonia—which are listed as the final cause of death.[69]

Use of health centers in Ghana dropped 85 percent following a 1986 change in policy that required a fee for service. In addition, greater access to drugs tends to encourage patients to pursue self-treatment, or treatment by traditional healers, and to avoid modern medicine. Despite these influences, however, malaria is still the primary reason that most Africans seek health care and a dominant cause of childhood mortality. Mortality rates during the early 1990s were approximately five per thousand each year, with annual growth rates of 7.3 percent for Zambia, 10.4 percent for Togo, 11 percent for Burkina Faso, and 21 percent for Rwanda. Zambian mortality rates for children were increasing at a rate of 5.2 percent per year.

North of the Sahara, the incidence of malaria is now quite low due to efficient control efforts in Algeria, Libya, and Morocco, where the dominant vector, *A. labranchiae,* remains susceptible to DDT. The completion of the trans-Saharan highway linking Northern Africa to highly malarious Western Africa, however, significantly increases the risk that resistant species of *Anopheles* and strains of *Plasmodia* will be dispersed.

Managing Surprise

In this chapter I have concentrated on the surprises of insect and parasite resistance to biocides and suggested that this resistance contributed to the persistence and evolution of the disease, especially in tropical parts of the world. But this is too simple a conclusion. Resistance emerged within a constellation of enormously complex social, medical, and ecological problems, many of which were poorly understood during the middle of the twentieth century. Among these problems, the limited knowledge of entomology, parasitology, and human immunology played a decisive role.

The twin problems of choosing appropriate pesticides and antimalarial drugs bear an uncanny resemblance. As resistance evolves, the choice of an appropriate substitute is complicated by other traits of the biocidal agent, such as persistence, toxicity, cost, and relative effectiveness. For both pesticides and drugs, experts worry that cross-resistance among chemically similar biocides might cause uncontrollable epidemics. Because of their concern, there have been efforts to reserve certain classes of pesticides and drugs for crises.

Malaria incidence and mortality continued unabated throughout the twentieth century because the relationships between the disease and settlement practices, housing technology, and rural development projects were continually misunderstood or disregarded. Agriculture, forestry, and mining tended to promote disease incidence and distribution, and mass migration, famine, and warfare have all been causally intertwined with malaria dispersion, especially in Africa.[70] Finally, the existence and quality of community health-care resources have proven crucial for the effective management of both epidemic and endemic malaria. By 1978, WHO based its malaria-control strategy on primary health care—a plan that depends on the development of basic infrastructures and therefore has achieved limited success in those rural areas with the highest incidence of the disease.[71] By articulating these problems I do not mean to criticize those who fought the disease, but rather to simply recognize malaria's extraordinary resilience, complexity, and variability across regions of the world.

To tell an adequate history of malaria would demand at least separate chapters on each of these subjects, an impossibility here. Instead I have chosen to explore the relationships between malaria incidence and pesticides. The modern pesticide industry, born in the 1940s, emerged into a world of immense suffering, and pesticide innovation during that period gave new hope that eradication was possible. But optimism quickly turned to disappointment as nature taught a harsh lesson on the evolution of biological resistance, one that appears to have created an addictive and perhaps endless demand for new biocides.

Misunderstanding the potential for parasites to rapidly evolve new methods for evading attack by the human immune system, or by antimalarial drugs, will likely increase world reliance on pesticides as the dominant vector-control strategy. This could cause a shift in disease management in those parts of the world—especially Africa—that have relied heavily on prophylactic drugs to combat parasites, rather than on the more expensive and hazardous approach of applying insecticides. It could simply justify the most deadly policy of all—complete neglect.

PART 2

Evolving Law

CHAPTER 4

Beyond Control Pesticide Law Before 1972

Debates over the health effects of pesticide residues date back to at least the turn of the century. The most commonly used pesticides in the early 1900s were such metals as copper acetoarsenite (Paris green) and lead arsenate, both of which are highly toxic to humans and persistent but relatively inexpensive. Farmers protested the absence of standards for purity, concentration, and labeling and demanded that the government protect their economic interests through regulation.[1] Congress responded in 1910 by passing the Insecticide Act, the first U.S. statute to govern pesticide use. The law was designed primarily to protect farmers from dangerous and ineffective products, and it demanded standards for the purity of each arsenical.[2] Still, the statute neither required product registration nor offered guidelines for protecting public health or the environment.[3]

Lead arsenate and calcium arsenate (which was introduced in 1917) were the pesticides most commonly used between 1917 and 1942, although their acute toxicity and environmental persistence were clearly a problem for farmers. As late as 1927, the chronic effects of lead and arsenic exposure from residues on food still had not been evaluated by the government.[4] Without cost-effective and less toxic alternatives, however, concerns over improvements in agricultural productivity dominated worries about public health.

Two new fluoride-based insecticides, cryolite and barium fluosilicate, seemed to offer a better alternative. Discovered in 1930, they appeared to have a chronic toxicity nearly twenty times lower than lead arsenate. These compounds were commonly used in 1933 by western fruit farmers, who were looking for new ways to avoid conflict with regulators over the levels of lead and arsenic residues on their crops.[5] Yet the anticipated broad substitution of the fluorine pesticides for lead arsenate never occurred; instead, both classes of pesticide created separate storms of controversy through the 1930s. Evidence gradually emerged that fluoride could mottle children's teeth even at residue levels as low as 1 part per million (ppm), and residues on fruits were common in the 5 ppm range. A tolerance was thus initially set at a level so low as to preclude use of the fluorides, but the secretary of agriculture quickly became the target of a letter-writing campaign by hundreds of outraged farmers.

The 1938 seizure by the federal government of a large shipment of California grapes that contained fluoride residues three times higher than the allowable level further inflamed the farmer protesters. Contributing to the other side of the debate, however, were two toxicologists from the Arizona Agricultural Experiment Station, Margaret and Howard Smith, who argued that because children's teeth develop more slowly than rats, children should be considered ten times more susceptible to fluorine poisoning than the lab animals.[6] The federal government responded to both pressures by doubling the allowable contamination limit. Farmers remained dissatisfied, while health officials believed that the limit was too high.[7]

Lead Tolerance

Efforts to set a lead contamination standard were discouraged until 1933 when a new lead-detection method became available. The first lead tolerance was set at 0.025 grains per pound in 1933—nearly two times the level suggested by an expert committee in 1926. Within five weeks, however, it was lowered to 0.014 by a new assistant secretary of agriculture known for his strict construction of the Food and Drugs Act. The tolerance reduction provoked an outcry from farmers. Within two months, it was again reset to 0.02, where it remained until 1938—when it was raised back to its original level of 0.025 grains per pound.

These reversals in regulation were driven partly by uncertainty about the toxic effects of lead in humans. Whereas the acute effects of both arsenic and lead were well recognized, their chronic effects were never carefully examined despite nearly half a century of federal approval of lead and arsenic residues in the food supply. The first thorough analyses of the chronic effects of lead on animals were planned by FDA in 1935. Initial results demonstrated chronic physiological damage in both rats and dogs, which caused farmers to demand "practical" human evidence rather than the "theoretical" evidence produced from laboratory animals. Soon, Depression-induced budgetary pressure was used as an excuse to prematurely abandon the studies.[8]

The Public Health Service finally initiated a study of 1,231 apple growers, packers, and their families near Wenatchee, Washington, in 1938 to estimate the effects of their chronic lead exposure. The study concluded that only seven workers manifested symptoms related to chronic exposure and absorption of lead arsenate, which was then termed "mimimal lead arsenate intoxication."[9]

The absence of a rapid detection method, combined with the lack of a reliable test for subclinical damage, frustrated attempts to set a tolerance for lead in food. Although the animal data demonstrated organ damage, human evidence of toxicity existed only for acute, sub-acute, and advanced chronic toxicity, which all resulted from high or moderate levels of exposure. The ability to identify damage, especially subtle neurological damage, from chronic, low-level exposure did not emerge until the development of sensitive cognitive testing techniques nearly forty years later.[10]

In 1940, the lead tolerance was increased to 0.05 grains per pound and the arsenic tolerance was increased to 0.025 grains per pound. Although the Food, Drug, and Cosmetic Act required a hearing prior to the adoption of a tolerance, the hearing was not held until 1950. The delay was justified first by World War II, and later by the wave of new chlorinated insecticides that rapidly replaced lead and arsenic—but posed additional questions concerning environmental fate and toxicity.[11]

The government's mismanagement of lead foreshadowed many later conflicts that dominated the attention of regulators during the second half of the twentieth century. Increased understanding of the toxicity and persistence of one class of pesticide had created a demand for less risky substitutes. But just as with current debates over these issues, greater understanding seldom emerged all at once or with great certainty. Agricultural interest groups therefore had ample opportunity to argue that allowable contamination levels should be raised, reducing the chance that their crops would be found to have excessive residues. The politics of tolerance-setting has been governed by the demands of growers—and later of chemical manufacturers—who have argued that crop yields, market shares, and even the stability of rural communities are threatened by "hypothetical" threats to human health. By contrast, health and environmental advocates have relied on advances in analytical detection technology, toxicology, and ecotoxicology to devise arguments favoring precautionary regulation.

Many believe that the American public and policymakers have a short attention span for technically complex problems.[12] The way in which Congress considered pesticides prior to World War II supports this view. Although two hearings were held prior to the passage of the 1910 Insecticide Act,[13] only two additional hearings were specifically devoted to pesticides between 1910 and 1945.[14] By contrast, nearly 300 separate Congressional hearings were held on pesticides during the past fifty years. Why was so little attention devoted to pesticide management before 1945? What caused such a radical shift in the attention of policymakers after World War II?

Two world wars and a severe economic depression overwhelmed any concern about environmental health, especially one in which the evidence of risk was so ambiguous. Further, this was the period of the Dust Bowl in the Midwest. Years of drought and soil mismanagement produced dust storms that blew millions of tons of topsoil from farmlands, and crop farming became impossible for many. Farmers in the 1930s faced the loss of crops, income, home, health, and family—risks that seemed far more immediate and important than invisible threats from chemical contamination.

At that time, pesticides represented a small proportion of overall agricultural investment, and the agrichemical industry itself was small and dependent upon relatively few compounds. Thus there was little public support for more stringent regulation of pesticide risks. If anything, government was far more interested in the design and adoption of new pesticide technologies that could propel rural development.

German and Japanese aggression finally fueled an unprecedented government-industry collaboration on pesticide research, primarily for the purpose of insect-borne disease control. As explained in chapter 2, the war was a testing ground for more efficient production, distribution, and application of pesticides. With the close of the war, these organizational and technological skills were turned to the problem of agricultural productivity, and pesticide use grew exponentially. The chemical industry was of central importance to the U.S. economy during the next several decades, and the production of new synthetic pesticides became highly profitable as international markets expanded. In 1945, for example, U.S. firms produced 10 million pounds of DDT with a market price of $1 per pound. In 1951, they produced 100 million pounds.[15] By the mid-1950s, the price had dropped to nearly 25 cents per pound. And between 1939 and 1954, the number of firms manufacturing insecticides and fungicides more than tripled.[16]

Postwar Reform and the Passage of FIFRA

By 1945, it was obvious that the original Insecticide Act provided little basis for effective government control of pesticides. Early in 1947, USDA estimated that nearly 25,000 products had been registered or licensed for use, and the numbers were growing quickly.[17] In fact, the government had great difficulty even tracking which pesticides and products were introduced into commerce.

The impression that the industry had moved beyond the control of government led Congress to adopt the Federal Insecticide, Rodenticide, and Fungicide Act (FIFRA) in 1947. This law dictated that "economic poisons" be licensed prior to their sale in interstate or international commerce.[18] It also required prominent display of warning labels for highly toxic pesticides, which included instructions for use. Manufacturers were compelled to color their powdered insecticides to prevent their being mistaken for foods such as sugar, flour, baking soda, and salt.[19] Authority to register pesticides was given to USDA, which could declare a product "misbranded" if it was found to induce harm to humans, animals, or vegetation even when properly used.

Thus the primary risk management strategy of the federal government during the postwar era was simply labeling. Despite expert knowledge of the bioaccumulation potential of chlorinated pesticides and their effects on non-target species of fish and wildlife, there was no discussion of limiting production or restricting use of pesticides. But labeling was not designed to inform users of the risks associated with the pesticide.[20] Instead, labels included instructions for use that implied that if directions were followed adverse effects could be avoided. This strategy circumvented the need for anyone to fully understand the health and ecological risks posed by compounds prior to registration. Dissatisfaction with the registration and labeling of federal law prompted four or five states to enact their own labeling statutes by 1947. Faced with the tremendous confusion of different labeling requirements among the states, manufacturers recognized the need for federal action.[21]

The potential effectiveness of the labeling strategy was questioned by the head of the Interstate Manufacturers' Association, who argued that requiring language such as "to be used according to instructions" offered the consumer little protection. It soon became apparent as well that although large interstate shipments required labeling under the act, once within the destination state the shipment could be broken into packages and sold without additional labeling. Further, the secretary of agriculture had no authority to deny registration if the pesticide was found to be especially dangerous despite precautionary labeling. Under these circumstances, the secretary could only register the pesticide "*under protest*."

This approach may have done far more to protect the entitlements of the pesticide manufacturers rather than either public health or environmental quality. It sheltered manufacturers from uncoordinated state regulations, and may simply have served to provide the public with a false sense of security that pesticide risks were being well contained by USDA. The reality was that USDA registered pesticides whenever asked.

The Delaney Hearings and the Miller Amendment

The passage of FIFRA did little to contain growing concerns about the environmental health effects of pesticides, which by 1950 had become a prominent public debate. Congressman James Delaney of New York held a series of hearings around the country to examine the nature of risks associated with food additives—including pesticides, which remained unregulated by the recently passed FIFRA. Much of the hearings' focus was directed toward

DDT and other chlorinated hydrocarbon pesticides, which were to enjoy continued registration for nearly three more decades.

By 1951, USDA had registered nearly 30,000 pesticide products. Further, DDT, one of the most widely used, was known to accumulate in human tissues, particularly in adipose tissue or fat and breast milk, and one of the primary sources of human exposure was suspected to be dairy products, especially cow's milk.[22] Pesticide manufacturers, USDA officials, farmers, and agricultural scientists, all of whom had dominated the Senate and House hearings that preceded the passage of FIFRA, felt that the law, passed in 1947, offered sufficient public protection.

The Food and Drug Administration, however, had a different view of these risks. FDA Commissioner Paul Dunbar argued that although wartime uses of DDT were appropriate given that troops were exposed to malaria and typhus, peacetime use of DDT warranted a more conservative standard. Dunbar concluded his testimony by calling for an amendment to FFDCA: "I feel that no new chemical or no chemical that is subject to any question as to safety should be employed until its possible injurious effect, both on an acute and on a long-time chronic basis, has been shown to be nonexistent. In other words, any chemical that is proposed for use ought to be proved in advance of distribution in a food product to be utterly and completely without the possibility of human injury."[23]

As an example of the need for tighter regulations, FDA presented evidence of liver damage in rats exposed at only 5 ppm of DDT in the diet, even though it was common for human body fat levels to exceed 10 ppm. In the same hearing, FDA presented a list of eight hundred compounds that were then classified as allowable food contaminants, including formaldehyde; cyanide; lead arsenate; chlorinated insecticides; the herbicides 2,4,5-T and 2,4-D; and mercury-based fungicides. FDA concluded that the twenty-three chlorinated insecticides "represent a definite health hazard."[24] Support for FDA's position came from another chemist, who found DDT residues in milk from Missouri, Oklahoma, Texas, and Wisconsin and from Robert Kehoe, director of the Kettering Laboratory at the School of Medicine of the University of Cincinnati, who testified that DDT could be detected in the tissues of virtually every American.

In an interesting preview of a debate that would reemerge in late 1980s over the growth regulator Alar, the Beech-Nut Packing Company complained in 1947 that pesticide residues on food crops forced them to institute a costly new detection program. Beech-Nut argued that there were "no regulations limiting the use of the organic spray residues on fruits and vegetables or any information as to what levels were safe for human consumption." The company found it nearly impossible to find residue-free produce, and reported that it had begun accepting fruits with detectable pesticide residues for processed baby food.

The concerns of the chemical industry, meanwhile, were well represented by the National Agricultural Chemical Association (NACA), which congratulated USDA for its efficient review and registration of 22,000 new pesticide products. When Lea Hitchner, NACA's executive secretary, was asked to comment on a statement by the American Medical Association calling for better testing and more careful review of the toxicity and residue fate of many pesticides already on the market, Hitchner flatly denied any such need, despite the fact that in 1950 USDA had only one toxicologist on its staff.[25]

Over the next twenty-two years, the Public Health Service became an important ally of the chemical industry in the fight to maintain DDT registrations. The chief of the toxicology branch, Wayland Hayes, was highly persuasive in his testimony on the low level of DDT toxicity. Hayes had studied patterns of exposure, accumulation, and the toxicity of DDT since its introduction in the United States in 1942. Those mixing the insecticide were found to have as much as 291 parts per million in their body fat with no discernible toxic effect.[26] Additionally, in 1947, twenty-eight people had been poisoned from accidentally eating biscuits made from flour contaminated with 10 percent DDT. The symptoms were vomiting, numbness, partial paralysis, mild convulsions, and hyperactive knee jerk reflex, but, as Hayes noted, all recovered quickly.[27] Hayes's only concession to those promoting more stringent regulation by FDA was to support the 1949 USDA decision to prohibit the use of DDT in cattle, particularly dairy cattle, because the Public Health Service had demonstrated in 1949 that feeding dairy cattle with alfalfa hay treated with DDT caused significant storage of the insecticide in fat and output in milk for nearly six months after feeding was stopped.[28] Hayes admitted that continued low-level ingestion of DDT from sources such as contaminated milk could cause it to accumulate in human fat beyond a safe level.

USDA presented an economic argument for its continued control over pesticides. In 1949, researchers there estimated that insects were responsible for $4 billion in crop losses even when insecticides were used and that it was extremely important to have several types of chemicals available for each pest problem. Accordingly, they had only issued their limited "registration under protest" on several occasions, and never to protect food from excessive pesticide residues.[29]

Conservation groups were notably absent from these hearings. Protection of public health from synthetic contaminants was not a concern of high priority. In the early 1950s, their attention was directed toward protection of wildlife and wilderness and the sustainable management of forests, water, and soil. Wildlife biologists during the 1950s—many of whom worked with the U.S. Fish and Wildlife Service—were the first to recognize the diversity of species adversely affected by chlorinated pesticides. One of these biologists was Rachel Carson, who would not capture wide public attention until 1962.

During the early 1950s, medical doctors were the pioneer "environmentalists" of the time and were regular participants in Senate and House hearings held to explore environmental contamination issues. For example, Charles Cameron, the scientific director of the American Cancer Society, testified that certain cancers appear to be related to the disruption of hormones.

> Mounting evidence strongly suggests that there are a number of causes of cancer—some (endogenous) operating within the body itself, as in the case of genetic factors and hormones—others operating from outside, as in the case of sunlight, X-rays, tar, and arsenic . . . at least 80,000,000 pounds of arsenicals, principally in the form of lead arsenate and copper arsenate, in the United States each year for the purposes of pesticide would appear to justify [chronic toxicity] studies. . . . A half dozen papers have appeared during the past five years indicating that the proportion of smokers among lung-cancer patients is higher than among the general population. . . . Is it the arsenic with which the leaf was sprayed? If I were forced to say what I thought was the cause of cancer, I would say it probably lies in an alteration of the steroid chemistry, or hormones, if you will, of our body.[30]

Although the estrogenic effects of chlorinated hydrocarbons were not directly addressed by Cameron, his speculation raised deeply rooted fears—fears that were echoed twelve years later by Rachel Carson and forty-five years later by others incredulous that EPA still does not require studies of estrogenic effects prior to licensing pesticides.[31]

Cameron was followed at the hearings by Harold Morris of the National Cancer Institute, who summarized the literature on chemical carcinogens between 1939 and 1947. By 1939, 696 compounds had been tested for one or more months, and of these, 169 (24.3 percent) produced cancer in one or more species of animals. Extending the analysis to 1947, Hartwell found that 1,329 compounds had been tested, and 322 (24.2 percent) were found to be carcinogenic. Most of the tests conducted were not designed to test carcinogenicity; instead they were general screens for chronic effects, which made the findings of carcinogenicity even more remarkable. Morris explained that the chemicals tested represented many different classes of compounds, and he demanded on the basis of this discovery that compounds that will likely remain as residues at the dinner table be tested for their chronic toxicity.

The Delaney hearings exposed the growing intensity of public fear about cancer in the early 1950s, as well as the possible role of environmental contaminants. Although concerns over the acute toxicity of DDT were fully deflated by testimony, especially that given by Hayes, the hearings painted a disturbing image of a national food supply laced with thousands of additives including pesticides, colorings, and preservatives. Americans had unquestionably been exposed to a complex mixture of compounds that were at least toxic to other species and that were inadequately tested for their influence on human health. Pesticides were far more suspect than other types of food additives, because they were used for their biocidal effects. But pesticides remained beyond the legal authority of USDA, and the agrichemical industry, farmers, and USDA enjoyed their freedom to choose pesticides that protected crop yields.

A Transfer of Regulatory Power

Beginning with the hearings in 1951, however, USDA's control over pesticides started to erode. Between 1952 and 1954, articles appeared in the popular scientific literature with titles such as: "DDT More Dangerous, Fat Accumulation Hints"; "Examination of Human Fat for the Presence of DDT"; "DDT Eaten in Every Meal but Amount Won't Harm"; "DDT, Miracle or Boomerang?"; "DDT Now Found in Human Body"; "DDT: Danger to Soil"; "Our Daily Poison"; and "Biological Deserts."[32]

In response to rising public fears, Congressman Arthur Miller of Nebraska drafted an amendment to the FFDCA that was eventually adopted by Congress in 1954. The amendment granted authority to the FDA to permit registration of a pesticide only if the manufacturer demonstrates that residue levels on food pose no danger to the public.[33] Residues were allowed to exist on raw agricultural commodities before they leave the farm gate, and the allowable contamination level, while protecting public health, must also recognize the need for "an adequate, wholesome, and economical food supply."[34] This dual standard—one which requires consideration of risks and benefits—has been the heart of the nation's pesticide-control strategy ever since.

The law allowed but did not compel FDA to consider benefits in addition to risks.[35] The amendment also had the effect of fragmenting the authority to regulate pesticides between USDA and FDA. USDA remained responsible for pesticide registration and was primarily interested in improved crop productivity. FDA, after 1954, held the authority to determine a safe level of food contamination for each pesticide.

FDA, however, did not anticipate the technical difficulties of judging safety. This task was enormous given the thousands of products already registered by USDA, especially because USDA had required neither toxicity nor environmental fate data from the manufacturers—both of which are necessary to evaluate potential human exposure. Collecting and evaluating toxicity evidence for thousands of registered compounds for different types of toxic effects such as carcinogenicity or reproductive effects proved to be a monumental job. Further, determining safe levels of exposure often required difficult judgments concerning ambiguous evidence of toxicity, which were derived primarily from animal studies.

Neither the Miller amendment nor the Delaney amendment shifted the burden of proof that a pesticide carried a significant potential for harm to humans, animals, or non-target species from USDA. Congress finally transferred this responsibility to manufacturers in 1964, yet USDA was still in charge of judging the quality of toxicity evidence presented by the industry. More importantly, they retained their "gatekeeping" responsibility over registrations, and by 1966, the USDA's pesticide registration division had licensed nearly sixty thousand separate products with only its infamous staff of one toxicologist to judge the sufficiency of health and safety claims.[36]

The Delaney Clause

The Miller amendment of 1954 required that FDA set tolerances to protect human health, *but neglected to give it the authority to demand industry tests of additives' safety prior to their use.* Thus the burden of proving health risks still lay with government. Eleven separate bills were introduced in 1956 to rectify this omission, a process that initiated a rancorous debate over whether then-existing food additives should be exempt from similar testing.

Congressman Delaney claimed that more than one-third of food additives then in use lacked satisfactory data to make judgments regarding their safety in food. Congressman Miller, however, supported the grandfather clause, if it applied to compounds "generally regarded as safe" (GRAS).[37] Predictably, the agrichemical industry supported Miller's position.[38]

Congress failed to act on any of these bills during 1956, and held additional hearings in 1957 and 1958 on seven bills that all proposed a similar policy of exempting additives from testing if they were listed as GRAS. One exception was Congressman Delaney's own bill, which included one additional provision, pertaining to carcinogenic substances: *"The Secretary shall not approve for use in food any chemical additive found to induce cancer in man, or, after tests, found to induce cancer in animals."*

To support the need for special protection from carcinogens, Delaney presented three specialists in cancer research to the committee. One was William Smith, a nutritionist who argued that substances that caused cancer in laboratory animals should be prohibited from the nation's food supply simply as a precautionary measure, even though differences in susceptibility between species may exist. A second expert presenting testimony was Francis Ray, director of the Cancer Research Laboratory at the University of Florida. Ray expressed concern over the long latency period—often more than thirty years—between exposure to a carcinogen and the onset of cancer. Although some compounds may be safe at extremely low doses, Ray explained, for others there is no safe level of exposure. Given these uncertain conditions and the virtual impossibility of isolating a cause thirty years following exposure to numerous carcinogens, Ray believed precaution was the only prudent policy.

Another important witness was William Hueper from the National Cancer Institute, who had spent more than fifteen years studying the carcinogenic effects of environmental contaminants. Hueper supported Delaney's clause by arguing that there is no reliable method for determining a "safe dose" of a carcinogen for humans. Even if there were, Hueper pointed out, it would be extremely difficult to control the amount of residues in the food supply to ensure that this dose was not exceeded given the diversity of dietary sources and variations in people's diets. Hueper was also concerned about the cumulative effect of numerous carcinogens already in the human environment and suggested that permitting synthetic carcinogens in the food supply would "increase the carcinogenic load of the consuming public."[39]

The Delaney proposal met strong criticism from the commissioner of FDA, who questioned the singling out of cancer and asserted that the statutory language alone would protect the public from all possible adverse effects. The House committee reinforced this view when it approved a bill requiring pretesting of food additives not "generally recognized as safe." The House report accompanying the bill explained: "Safety requires proof of a reasonable certainty that no harm will result from the proposed use of an additive. It does not—and cannot—require proof beyond any possible doubt that no harm will result under any conceivable circumstance."[40]

Manufacturers had therefore won FDA support to grandfather compounds already in use—if they were "generally regarded as safe"; the manufacturers and FDA had succeeded in deleting the Delaney clause from the reported bill. But the fight was not over. Delaney had already demonstrated that consensus on safety was commonly achieved without sufficient test data, and that the lack of evidence was considered by FDA to indicate the absence of any adverse effect. On the House floor, Delaney reinstated his provision in the final language, and it passed on August 13, 1958. Five days later, the Senate Committee on Labor and Public Welfare reported the House bill to the full Senate, which adopted the Delaney language and commended Congressman Delaney: "We applaud Congressman Delaney for having taken this, as he has taken every other opportunity, to focus our attention on the cancer producing potentialities of various substances, but we want the record to show that in our opinion the bill is aimed at preventing the addition to the food our people eat of any substances the ingestion of which reasonable people would expect to produce not just cancer but any disease or disability. In short, we believe the bill reads and means the same with or without the inclusion of the clause referred to. This is also the view of the FDA."[41]

Thus the Senate used FDA's own logic—that the general safety provisions of the bill were designed to protect against any adverse effect—to choose a policy actively opposed by FDA and HEW. Members of the Senate concluded that the Delaney clause did not substantively change the law, and took the opportunity to support a House colleague in his fight against environmental carcinogens. As it turned out, everyone underestimated the severity of the Delaney clause, and its potency led regulators to avoid its strict application (a history described in chapter 6).

The Science and Politics of Silent Spring

Before 1962, technical knowledge of the risks of pesticides was confined largely to scientists, especially entomologists, toxicologists, and doctors employed most often within industry or government. This understanding of hazard occasionally reached public hearing records and, less frequently, the popular press. Yet few laypeople could interpret the expert technical language, and no one ventured to forecast the effects of releasing billions of pounds of pesticides into the global environment. By 1962 nearly fifty-five thousand pesticide products had been registered by USDA, and since 1955 FDA had issued thousands of food tolerances, which set residue limits for foods. Now both agencies had a stake in defending their records of registration and standard-setting.

During the summer of 1962, Rachel Carson published three articles in the *New Yorker* magazine presenting the heart of an argument on pesticide risks that would appear in *Silent Spring* later that year. She compared the threat of environmental destruction from chemical contaminants with the risks of nuclear war. She marshalled an impressive body of evidence, primarily by synthesizing and interpreting the work of others, and claimed that pesticides had been used carelessly, with "little or no advance investigation of their effect on soil, water, wildlife, and man himself." She claimed that humans were exposed to untested compounds in every meal, and that the government officials responsible for their management had acted neither with regard for human health nor with "concern for the integrity of the natural world that supports all life."

Carson's language was emotional, but she was a gifted communicator of technical arguments. Although she was trained as a wildlife biologist, she was careful to rely on evidence produced by others with more expertise, especially in her claims of risk to human health. She acted instead as a human ecologist, constructing an image of health risk caused by contaminants flowing through complex ecosystems. She forecast that pesticides indiscriminately dispersed into the global environment would haunt us long into the future by disrupting health and ecological processes fundamental to life on earth.

In many ways, the debate over atomic weapons testing made public comprehension of the ecological and health implications of pesticides possible. The detection of strontium 90 in the nation's milk supply was one of the first cases of global pollution to be understood by the public. This understanding was broadened by Adlai Stevenson's 1956 presidential campaign, in which he made the control of atomic weapons and associated fallout a prominent campaign issue.[42]

Pesticides, Carson argued, connect society to its surrounding environment in an unexpected and deeply disturbing way: through its food supply. The intimacy of the connection was driven home by Carson as she explained how DDT accumulated in food chains and human fat and was excreted in human milk. What was worse, she revealed, the government had known about this residue fate for more than a decade. She also recognized the estrogenic properties of DDT as shown by the reproductive failure of fish and wildlife, and wondered why humans should be immune to these effects.

Carson's elegant narrative of the persistence and accumulation of DDT taught a broad cross-section of the public much about the young discipline of ecology as a body of knowledge concerning connections in nature. The traditional disciplines of academic inquiry like chemistry, biology, and geology and the more professional disciplines such as entomology, agricultural science, forestry, and public health had all failed to warn of the potential for global distribution of contaminants and their effects on the health of people and their environments. Each discipline or profession seemed to have broken the environment into highly specialized compartments or to have focused too narrowly on problems of production. Ecology provided Carson with a more systematic and holistic perspective, which she used to warn an unsuspecting public of pesticide risks.

During the spring of 1963, the Senate held ten days of hearings on the "Interagency Coordination of Environmental Hazards." Because Carson's dominant thesis was that there was no control over the use and dispersal of toxic pesticides, it seemed appropriate that the Senate Committee on Government Operations consider the weaknesses of federal regulatory efforts.

These hearings were different from those held to consider the adoption of FIFRA sixteen years earlier in several important ways. The most striking distinction was that chemical manufacturers and their supporters within USDA had, by 1962, been placed in a defensive position. They had earlier claimed to be managing the acute risks of pesticides carefully and asserted that the evidence of chronic risks was hypothetical. Although the Miller Amendment and the Delaney Amendment to FFDCA compelled manufacturers to provide more precise estimates of health risks, crude detection technology, the absence of credible toxicity data, and primitive methods for estimating food and water intake severely limited the quality of these estimates.

With a gentle but firm demeanor and substantial documentation, Carson challenged the position of the USDA and the manufacturers with stories about the poisoning of fish, eagles, falcons, and farmworkers. She chronicled a legacy of errors in judgment by those charged with the public trust to protect health and environmental quality. Underlying her charges lay a deeply held conviction that "we in this generation must come to terms with nature"—meaning that we had to become responsible for the complex array of poisons we were releasing to the environment.[43] She felt no need to present the opposing case—that pesticides carried substantial benefits—because the agrichemical companies expended millions of dollars on advertising to be certain that their message was heard around the world. Instead, she articulated the frailty of life in an age of pesticides and the arrogance of those who failed to understand and respect it.

Carson's work was met by a storm of criticism from those who made pesticides and those in government who had licensed their use. Monsanto Chemical Company, which began producing DDT in 1944, published its own fable to parody the introduction to *Silent Spring,* one that presented an apocalyptic vision of insects overrunning a world without pesticides.[44] The attack of a chemical industry spokesperson, Robert White-Stevens, was more typical, but equally scathing:

> The major claims of Miss Carson are gross distortions of the actual facts, completely unsupported by scientific, experimental evidence, and general practical experience in the field. Her suggestion that pesticides are in fact biocides destroying all life is obviously absurd in the light of the fact that without selective biologicals these compounds would be completely useless. The real threat then to the survival of man is not chemical but biological, in the shape of hordes of insects that can denude our forests, sweep over our crop lands, ravage our food supply and leave in their wake a train of destitution and hunger, conveying to an undernourished population the major diseases (and) scourges of

> mankind. If man were to faithfully follow the teachings of Miss Carson, we would return to the dark ages, and the insects and diseases and vermin would once again inherit the earth.[45]

In the wake of Carson's book, the Department of Agriculture struggled to emerge from a defensive posture. Between 1947 and 1963, USDA had registered nearly fifty-five thousand separate products containing pesticides. Of these, only twenty-three were registered "under protest" due to health concerns, and they too were allowed to be sold to an unsuspecting public.[46] In California alone, more than 12.2 million acres of cropland were being sprayed each year, and by 1960 three thousand poisonings of children occurred there annually.[47]

Secretary Freeman's response was to articulate the benefits associated with pesticide use by painting a picture of a society without pesticides, the opposite of the image created by Carson as she opens her book with the fable of a silent spring.

> Perhaps one way to indicate just how important these benefits really are is to point out what would happen if we did not have them. Production of commercial quantities of many of our common vegetables would be drastically reduced, including corn, tomatoes, and lima beans. . . . Commercial production of apples would be impossible. Peaches and cherries would almost disappear from our markets. A number of diseases that infect grapes, cranberries, and raspberries would drive these fruits off the market. Commercial production of strawberries, peaches, and citrus would be impractical. . . . The production of eggs, chickens, and other poultry in the South and Southwest would no longer be profitable. Economic production of beef in the South would be virtually impossible. . . . Our total supplies of meat and milk could be drastically reduced. . . . The percentage of stored food condemned for spoilage or other damage would spiral upward with consequent reduction of supply and higher prices to the consumer. . . . Only families in the higher economic brackets would be able to enjoy many of the most nutritious foods now available to all. . . . The lowering of national dietary standards in itself would be a severe restriction to public health. Economic repercussions—affecting not only the farmer but the processor, retailer, and consumer—would be equally serious.[48]

It is curious that in the Delaney hearings of 1951 and the "Carson" hearings of 1963, claims such as these went unchallenged whereas claims of health or ecological risk were quickly attacked by manufacturers and USDA officials. Freeman was warmly thanked for his comments, which were characterized by Senators Ribicoff, Javits, and Gruening as "well-planned, thoughtful, knowledgeable, and splendid." Hyperbole and sweeping generalizations such as these were used to justify the unrestricted use of pesticides, rather than to encourage a strategy of risk minimization using available chemical and nonchemical options.[49]

Growing public distrust of USDA was seized upon as an opportunity by the Department of Health, Education, and Welfare, which housed FDA. The secretary testified that animal evidence of chronic toxicity raised the suspicion of the long-term effects on humans, and that expert understanding of human toxicity as well as pesticide transport and fate in the environment was primitive. "The long-term problem before the Nation, in our opinion," he announced, "is whether or not the continued use of pesticides may bring in the future contamination of the environment to levels that would be damaging to the health of people and other forms of life. All of these points, however, indicate vividly that we need much more knowledge. . . . We need all of this because we can base our actions only on what we know."[50]

Secretary Calabrezze stated further that "we have no evidence as of now that the food supply is dangerous." When linked with the statement above claiming that we know little of the toxicity of pesticides, Calabrezze's claim that we faced an absence of evidence of danger is very different than a claim that we know the food supply to be safe. Although the distinction is subtle, it is important for understanding the emotional appeal of Carson's argument. The absence of certain evidence of risk had led USDA to avoid regulation. Following the release of *Silent Spring,* the public was angry that ambiguous evidence was consistently interpreted in favor of industry. The burden of proof of damage appeared to lie with the public or government, and without resources to conduct studies, registrations and tolerances were simply issued. This policy, Carson argued, had turned us into a nation of industry guinea pigs.[51]

With responsibility for managing pesticides fractured between USDA, which was responsible for registration, and FDA, which was in charge of setting tolerances, Congress had set the stage for continuing conflict between those claiming benefit and those claiming health risks. But these disagreements masked the fact that estimating benefits or risks was simply beyond the resources or expertise of either agency.

At the same time, *Silent Spring* engendered enormous public sympathy. Thousands of letters were sent to newspapers around the world, supporting her conclusions and her courage. In addition, many in the scientific community sustained her claims. Loren Eisley, then a professor at the University of Pennsylvania, declared the book "a devastating, heavily documented attack upon human carelessness, greed and irresponsibility . . . without parallel in medical history."[52] And Julian Huxley, the renowned British biologist, celebrated the book as one that highlighted "the conflict between the present and future, between immediate and partial interests and the continuing interests of the entire human species."

The debate also caught the attention of President Kennedy, who during a press conference on August 29, 1962, promised a full investigation of Carson's claims. President Kennedy appointed his science adviser, Jerome Weisner, to lead a panel of experts in judging the seriousness of the problems suggested by Carson. The report of Weisner's Science Advisory Committee opened with a careful review of the enormous benefits pesticides had conferred on societies around the world, especially malaria reduction and crop protection. Yet the committee concluded that pesticides constituted only one class of a larger set of contaminants—which included air pollutants, water pollutants, and radioactive particles from nuclear-weapons testing—that pose similar management problems for two dominant reasons. First, they are present in the environment in extremely low concentrations that are normally difficult to detect; and second, human exposure is likely to be intermittent. Together, these two characteristics make it extremely difficult to judge the risks they pose to humans.[53] The committee called for eliminating persistent toxic pesticides, particularly those that accumulate in human tissues, and advocated a comprehensive research program to understand more precisely the toxic characteristics of other compounds.

The committee collected statistics that captured the media's attention and the public's imagination. In 1962, over 350 million pounds of insecticides and nearly the same amount of herbicides were applied. An additional 45 million pounds of insecticides were sprayed around homes and urban areas. Nearly one acre out of every twelve in the United States was treated by either an insecticide or an herbicide, and not uncommonly both were used.[54]

Ultimately, the committee supported Carson's claims that we were conducting a grand-scale experiment in ecology and public health without understanding much of the long-term consequences of chronic low-level contamination by hundreds if not thousands of toxins. The chairman even agreed with Carson that pesticides may pose a more significant risk than radioactive compounds: "Review of pesticides brings into focus their great merits while suggesting that there are apparent risks. This is the nature of the dilemma that confronts the Nation. . . . In the end society must decide (on the level of desired control), and to do so it must obtain adequate information on which to base its judgments. The decision is an uncomfortable one which can never be final but must be constantly in flux as circumstances change and knowledge increases."[55]

In a 1962 Christmas letter by Carson published in the *New Yorker,* she reflected that her intention was less to focus the public's eye on pesticides than to reveal a pattern of contamination that was largely uncontrolled by either industry or government. Her letter was well received. The attack on Carson's argument and credibility launched by industry and some government agencies had been seriously deflated by the Advisory Committee report. The gentle, tenacious, and courageous woman who repeatedly stood before hostile Senators and Congressmen and leveled her attack on the global dispersal of an untested but poisonous technology had fundamentally changed the public's understanding of agriculture.

Rachel Carson died of cancer in April 1964. In that year, Secretary of Agriculture Orville Freeman formed a task force to review the practices of USDA in regulating pesticides. The committee concluded their work in 1965, and by then nearly sixty thousand formulations had been registered by the department. The task force concluded:

> In reviewing the Registration procedures, the Task Force gained the impression that products have been registered on the basis of professional judgment in lieu of adequate data with no record being made of the basis for action. . . .
>
> The most important aspect of registering a pesticide chemical is the criteria used in determining safety and effectiveness. The Act and Regulations require that the registrant must provide data to support the registration and the Division must decide what criteria to use in accepting or rejecting the application. . . .
>
> While in many instances the judgments may have been good, they have nevertheless been much too arbitrary. Decisions have been made on the basis of personal knowledge possessed by the Section Head, but many of these have not been documented with adequate experimental data. Inspection of some documents has shown no appropriate information on translocation, persistence, fate, crop safety and only a very meager amount of toxicological data. The Division does not have a policy of requesting new data in support of renewal applications, even though the published literature has indicated the need for further review. . . .
>
> The reason for this apparent lack of a critical review of all renewals as well as new applications has been attributed to a shortage of qualified personnel trained to evaluate safety and effectiveness. We feel that it is more likely due to having no established set of requirements that must be met before registration can be granted.[56]

The task force recognized that USDA's primary concern in pesticide regulation had been product effectiveness and that they had become overwhelmed by the rapid innovation by the chemical industry. In their opinion, the absence of scientific evidence of hazard, or at least the equivocal nature of the evidence, supported the USDA's casual consideration of health, safety, and environmental quality.

A Growing Tension: Health Security versus Economic Security

During the 1940s, pesticide technology enjoyed a highly favorable public image, clarified by the effects of insect-borne disease reduction and crop protection. These benefits of health and economic security legitimated a federal control strategy that in 1947 simply required product registration by USDA. Expert and public concern over food safety gradually led to changes in the decision criteria for setting limits on food contamination by pesticides. These concerns were reflected by the Miller Amendment to the FFDCA in 1954, which formally required consideration of public health and safety in setting tolerances, as well as the Delaney Amendment, which followed in 1958. Although the Delaney clause prohibited pesticide residues under certain highly restricted circumstances, the standard almost always applied was one requiring that economic concerns be balanced against public health. Given the poor understanding of pesticide use patterns, residue movement and fate, and toxicity, it was relatively simple for USDA to conclude that evidence was insufficient to justify strict regulation. Weighing the ambiguous evidence of risk against the relatively clear evidence of benefit created a pattern in which tens of thousands of registrations and tolerances were issued. How did this complex set of regulations—and the use entitlements they conferred—respond to increasing evidence of health and ecological risks during the decades that followed? This is the question to which I now turn.

CHAPTER 5

EPA as the Gatekeeper of Risk

Rachel Carson's fundamental claim in *Silent Spring* was that pesticide use had grown well beyond the control of government. She argued that the law governing pesticide registration, in force since 1947, was far too weak to ensure protection of environmental health. With eloquent but simple prose, she expressed moral outrage at the behavior of USDA, the chief administrative agency responsible for regulating pesticides, which appeared to register new pesticides whenever asked by industry. How could USDA be trusted to regulate the industry it sought to promote and protect?

As mentioned in chapter 4, the acrimonious and very public debates that followed the release of *Silent Spring* led President Kennedy to have his Science Advisory Committee review Carson's work. This committee concluded that pesticides may pose a more serious threat to environmental health than radioactive fallout and largely supported Carson's claims. Yet Kennedy's brief administration was dominated if not overwhelmed by domestic issues such as civil rights, the expansion of communism, the Cuban missile crisis, and the escalation of the Vietnam War. The result was that no reform initiative emerged from the executive branch.

Instead, the problem moved to a more sympathetic arena, a Senate committee chaired by Senator Abraham Ribicoff of Connecticut. Ribicoff held an extended set of hearings on pesticide reform in 1963 and early 1964 that set the stage for the dramatic claims and counterclaims that followed. Health advocates challenged traditional supporters of agriculture and the chemical technology that had propelled the astounding improvements in crop yields during the 1950s and early 1960s.

Despite the intensity of feeling on both sides of the debate, only modest changes to FIFRA resulted from the hearings. First, USDA was given the authority to refuse registration if public health or safety was jeopardized and to cancel registrations in light of new evidence of hazard. Second, the burden of proof was shifted from USDA to the registrant to demonstrate product safety and effectiveness.[1] Yet even after these changes, the dominant legal strategy for managing pesticides—risk communication through labeling—remained, leaving the most prominent defects of FIFRA intact through the 1960s. These limitations included product labeling as the primary risk management strategy, limited provisions for public participation, and granted broad administrative discretion to the secretary to define when a pesticide posed an "imminent hazard to the public," a finding necessary for product cancellation. Each of these problems is considered briefly below.[2]

We saw in chapter 4 how FIFRA's reliance on labeling was based on the premise that a statement of instructions for effective use, together with precautionary language warning of risks, was sufficient to protect the public from pesticide dangers. The strategy failed for several different reasons: not all users were literate enough to comprehend the directions; even users who understood the warnings were often careless; and the government failed to understand that some pesticides persisted or migrated from their point of application while maintaining their toxicity.

In one case, the Stearns Electric Paste Company sued EPA when it tried to cancel the registration for a roach and rat killer that caused numerous deaths of children and adults. EPA claimed that cancellation was justified by a balancing of risks and benefits. The court, however, concluded that if the label had been followed the injuries would have been avoided. If the product could be used safely, EPA had no statutory basis for removing the registration. The court contrasted the roach and rat poison to DDT, which, it claimed, carried the potential for harm even if used as directed by labels. Thus, under FIFRA, EPA was constrained from prohibiting the use of even extremely dangerous compounds, as long as the label specified an appropriate precautionary message.[3]

Before 1972, FIFRA made little provision for extra-governmental groups or individuals to become involved in pesticide regulatory decisions. This exclusionary practice was contested, however, over the regulation of DDT. The Environmental Defense Fund sued the secretary of agriculture, and then the administrator of EPA after it was created in 1970, to force cancellation of all uses of DDT in 1970.[4] EDF et al. contended that EPA's *inaction* prohibited them from participating in decisions regarding registration. EPA countered that because DDT did not carry an imminent public hazard—and a cancellation notice had not been issued—neither the court nor EDF had the right to revisit the decision to refuse cancellation. But the D.C. Circuit Court disagreed, claiming that once a "substantial question concerning safety" existed the intent of the law was that government should issue a cancellation notice and hold hearings where public participation would be solicited. In addition, the burden of proof of safety would then shift to the producer. The court therefore ordered the administrator to issue cancellation orders, which merely required him to hold hearings on the possibility of prohibiting the use of DDT. The enormity of effort required by EDF to simply gain participatory rights pointed to a fundamental defect in FIFRA, which discouraged competing interpretations of risk.[5]

Another weakness of FIFRA was its ambiguity in defining "imminent hazard," a finding necessary to cancel a pesticide product's registration.[6] If the secretary of agriculture or EPA administrator did not articulate criteria for declaring an imminent hazard, no one could challenge the registration decisions or the failure of EPA—or, before 1970, USDA—to issue cancellation notices. In *EDF v. Ruckelshaus,* the court required that an imminent hazard be defined by a "substantial likelihood that serious harm will be experienced." The court also required that the standard be clearly articulated to permit public review and that it be applied uniformly.[7]

The Environmental Defense Fund petitioned FDA in 1968 to reduce all food tolerances for two chlorinated hydrocarbon insecticides—aldrin and dieldrin—to zero following evidence from a Shell Chemical Company study that suggested the pesticides caused cancer in mice. These compounds had been in common use for nearly twenty-five years. The EDF request was quickly denied by FDA. The day after the Environmental Protection Agency was born, EDF asked the agency to cancel and suspend all uses of aldrin and dieldrin. Administrator Ruckelshaus issued a notice of intent to cancel all registrations; the firms holding the registrations, however, requested a review of the data by NAS, a right they held under FIFRA. This review was completed approximately two years later, and concluded that several uses, including its dominant use as a corn insecticide, appeared to pose no threat to human health.[8]

EDF next turned to the D.C. Circuit Court in 1972 requesting that EPA suspend the uses of aldrin and dieldrin. Both compounds had been earlier highlighted by President Kennedy's Science Advisory Committee in 1963 as "persistent toxic pesticides" needing careful review. EDF's request was remanded to EPA, which was asked to explain why it had not issued an order for suspension. The agency responded that it would treat each suspension case separately based upon a balance of anticipated risks and benefits. Applying this criterion to aldrin and dieldrin, EPA claimed that their risks were slight but their benefits were substantial. The court then required EPA to clarify what it meant by benefits.

Cancellation hearings for aldrin and dieldrin began in August 1973, and within twelve months thirty-five thousand pages of testimony, which took up nearly sixty feet of shelf space, were accumulated. Russell Train, EPA administrator at the time, announced that his interpretation of new information had persuaded him that the compounds posed an "imminent hazard to man and the environment." An expedited suspension hearing was held in fifteen days, followed by a judge's recommendation to suspend all uses. This proposal was agreed to by Train, appealed by Shell Chemical Company, but finally upheld by the D.C. Circuit Court in April 1975. Ambiguity in the statutory definition of "imminent hazard" had become a focal point for conflict between industry and environmental groups.

The Birth of EPA and the Legacy of USDA

Richard Nixon was extraordinarily sensitive to the breadth of public sentiment favoring environmental protection during the early 1970s. Nixon and his staff understood that the desire for improved environmental quality cut across income, ethnic, and party boundaries. Although his contributions to environmental law and policy are often unrecognized, one of Nixon's more important domestic accomplishments was the creation of EPA on December 2, 1970.[9] EPA was created by consolidating responsibilities and personnel from thirteen different federal agencies, including divisions within the Departments of Interior; Agriculture; and Health, Education, and Welfare. It emerged into a world of intense debate over the meaning of environmental quality and the magnitude of risks faced from pollution, including pesticides. EPA inherited staff and pesticide lawsuits from USDA, and the judicial opinions that flowed from those cases played a formative role in shaping EPA's pesticide policies.

EPA's ban of DDT brought pesticides into the public eye during 1971, and laid a foundation for the first significant reform of FIFRA in twenty-five years. The increasing activism by courts in the early 1970s reflected public dissatisfaction with the deficiencies of FIFRA. Yet judicial direction was offered only in piecemeal fashion, as specific issues were raised for adjudication.[10] Several states enacted pesticide control laws, creating a frightening vision of fifty unique state regulatory programs to industry.[11]

Revised legislation introduced by President Nixon resulted in the passage of the Federal Environmental Pesticides Control Act (FEPCA) in 1972.[12] FEPCA redirected the mission of the federal government from accuracy in product labeling to the protection of public health and the environment. Whereas FIFRA required registration once a product was found to be effective and safe if used as directed, FEPCA required that additional standards be met before registration, cancellation, or suspension of registration. Manufacturers now needed to demonstrate that a pesticide would perform its "intended function" and, "when used in accordance with widespread and commonly accepted practice," would not cause "unreasonable adverse effects on the environment." This is a pivotal phrase, further defined as "any unreasonable risk to man or the environment, taking into account the economic, social, and environmental costs and benefits of the use of any pesticide."[13] FEPCA required that these standards be applied to compounds already registered, posing an enormous risk assessment problem for EPA.

Key terms in the statute such as risk, cost, and benefit were not clearly defined, and since 1972, conflict over appropriate ways of measuring risk and its distribution have dominated deliberations over pesticide registration and tolerances. Pesticides already registered were not threatened unless new evidence of risk was discovered by EPA. Benefits for these compounds are presumed simply by the fact that the manufacturer seeks continued registration, often to serve a well-established market. Thus regulatory resources have been focused far less on benefits than on risks and the costs of risk avoidance.

No guidance was provided by Congress to EPA to judge the "reasonableness" of risks. The phrase implies that the judgment is somehow related to a balancing of costs and benefits. Translating pesticide use, or its prohibition, into a net estimate of social welfare is a problem of enormous scale. Little agreement exists over types of costs and benefits to be accounted, and how they should be estimated. In considering benefits, EPA must consider the effect of its decision on "production and prices of agricultural commodities, retail food prices, and otherwise on the agricultural economy."[14]

Even less consensus has evolved over the need to understand distributional patterns of costs and benefits, despite their obvious centrality to the issue of judging the reasonableness of a regulatory decision. Although the court attempted to guide EPA by demanding that it consider the "magnitude of damage" and its "probability of occurrence," probabilities have rarely been assigned to formal estimates of cost or benefit. The absence of direction on these matters, either from Congress or the judiciary, has provided EPA with broad discretion to change the rules of analysis and criteria for choice regarding pesticide registrations and tolerances. In short, EPA can decide to register a pesticide that poses significant risks if it finds the benefits to be more important.

Before 1972, pesticides were simply registered for specific uses, whereas the new law created categories of use: general and restricted. Pesticides were permitted for general use if they were unlikely to cause adverse effects when used properly. Some pesticides were more dangerous, however, and could cause "unreasonable adverse effects" even if used according to label instructions. After FEPCA, these compounds could be used only "under direct supervision of a certified applicator." Yet a closer reading reveals that any "competent" person acting under the control of a "certified applicator"—who need not be present when the pesticide is actually applied—is allowed under the law.[15] Although the original intent of the statute was to control extremely hazardous substances by training applicators in proper methods of handling, the applicator can delegate this authority to an untrained person. The health and safety of the public thus rests on whether the applicator, trained or not, complies with the special precautionary measures provided with the restricted compound. Without extensive monitoring and enforcement by the states, "unreasonable adverse effects" seemed not only possible, but probable.[16] Not surprisingly, the agrichemical industry supported the 1972 amendments largely because of the certified applicator provision, which has been applied to a limited number of compounds and has thereby taken regulatory pressure off the majority of others.[17] Also, the statute left significant room for debate over which compounds were to be classified as restricted-use. By 1975, the requirements were further relaxed when EPA was forbidden to test an applicator's knowledge of a pesticide's potential to injure health or the environment.[18]

In 1970, EPA faced the responsibility to reregister nearly sixty thousand already registered products. Again, for a compound to be reregistered under FEPCA, it had to have "no unreasonable adverse effect on the environment" when used in accordance with the product label. EPA was given the authority to issue cancellation notices to manufacturers, which meant only that it would begin a thorough review of all available data about a chemical's toxicological and ecological effects. Reviews, then as now, were often delayed by poor quality data, and often took years to complete if the manufacturer challenged the order and was granted a public hearing or a special review by the scientific advisory panel. Although the EPA administrator would make the final decision, the company would be free to produce and distribute the compound during the entire review period.

The new law, however, did clarify standards for cancellation and suspension. Under FIFRA, suspension had been reserved for compounds that posed an "imminent hazard to the public," and as described above, this phrase was left undefined. FEPCA offered modest clarification by defining "imminent hazard" as "a situation which exists when the continued use of a pesticide during the time required for cancellation proceedings would be likely to result in unreasonable adverse effects on the environment or will involve unreasonable hazard to the survival of a species declared endangered by the Secretary of the Interior."[19]

Since the 1970s, then, if a pesticide poses an "imminent hazard" to humans or the environment, EPA can suspend the registration.[20] The suspension can take effect without notice to the manufacturer and can be invoked at any time during cancellation proceedings. The registrant is afforded a five-day period during which a hearing may be requested; otherwise the order takes effect. Unfortunately, however, the distinction between a situation warranting suspension and one justifying cancellation has remained unclear. One court suggested that a suspension is justified if there "is a substantial likelihood that serious harm will be experienced during the three or four months required in any realistic projection of the administrative suspension process."[21] "Imminent hazard," another court directed, is a serious threat to public health, even if the effect will not be recognized for several years.[22]

EPA first used its emergency suspension power in 1979 when it halted the sale and certain uses of 2,4,5-T and Silvex because of contamination by dioxins. In *Dow Chemical Co. v. Blum,* a Michigan district court supported EPA's suspension, saying that the EPA administrator was required to examine five factors: (1) the seriousness of the potential harm, (2) the immediacy of this risk, (3) the probability that the threatened harm would result, (4) the public benefits of continued use of the compound, and (5) the quality of data available to the administrator at the time of the decision.[23]

Interestingly enough, once a compound is canceled or suspended due to concerns over human health or ecological effects, EPA normally has allowed the existing supplies to be used without recall.[24] Between 1972 and 1988, this allowance was encouraged by FIFRA's indemnification clause, which required EPA to purchase, collect, and dispose of any recalled products. Mercury-containing pesticides, for example, were not recalled under the convoluted logic that they would be uniformly dispersed across the nation and thus lead to a relatively low "average" exposure and risk. Further, local sales and use of existing pesticide stocks were allowed for registrations suspended before the 1972 amendments—even if an imminent hazard existed—because intrastate commerce was not regulated. Thus if EPA chose to issue a cancellation or suspension order, FEPCA authorized far more control over patterns of use.

Because the vast majority of pesticides were registered by USDA under criteria established by FIFRA in 1947, EPA's task has largely been to obtain current, high-quality evidence about residue fate and toxicity for making reregistration decisions. These data are also vitally important for determining whether a compound should be classified as general or restricted-use. EPA's role has therefore been to manage the scientific process, judge the potential for unreasonable hazard, and then adjust existing use entitlements.

As mentioned earlier, one of the most controversial sections of the FEPCA amendments concerned indemnification, a provision requiring the federal government to compensate manufacturers, applicators, and farmers for stocks of pesticides that were either canceled or suspended.[25] This indemnification provision was a necessary condition for industry's support of the 1972 amendments, but environmentalists were furious at the prospect of taxpayer's dollars compensating for inadequate health and safety testing—and thereby possibly encouraging the introduction of risky compounds.[26] The manufacturers argued that this insurance was necessary for them to invest in research of new and safer pesticides. Environmentalists, however, believed the costs of indemnification, borne directly by the Office of Pesticide Programs within EPA, could deter federal cancellation or suspension of pesticides suspected of posing unacceptable risks.

Indemnity persisted from 1972 to 1988, and EPA has paid $20 million to the manufacturers of 2,4,5-T, canceled in 1979, and to the makers of ethylene dibromide (EDB), canceled in 1984. Not only did EPA pay for existing stocks of EDB, they also "owned" these stocks and paid for their transfer from leaking steel drums to stainless-steel railroad cars. EPA then paid for high-temperature incineration, required by the Resource Conservation and Recovery Act. An additional $40 million has been spent for the cancellation of dinoseb.[27]

The indemnity clause was changed by Congress in 1988, ending financial protection for manufacturers—but continuing to shelter farmers and applicators. These "end users" are now eligible for compensation from the regular federal Judgment Fund administered by the Treasury Department, rather than from EPA's pesticide-program regulatory budget.[28]

This history suggests that manufacturers were treating registrations and tolerances as use entitlements or property rights. Because entitlements were originally issued based upon a government pronouncement of safety, what should happen to them if the government later concludes that the product is damaging health or environmental quality? Holders of the entitlements argued that, like any other private property, use rights should be protected by the Fifth Amendment to the U.S. Constitution, which requires compensation for any such "taking." Consumer and environmental groups, however, expressed outrage that the government should bear the costs of product collection, management, and disposal.

A second important property-rights conflict arose over control of data on chemical formulae, manufacturing processes, and toxicity data. Public availability of these data has been contested by industry for decades on the grounds that if they become public, competitors could use them to justify proposed registrations of similar products. Also, controlling access to primary data helped manufacturers maintain some modicum of control over judgments about risk. The 1972 amendments to FIFRA require that trade secrets be kept confidential as a form of private property deserving Constitutional protection.[29] This requirement was interpreted by EPA to apply only to limited sets of data, primarily pesticide formulae and manufacturing processes, not data relevant to judgments of a pesticide's hazard or efficacy. In 1978 Congress clarified the property claim by formalizing EPA's policy of permitting confidentiality only for chemical formulae and manufacturing processes.

Reregistration: Delay in Pursuit of Certainty

Between 1964 and 1994, pesticide use in the United States doubled from 500 million to over 1 billion pounds per year. In 1983 there were approximately fifty thousand separate compounds registered by USDA; this number was reduced to approximately twenty-one thousand compounds, including 860 active ingredients, by 1993.[30] Nearly nineteen thousand of these products need to be reregistered by EPA, because the quality of data supporting the original claims of environmental health and safety were suspect. Available data for these compounds tended to support claims of effectiveness in managing pests, rather than to demonstrate the absence of their adverse health or ecological effects.

Our understanding of the risks posed by pesticides is continually changing for several reasons: (1) the increasing sensitivity of chemical detection technology, which has revealed that environmental contamination is more widespread than earlier believed, (2) more stringent and sensitive toxicity testing requirements, which have expanded the universe of chemicals believed to pose significant risk to human health, and (3) the improving quality of epidemiological evidence regarding the incidence and distribution of environmentally induced disease. Generally, where experts have bothered to look for more persistent pesticides in the environment and in our bodies, they have been found. And when older pesticides are tested for toxic effects according to state-of-the-art protocols, new hazards are discovered.

When EPA was created in 1970, it inherited not only the staff of the USDA Economic Research Service previously responsible for pesticide registration, but also the truckloads of poorly organized data submitted by manufacturers to support registrations. These data primarily demonstrated the effectiveness and efficiency of the chemical products in controlling pests. Although toxicity data was necessary for USDA to justify precautionary statements on labels, it was extremely variable in quality and often more than a decade old. In addition, the absence of comparable environmental fate and evidence across different pesticides made the practice of substituting one risk for another, as farmers choose among registered pest-control options, nearly impossible to control. Instead, evidence has been produced compound by compound, according to a staggered schedule often spanning more than a decade for each one.

The reregistration requirement quickly overwhelmed EPA's limited resources. Despite the similarity in chemical structure among many compounds, individual manufacturers were successful in preventing the data they submitted from being used by EPA to judge the risks and support registration of a competitor's product. Although Congress had initially required that reregistration be completed by 1976, it later extended the deadline to 1977. In 1978 Congress removed the deadline from the law altogether due to uncertainty associated with data quality and availability of agency resources for data review. By 1980 the agency was not even close to finishing its mission, so it tried a new tack. It directed its scarce attention toward the review of some six hundred active ingredients, rather than the fifty thousand then-registered compounds. The active ingredients were grouped into "cases," and "registration standards" were issued for the cases. These standards specified the data that must be submitted to support continued registration of the older suspect compounds. Initial data submissions by manufacturers often elicited additional data requests, which further delayed the final decision, all while the product remained on the market.[31] High-volume pesticides and those used in food received priority attention, and by 1988, EPA had completed 194 registration standards covering 350 active ingredients, representing 85 to 90 percent of the total weight of pesticides used in the United States at the time.[32] Because the toxicity of pesticides is not necessarily related to weight or volume, these figures do not provide a clear impression of effective risk management.

In 1976 EPA discovered that the Industrial Bio-Test Laboratory (IBT) had falsified data submitted to support over two hundred pesticide registrations. Many of the suspect studies were chronic toxicity tests, such as two-year chronic-feeding cancer studies that commonly take five years to contract, conduct, interpret, and validate. Given the lethargic pace of the data production, combined with EPA's practice of reviewing pesticides one at a time, the scandal conservatively set the registration program back by a decade and, by suspending the regulatory animation of EPA, extended the registration life of numerous compounds.[33] In response to the IBT case, EPA created a laboratory audit program that conducted six inspections in 1978, thirty-three in 1979, seventeen in 1980, and four in 1981. By 1983, however, only one full-time position was allocated to the effort. The effectiveness of any public-risk management effort rests on the quality of toxicity evidence. Because the U.S. program relies on private-sector production of this information, quality controls such as laboratory audits are essential to its potential success.[34]

In 1978, in addition to the Congressional reprieve from meeting the reregistration deadline, EPA was given the authority to issue "conditional" registration for a pesticide even if data on human health risks had not been provided for its review, as long as the "use of the pesticide is in the public interest." Much of this problem was caused by EPA's ever-expanding data requirements for continued registration. These demands included data on pesticide residue distribution and fate in the environment; toxicology, including different studies on carcinogenicity, mutagenicity, neurotoxicity, and teratogenicity; and ecological effects such as bioaccumulation potential, persistence, and effects on unintended targets such as fish, wildlife, and even desirable insects like honeybees. As the disciplines of toxicology, epidemiology, ecology, and ecotoxicology evolved, the standards of study quality also developed. Single chemical reviews often spanned such long periods that studies initially judged to be of sufficient quality were later found to be inadequate. By 1983, an NAS committee reviewed the quality of health effects data on pesticides and concluded that only 10 percent of registered compounds had data of sufficient quality to conduct a complete assessment of their risks to human health.[35]

Coordinating the production of health and ecological effects data for over six hundred active ingredients, each requiring reregistration every five years, became a recurring management nightmare for EPA. The problem was especially acute during the early 1980s when the agency requested cuts in staff just as the manufacturers were delivering enormous numbers of new studies supporting their registration requests. While the assistant administrator for toxic substances and pesticides was requesting budgetary cuts before a reluctant Congress, he was also compressing pesticide review periods. Regulatory reform during the administrations of Presidents Reagan and Bush was often described as the pursuit of efficiency. This translated into reviewing chemical risks more quickly and with fewer staff.[36]

Further, in addition to the problem of ever-expanding data requirements, advances in analytical chemistry have made pesticide detection possible at very low levels. Insensitive detection tests often had led regulators to conclude that residues had completely dissipated. More recent, sensitive tests may find residues that, if toxic enough, could lead to significant public health risks even if found only at very low levels. EPA would then be faced with the difficult prospect of canceling or suspending a compound with an established and valuable market.

In 1991, EPA used nine separate database management systems to track information required by the pesticide registration program. Those responsible for its coordination remarked: "They need to be integrated . . . we have a strategy for that . . . [though] not yet approved."[37] The data required for submission are now so costly to produce that manufacturers argue that they provide an incentive to continue production of older, riskier compounds rather than suffer the costs of bringing a new, less risky one to market.[38] Data on the chemical composition and characteristics of active and inert ingredients are required, along with limits on the percentage of each ingredient in the final product. Environmental fate data are required to estimate the distribution and effects of chemical residues in food, water, soil, and air. The degree of hazard posed to humans must be estimated by conducting tests for acute, subchronic, and chronic health effects.

Once these data are submitted to EPA, its attention is devoted to judging whether the compound poses an unreasonable adverse effect on the environment. To reach a conclusion, EPA weighs the evidence against six separate "risk criteria": (1) acute toxicity, (2) chronic or delayed acute toxicity, including carcinogenic, mutagenic, fetotoxic, and teratogenic potentials, (3) the potential for residues to induce toxic effects in nontarget organisms, (4) the possible effects on threatened or endangered species, (5) the possible effects on the habitats of threatened or endangered species, and (6) whether the compound might pose other types of risk to humans or the environment.[39]

If any of these risk criteria are triggered by preliminary evidence, EPA informs the manufacturer of its conclusion that the product should not be registered and that a more intensive analysis, known as "Special Review," will be conducted.[40] After it completes its risk-benefit assessment, EPA is required by law to explore ways of reducing significant hazards before it chooses not to register a pesticide. Conditions for registration designed to diminish risks could include restricting uses from certain highly consumed crops; prohibiting uses in sensitive ecological zones such as recharge areas for drinking-water aquifers or habitats for endangered species; increasing the time before which farmworkers can reenter fields after spraying; or requiring that applicators be better trained and more effectively protected.[41]

As we have seen, amendments to FIFRA in 1978, 1980, and 1988 established deadlines for the production of data that industry and EPA have consistently failed to meet. By 1988, Congress reached its limit of patience and added a requirement that data submitted before 1970 to justify registration for a chemical could no longer be used to support its reregistration. The burden of proof regarding safety was clearly directed to the manufacturer, which was given a maximum of four years to produce the required data.[42] If the deadline was not met, the registration would automatically expire.

Emergency Exemptions and Special Local Needs

EPA has the authority to allow the "emergency" use of pesticides that have not gone through the extensive analysis described above.[43] These exemptions are intended to allow response to public health emergencies, to minimize or prevent significant economic losses to farmers, and to quarantine pests recently imported to the United States. Since 1978, EPA and the states have issued over four thousand emergency exemptions, many of which are reapproved year after year. In one instance, the "emergency" has lasted for twelve years.[44] In 1983, the director of the Office of Pesticide Programs noted before a House hearing that many requests for exemptions were for compounds undergoing Special Review because of suspected health or ecological effects or due to grossly inadequate data.[45]

The appointment of Ann Gorsuch to lead EPA in 1981 signalled a change in mission for the pesticide program. Between 1980 and 1983, the pesticide program staff was reduced from 760 to 540, whereas the approval of emergency exemptions issued under section 18 of FIFRA jumped from 180 in 1978 to 750 in 1982.[46] After 1980, EPA granted numerous exemptions for chlorinated compounds, even the previously banned DDT and dieldrin—among dozens of others that later were more stringently regulated to control the risks they posed to public health.[47]

Another exemption that also sidesteps conventional health and safety protections is the "special local need" registration. This provision allows pesticides already federally registered to be used for additional uses authorized by states. If benomyl, for example, were registered for use on apples by EPA, California could authorize its use on additional crops. Through the end of 1982, EPA had issued 8,650 of these registrations, each of which covered an average of twelve crop-pest combinations. Thus nearly one hundred thousand separate uses not covered by federal registration were authorized by this exemption program.[48] Requirements for permits became streamlined, and pesticide manufacturers, growers, and state agencies learned that the tolerance-setting requirements of FIFRA could be circumvented by declaring that "special local needs" existed.[49] The House Committee on Agriculture questioned the consequences of these decisions, especially because state regulatory staffs were commonly too small and insufficiently trained in pesticide toxicology and ecotoxicology. Also, the data requirements of state programs were commonly less rigorous than those used for determining federal registration.[50]

The assistant administrator for toxic substances and pesticides, John Todhunter, also made it clear during the early years of the Reagan administration that new emphasis would be given to economic benefits of pesticide use when balanced against uncertain risks. His statements confirmed work already under way to isolate and revoke "burdensome, unnecessary, or counterproductive federal regulations," efforts proposed by Vice President Bush's Regulatory Relief Task Force.[51]

In August 1981, Bush announced that the pesticide registration program had been targeted for the task force's special review. Industry had three major complaints. First, the agency took too long to review registration and tolerance requests. Second, the larger manufacturers desired greater "exclusive use" protection for data submitted for EPA review. Third, the industry opposed promulgation of data requirements as regulations. They argued that because risk estimates were changing continually it was virtually impossible for regulations to keep current with advances in science.

EPA's internal response to the Bush task force review set the tone for pesticide regulatory policy for several crucial years. First, EPA promised "regulatory relief," which it defined as a softening of adversarial relations with industry and a streamlining of reviews of health and environmental effects.

> Voluntary compliance by the industry has allowed efficiencies not possible under rulemaking procedures which had been the predominant approach. Unnecessary regulations have been eliminated or amended to allow exemptions, expansion of waivers, self-regulation, and greater flexibility with resulting improvements in protection of health and the environment through more timely decisionmaking. Testing protocols have been removed from rulemaking status to permit recognition and utilization of the scientific state of the art.[52] The exceedingly lengthy and expensive Rebuttable Presumption Against Registration Process has been streamlined and a change from an adversarial approach to negotiation has been implemented.[53]

Part of EPA's response was a new emphasis on negotiated settlements to avoid litigation. "Specifically, we are planning to increase industry involvement with pesticides staff in non-adversarial preregistration conferences, and new chemical and major use decision conferences. . . . We are also considering options for deregulation. . . . Ideas under consideration include exempting from registration or possibly from FIFRA altogether, specific products which contain natural non-toxic components, [and] pesticide combinations which could be regulated by another agency . . . reducing requirements for registrants who amend labels, and expanding use patterns on labels."[54]

Although the registration standard made EPA's requests for additional information more uniform, by all accounts the pace of reregistration was still glacial. In 1987, the General Accounting Office estimated that the reregistration program would not be completed until 2024.[55] By 1988, Congress had lost patience. They amended FIFRA yet again by specifying a new reregistration process that must occur in five phases and be completed by 1997. In the first phase, pesticides were grouped into four lists, A through D, in order of review priority. The second is essentially a review of data previously required to support a pesticide's registration. The third stage requires submission of summaries of data provided in phase two, along with any additional data EPA deems necessary to complete the review. Congress was clearly interested in distinguishing between previously submitted and newly required data. It required a summary of previously submitted data to support registration along with a summary of any studies that the applicant believes *adequate* to support continued registration. The question of adequacy is very important to the ultimate decision, because a finding of inadequacy for a chronic-effects study can cause a four- to five-year reregistration delay at a cost to the manufacturer of more than $5 million. A delay in the name of improved scientific evidence seems rational, yet it clearly has served the manufacturer's interest by protecting use entitlements during the review. If EPA eventually decides to cancel a pesticide due to excessive health risk, the public's interest has clearly been poorly served if the pesticide has been in use during the agency's examination.

Previous registration often occurred after EPA reviewed *interpretations* of raw data. EPA, however, has become increasingly sensitive to ways that subtle differences in interpretation may inflate or deflate risks. Summary statistics such as averages, medians, and outer percentiles can often mask unusual distributions in estimates of pesticide residues in the environment and of human exposures. Requiring access to raw data permits EPA to interpret quality and variance in ways that might be relevant to crafting conditions for continued registration, or cancellation.[56] Further, there is a regulatory "hammer" associated with the third phase that requires that EPA cancel a pesticide's registration without a hearing if the manufacturer fails to submit required data within the time period prescribed.

During the fourth phase, EPA reviews the data submitted and identifies any data gaps. Additional delays may occur during this period depending on how difficult it is to complete the data set. In the fifth and final phase, which begins when EPA concludes that submitted data are sufficient, EPA has one year within which it must make the reregistration decision.

By March 1992, only two of the nineteen thousand older pesticides had been reregistered.[57] By 1993, following a flood of supporting data submitted to the agency, this figure had still reached only 250. Moreover, in 1993 the General Accounting Office found that most of the reregistrations completed were not for "high priority" food-use pesticides. The reasons for the delay are complex, but are largely related to difficulties encountered in judging the quality of data. Of the studies submitted to it under the 1988 requirements, the agency found the quality of 45 percent "unacceptable," and by late 1993 EPA anticipated that reregistration would not be complete until 2006, thirty-four years after it was begun.[58] The agency claims that the delay was caused by its late discovery in 1990 that more than half of the highest-priority pesticide studies required a substantial amount of additional information. The General Accounting Office recommended further Congressional action to require that EPA concentrate its scarce regulatory resources on the highest priority food-use pesticides.[59]

Although Congress, EPA, and manufacturers have demanded better data concerning the health and ecological risks associated with pesticides, no one anticipated the difficulty EPA would have in interpreting data once it was submitted. An especially important effect was one of stimulating additional questions about a chemical's potential fate in the environment and its possible toxic effects on humans and nontargeted species. This Pandora's box of questions has delayed the agency's reviews; meanwhile, however, use rights continue for most compounds originally registered.

The recurring increases in data requirements, and the increasingly technical character of these data, came about primarily during the 1980s. During this period of antiregulatory fervor under the Reagan and Bush administrations, reductions in EPA's regulatory budget were perhaps inevitable.[60] The staff of the pesticide program reached its peak in 1980 at 829 full-time employees, fell to 555 by 1985, and did not rebound until 1992, largely due to the commitment of William Reilly to improved quality of science within the agency as well as to a sympathetic Congress.[61] The question frequently heard in the halls of the human health effects division of the pesticide program was "Who is going to review and interpret all of these data once they arrive?"

EPA's inability to handle all these data was demonstrated in 1991 when a train derailed in Northern California, spilling the herbicide metam sodium into the Sacramento River, which feeds Lake Shasta. Although the manufacturer had submitted data to EPA four years earlier demonstrating that metam sodium caused birth defects, EPA did not know that it had this study, had not reviewed it, and therefore was incapable of providing appropriate warnings to pregnant women who might drink from the river, which supplies public drinking water.[62]

Another example is EPA's management of ground and drinking water contaminated by pesticides. Whereas 40 percent of the overall U.S. population depends on underground supplies for their drinking water, 90 percent of rural residents count on this resource. Pesticide contamination of groundwater was not believed to be a problem until two compounds were discovered in one 1979 groundwater study. By 1985, EPA had found sixteen pesticides in groundwater, and by 1988, forty-six, all from normal agricultural use. Contaminated drinking water may carry substantial human health risks because so much is consumed, and it may pose particular hazards to pregnant women, infants, and young children. EPA waited five years before it even reviewed submitted studies of toxicity and leaching potential for some of the sixteen pesticides detected in 1985.

Risks and Rights in Pesticide Management

Modern efforts to control pesticides may be best understood as a collision between manufacturers trying to protect pesticide use entitlements and public-interest groups seeking to reduce risk and protect environmental health. Registration gives farmers license to use pesticides, and tolerance-setting allows food to be contaminated to a specified maximum level. Most of these rights were created in the absence of clear evidence that pesticide risks were significant and difficult to control. Thus government normally concluded that if products were used in accordance with labeling precautions, significant risks would be avoidable.

In many cases, registration was followed by rapid growth of domestic and international pesticide markets, a process that inflated the value of the original entitlement to the manufacturer. The release of pesticides into the environment also created a monitoring and enforcement problem of overwhelming proportions, because it is necessary to know where residues end up to manage exposure and risk.

The 1972 FEPCA amendments threatened these entitlements in several ways. First, Congress required that EPA consider public health and environmental risk directly and balance these effects against benefits. Second, scientific evidence appeared to Congress to be changing rapidly—hence its requirement for five-year reviews of every registration. As the reregistration program evolved during the late 1970s and especially during the 1980s, the conflict between claims of risk and claims of property intensified.

Although Congressional leaders may have had the best of intentions, they had little understanding of the scale of the analytical burden they imposed on EPA. The young agency faced a mandate to review the environmental health and safety of tens of thousands of products, formerly registered by USDA in the absence of sufficient quality data. Given the workload estimates and promises made before Congress, it is probably fair to say that EPA administrators did not fully comprehend their predicament for the first decade of the agency's life. In addition to the impoverished database, the statutory ambiguity of key terms such as "unreasonable," "risk," "cost," and "benefit" conferred tremendous administrative discretion on EPA as well as an analytical task of immense proportions. How was EPA to estimate risks, costs, and benefits? How should it translate risks into costs so that they may be meaningfully compared with benefits? When were risks so significant that benefits should be disregarded, and product registrations forbidden?

Even when Congress allowed EPA to review registrations by "active ingredient" via the 1978 amendments, the list still included six hundred compounds. The agency's request for additional data necessary to judge health and safety effects was not completed until 1988. The agency will not even receive all of these data until some time early in the twenty-first century. Given the time required to review these data, however, and the high probability that standards for the quality of these studies will continue to rise, more data will likely be requested, and the pattern of delay in the pursuit of certainty will continue. The agency has mistakenly believed that there is an end to the process. Instead, the agency lies at the intersection of two highly dynamic and endless social processes—law and science—where decision rules have remained ambiguous.

The Congressional strategy of reregistration was built on a logic that later evidence of significant risk would lead EPA to modify pesticide-use entitlements. In this way, it was thought, risks could be reduced to acceptable levels while preserving as large a proportion of the benefits as possible. EPA's scarce analytical resources have been directed toward understanding risks of pesticides, but meanwhile a pesticide's benefits have been deemed significant simply by the fact that corporations have requested registration and are willing to bear the costs of data production and legal wrangling.[63]

Much of this chapter has been devoted to showing how debates about the quality and certainty of claims of risk have become a central strategy to protect entitlements for pesticide use. If new claims of risk threatened existing private entitlements, stakeholders have been quick to expose any uncertainty surrounding estimates of future damage. Because these forecasts are always uncertain—and therefore easy to criticize—they normally provided a basis for delaying prohibitive decisions while more certain evidence was sought. Predictably, manufacturers have demanded a high level of certainty before accepting a reduction in entitlements—especially an uncompensated reduction—and their demands have resulted in lengthy regulatory delays.

EPA has had different incentives that have also promoted regulatory delay. These pressures have included a very limited budget (given the scale of its responsibility); the need to cultivate its image as a careful and unbiased scientific body; its inheritance from USDA of a woefully inadequate database on the environmental health effects of thousands of products; and the necessity to carefully control its decisionmaking agenda to avoid too many controversial choices at any one time. Thus both key actors in the regulatory process—the manufacturers and EPA—have had powerful but different reasons to defer more prohibitive decisions.

Key questions emerge from this analysis, however—questions that center on the effects on public health and environmental quality of these bureaucratic pressures and counterpressures. If uncertainty in risk assessment had the effect of protecting pesticide-use registrations, how did it influence the setting of limits for food contamination by pesticide residues? Are tolerances for pesticide residues in foods set to protect the public health from significant risks? I address these questions next.

CHAPTER 6

Risk Assessment and Tolerance Setting

The Delaney Paradox

In 1992, Senator Edward Kennedy sent a letter to William Reilly, then administrator of EPA, requesting more information about pesticide residues legally allowed to persist on foods. What Kennedy wanted to know was whether the legal tolerances adequately protect public health from the toxic effects of pesticides. Linda Fisher, then an assistant administrator under Reilly, responded that the residue limits allowed by law often exceeded a "safe" exposure level for at least sixty of 325 food-use pesticides, if average dietary exposures were assumed. Fisher's response indicated that if childhood patterns of food intake and exposure were used to answer the question, many more compounds would be implicated. Reflecting on the list provided to Kennedy, Fisher commented: "As you can see, infants and children are the two subgroups that typically receive the most exposure to pesticide residues in the diet as a percentage of body weight."[1]

EPA's admission in 1992 that the complex network of tolerances was not health protective, especially for infants and children, raised important questions concerning how the tolerance system evolved and how unresponsive it has been to rapidly evolving toxicological evidence that residues may pose a significant hazard, especially to young people.[2] Before addressing these questions, however, it is necessary to understand the convoluted legal architecture that has shaped the tolerance system.

The Legal Context

The most important event in regulating a pesticide is the act of registration, which is governed by FIFRA and described more fully in chapter 5.[3] After a pesticide is registered, the primary method of controlling human exposure from food residues is tolerance-setting, which establishes a legal limit on the amount of a residue that can remain on a food. Nearly 8,500 tolerances were set by FDA before 1970, and FIFRA requires their periodic review by EPA. Each tolerance specifies maximum allowable residues of nearly 325 pesticides on 675 different types of food. Tolerances need not be set for the inactive pesticide ingredients, which may also persist as food residues.

Tolerances are required to be established by the Federal Food, Drug, and Cosmetic Act (FFDCA).[4] Prior to 1972, the secretary of Health, Education, and Welfare had this authority and acted through FDA, one of its divisions. FDA would then confer with the Department of Agriculture, which was responsible for registering the use of a pesticide on a specific crop. So two approvals were required from the federal government before a compound could be sold and used on any crop—a registration from USDA, and a tolerance from FDA.

This fractured administrative responsibility caused considerable conflict between USDA and FDA. Because USDA had traditionally registered pesticides with the expectation that tolerances would be set to ensure only safe human exposures, there was a clear need for interagency coordination to estimate probable residue levels in foods and associated risks. Yet coordination was inhibited by the fundamental differences in purpose between the two bureaucracies—USDA had the mission of improving crop yields, whereas FDA was responsible for ensuring food safety. By 1969, nearly sixty thousand separate products were registered by USDA, and removal of any registration required that FDA produce evidence of hazard to health or the environment—an overwhelming burden. In that year, the Mrak Commission recommended that this responsibility be shifted to the manufacturer to demonstrate safety and that registration be renewed every five years.[5]

In 1970, both registration and tolerance-setting responsibilities were consolidated within EPA. By 1972, FIFRA was strengthened to require that registration be contingent upon EPA finding no "unreasonable adverse effects on the environment, taking into account the economic, social, and environmental costs and benefits of the use of any pesticide."[6] When deciding whether to cancel or suspend a registration, EPA is also directed to consider the effect of their decision on "production and prices of agricultural commodities, retail food prices, and otherwise on the agricultural economy."[7] Although the 1972 amendments are commonly described as having strengthened the environmental health considerations of FIFRA, EPA retained substantial discretion to judge what is reasonable and adverse and to balance costs and benefits. Thus, even if EPA found health risks to be significant, it could issue or continue a registration if it concluded that benefits were substantial.

The Food, Drug, and Cosmetic Act

Although pesticide registration is governed primarily by FIFRA, tolerance-setting—establishing allowable residue limits in foods for specific pesticides—is governed by two sections of FFDCA.[8] The first section, §408 or "the Miller Amendment," was adopted by Congress in 1954 and directs that residue limits be established for "raw agricultural commodities" such as fresh fruits, vegetables, or milk. Residues that "induce cancer" in laboratory animals or humans are considered to be food additives under certain restricted circumstances and are not permitted for use on food crops or animals. Each of these provisions is described briefly below.

Following the rapid expansion of the pesticide industry during the decade following World War II, especially the growth in use of highly persistent chlorinated insecticides, Congress focused on the potential health effects of pesticide residues in the food supply. In a departure from FIFRA, which simply required registration and labeling, the FFDCA was amended in 1954 to permit registration of pesticides only if the manufacturer presented data demonstrating that residue levels on food crops posed no danger to public health.[9] Yet public health was not the only criterion for tolerance-setting; FDA was required to recognize the need for "an adequate, wholesome, and economical food supply."[10] This dual standard, necessitating the protection of both public and economic health, has become the essence of the nation's pesticide control strategy, as structured both by FIFRA and FFDCA.

Through the Miller Amendment, FDA secured broad authority to issue tolerances for each pesticide-crop combination. The law allowed but did not compel FDA to consider benefits in addition to risks.[11] Once the residue limit or tolerance has been set, any food that contains higher than allowable levels is "unsafe," "adulterated," and therefore unlawful. The amendment also began a process of fragmenting administrative authority to control pesticides between USDA, the Department of Health, Education, and Welfare, and the Department of the Interior (especially the Fish and Wildlife Service). Whereas USDA remained responsible for pesticide registration between 1954 and 1970, when EPA was created, FDA held the authority to determine a safe level of food contamination for each pesticide and crop.

Pesticide Residues in Processed Foods (§409)

One of the most enduring debates over pesticides concerns their potential to induce cancer. EPA is required to judge the potential for every pesticide to cause a wide array of health problems, including cancer, neurological damage, reproductive failure, birth defects, mutagenic effects, and fetotoxic effects. Despite this list of concerns, EPA's attention during pesticide regulation has been primarily directed toward judging the link between pesticides and cancer. The disproportionate attention afforded cancer appears to have several causes. The first has simply been the increase in cancer incidence in the U.S. population—more than 25 percent of U.S. citizens will experience some form of cancer in their lifetime, and the public has clearly been alarmed about the potential link between environmental contamination and the disease. The second reason is the legal requirement that prohibits FDA from approving any food additive (including pesticides under certain circumstances) shown to induce cancer in animals or humans. This prohibition is part of the Delaney clause, which is contained within the general safety clause of the FFDCA, §409.

This clause requires that a manufacturer demonstrate with reasonable certainty that any proposed food additive will cause no harm to consumers.[12] Thus food additives found to induce cancer, including pesticides under some circumstances, are prohibited, a policy effected by either prohibiting registration or setting a tolerance for allowed residues to zero. This "zero risk" standard lies in stark contrast to the standard for pesticide residues on raw agricultural commodities, which directs the administrator to balance the need to protect public health with the need to provide the public with an "adequate and wholesome food supply."

When Congress adopted the Delaney clause in 1958, it foresaw a serious conflict in the application of the two standards. Pesticide residues on raw foods, subject to the risk-benefit standard, often migrate to processed food, subject to the zero-risk standard. Strict application of the this standard would have effectively canceled some tolerances, or residue limits, on raw commodities based on their concentration during processing. These conditions, although understandable, are hard to justify. Congress included a "flow-through" provision in §402 that allows a residue to remain in processed food at the same concentration allowed on raw foods. If the pesticide in question is found to "induce cancer," and if the processed food residue level (say in apple juice) exceeds the allowable level on the raw commodity (in this case, fresh apples), then the pesticide becomes a food additive governed by the Delaney amendment, and the food is declared adulterated and unlawful.[13] If residues do not concentrate during processing, then the risk-benefit balancing standard, which provides for more discretion in tolerance setting, is applied.

The term "tolerance" is confusing because it implies some biological tolerance to a pesticide. Instead, it is a maximum allowable residue level in food, set with one eye cast toward public health protection and the other directed toward crop protection. Historically, tolerances on raw foods have been set to legalize anticipated residue levels of a pesticide after it is applied in field tests at a concentration necessary to accomplish intended pest control. These field tests were normally conducted under conditions likely to cause residue persistence. These circumstances mean that the allowable contamination ceilings were originally set with a primary concern for field effectiveness.

Raw food tolerances set to legalize high concentrations of residues that *might* occur under extreme field application conditions have repeatedly been criticized by environmentalists for their failure to protect public health. FDA—responsible for monitoring residues in the food supply—claims yearly that the food supply is safe based upon the low incidence of cases in which residue levels exceed legal tolerances. Health advocates, however, argue that this outcome is like setting the highway speed limit at five hundred miles per hour and claiming the roadways are safe because people generally drive only at two hundred miles per hour.[14] Also, this high ceiling allows for significant residue concentration during processing. Any petitioner for a raw food tolerance must consider the potential for a residue to concentrate during processes such as juicing, milling, or drying. One obvious strategy to avoid the Delaney prohibition has been for manufacturers to request raw-food tolerances high enough to allow for concentration of residues. If carcinogenic residues in processed foods do not exceed the generously high raw-food tolerances, the Delaney prohibition is avoided and both the parent and processed foods are protected from being impounded as "adulterated."[15]

Although the proportion of the nation's food supply made up of processed foods is expanding rapidly, the database on the effects of processing has astonishingly few samples per food, and is often years if not decades old. A National Research Council Committee concluded in 1993 that "a comprehensive study of the effects of processing on food residues is badly needed."[16] The absence of evidence of residue concentration during food processing is commonly used to explain the relative abundance of raw food tolerances compared with processed food tolerances. The Delaney guillotine, which is supposed to fall not only on the processed food tolerance but also on the "parent" raw food tolerance, is thereby avoided.[17]

Routes for Escaping the Delaney Prohibition

Although the Delaney clause appears to be quite clear in its prohibition against food additives that "induce cancer," the history of its implementation by both FDA and EPA demonstrates the enormous role that scientific uncertainty has played in federal attempts to escape from its strict interpretation.[18]

The Delaney clause was passed in 1958 and the color additive amendments in 1960. The initial expectation was that few compounds (pesticides, colors, or other additives) would be affected.[19] Two trends dashed these hopes. First, the universe of compounds meeting the definition of food and color additives continually expanded as analytical testing became increasingly sensitive. Second, as more compounds were tested, more were found to induce cancer, especially if they were administered to laboratory animals at high doses. Both FDA and EPA increasingly found that older compounds, previously believed to be noncarcinogenic, induced cancer in laboratory animals.

As a result of these developments, a different logic emerged within the agency tied to advancements in our understanding of how cancer evolves. During the early 1980s, considerable progress was made by toxicologists in understanding the relationship between dose and tumor response. Animal tests are inherently problematic for estimating human cancer risk, not only because of the inference required to suspect the same effect in humans, but also because of the magnitude of dosing often required to demonstrate a statistically significant difference in effects between control and treated groups. Inducing tumors by feeding animals extremely high doses of a chemical provides toxicologists with a basis for "extrapolating" the tumor incidence from the higher animal doses to the lower doses commonly experienced by humans.

The concept of dose-response extrapolation became a cornerstone for judging the relative potency of cancer-inducing agents during the 1980s.[20] EPA maintained a list of pesticides that caused cancer in laboratory animals, each of which was given a number indicating the compound's potency.[21] As more pesticides were tested, more were added to EPA's list of carcinogens, raising the possibility that the Delaney clause might apply to residues concentrating in processed foods.

Once toxicologists have determined that cancerous lesions are in fact occurring at a statistically higher incidence in treated animals, controversy then focuses on the method or "model" chosen to estimate human-tumor incidence based upon the animal data. Animal tests are structured to compare tumor incidence between exposed and unexposed groups and to identify if the tumor incidence is related to the magnitude of the dose or exposure. Thus animals are commonly segregated into high-, moderate-, and low-exposure groups, after which their effects are compared.

Tremendous controversy has surrounded the choice of methods to forecast human cancer risks based upon animal evidence. If a dose-response relation is found, how should test results be interpreted when human exposure would normally be far lower than the lowest dose that caused cancer in animals? EPA has normally assumed that the dose response relation is linear and that any exposure, even the most minute, increases the probability of tumor incidence.

FDA rarely relied upon the Delaney clause to ban a substance from the food supply, and during the 1960s few decisions even required FDA's interpretation. The administration was able to sidestep this rule because of both the very limited definition of a "food additive" and the several explicit exceptions provided in the statute. Pesticides, as described above, become food additives only under very restricted circumstances. Thus pesticide residues in or on raw agricultural commodities are specifically exempt from the prohibition. Only residues from pesticides that induce cancer and that concentrate during processing are caught in the Delaney trap. Another exception applies to drug residues that can flow through livestock to contaminate meat, dairy products, and eggs. A third category includes compounds that have been approved by FDA or USDA prior to 1958.[22]

Carcinogenic food additives may also make their way into the human food supply via contaminated animal feed. Before 1958, FDA had approved numerous hormones to promote livestock growth, including diethylstilbestrol (DES). DES was administered via animal feed, and was known to cause cancer in animals in the 1950s.[23] The Delaney clause is clear in its prohibition of carcinogenic additives to human food or to animal feeds, but FDA then believed that the hormone could not be detected in the human food supply. Following the 1958 Food Additives Amendment, FDA stopped approving new petitions for DES use; existing producers, however, were allowed to continue its distribution. Congress yielded to drug-industry pressure and amended the FFDCA in 1962 with the "DES proviso." This escape route does not prohibit FDA from approving a carcinogenic additive, as long as no residue of the additive is found in human food using an approved method of chemical detection.[24]

This "best available method" strategy amounted to a legal recipe for constant change in the terms for approval, because the sensitivity of detection technology is continually advancing. For example, prior to 1971 DES could not be detected in beef liver if the administration of DES was stopped forty-eight hours before slaughter. A new detection method developed in 1971, however, identified DES residues beyond the forty-eight-hour window, and administration was then stopped seven days before slaughter. By 1972, FDA was able to find residues persisting beyond seven days. A new detection method, which attached radioactive isotopes to DES, found that after the hormone was implanted residues were detectable for at least 120 days. Finally, in 1979, FDA revoked all new animal applications of DES with the Delaney clause, arguing that no current analytical method of detection was sufficient.[25]

In 1970, FDA required that anyone wishing approval for a carcinogenic additive must supply an analytical method capable of detecting the residue to a level of 2 parts per billion (ppb).[26] Yet requiring a uniform sensitivity among methods of detection ignored the fact that different compounds have different toxic potencies at the same concentration. Whereas one compound might be quite safe at 1 ppb in the diet, another could be quite hazardous.

Residues of some highly toxic substances, for example, may pose significant risks at concentrations less than 2 ppb. This strategy also ignores variance in human exposure caused by wide differences in dietary habits. These patterns are considered more fully in chapter 9.

FDA recognized this dependency of risk on detection limits, and proposed a solution that would guide its policy toward carcinogenic animal drugs and feed additives for nearly two decades. The approach became known as the "sensitivity of method" (SOM) policy, and it required that residue-detection sensitivity be tied to the carcinogenic potency of each compound. Thus, for any compound to be considered carcinogenic, the residue-detection test must be sensitive enough to find residues at a level that poses a significant human cancer risk. In 1973, FDA defined "significant risk" as an excess lifetime risk of one in one million; in 1977 it revised this level to one in 100 million, which essentially required a hundredfold increase in detection sensitivity.

This "sensitivity of method" policy is a watershed in the history of U.S. attempts to regulate environmental carcinogens, because it ties the prohibitive power of the Delaney clause to quantitative risk assessment methods. Animal test data are used to judge the toxic potency of a compound and to identify, if possible, a dose-response relationship. Human exposure is then estimated and combined with the potency estimate to project human risk. To a layperson, these estimates may appear to be sound scientific evidence. In reality, considerable scientific uncertainty embedded in the projections of risk provides FDA with a wide berth of discretion in judging whether any compound "induces cancer." Uncertainty may result from differing interpretations of tissue and lesion samples, the quality of a study's design, or the statistical significance of results. Many additional questions are related to the conclusion that a compound "induces cancer": "FDA effectively has authority to decide what evidence will support or dictate a positive finding. Will it be satisfied with a positive result in one sex of one species? In both sexes of a single species? Only in two species? Are benign tumors to count? Only if they accompany malignant tumors?"

The method also demanded an estimate of the range of human exposure, which a later discussion will show has been grossly underestimated by both FDA and EPA for pesticide residues. By linking the required detection sensitivity to quantitative estimates of risk, FDA believed it was controlling for variance in potency among carcinogens. What it failed to recognize was that for any "additive" the one in 1 million risk threshold could be calculated in an infinite number of ways, depending upon choice of animal study, interpretation of tissue samples, animal-to-human extrapolation methods, residue data, and food intake data. The ultimate effect of the policy has been to permit residues of additives in the food supply if they pose "less than significant risks." The flat prohibitory language of Delaney had been circumvented yet again.

The last attempt by FDA to strictly enforce the Delaney clause came in 1977 when it attempted to ban saccharin, an artificial sweetener. A series of animal studies during the 1970s demonstrated that saccharin induced malignant tumors in rats. The "sensitivity of method" policy did not apply because saccharin is added directly to human food, and the constituents policy was not relevant because pure saccharin was proved to be cancer-causing in a Canadian study. When FDA conducted its quantitative risk assessment, it concluded that the sweetener posed an excess risk of one tumor among ten thousand exposed,[27] a result far higher than its recently articulated threshold of significant excess risk of one tumor among one million people exposed.[28] Credible escape routes from a strict interpretation of Delaney appeared shut, and in 1977 FDA proposed to ban saccharin.[29]

The public outcry that ensued was enormous, however, and it was voiced especially by diabetics and those with other health-related weight control problems. In the same year, Congress enacted the Saccharin Study and Labeling Act,[30] which stalled the FDA ban for two years, required product labeling, and requested that NAS study the problem.[31] Congress reauthorized the restraint on four additional occasions, allowing saccharin to remain in the marketplace.[32]

FDA's more intensive testing of color additives during the 1970s demonstrated the carcinogenicity of the colors themselves, but unexpectedly showed that trace contaminants associated with the colors also caused malignant tumors in animal studies. These trace contaminants became known as "constituents" of additives, and presented the FDA with cases in which the additive—or color—was not carcinogenic, but the trace contaminant was. FDA found this problem while reviewing Green No. 6, which contained the contaminant n-toluene, a known animal carcinogen. FDA decided again to rely upon quantitative risk assessment to judge whether the potential for harm was significant, and its conclusion was driven primarily by low estimates of human exposure. FDA forecast excess human cancer risks ranging between one in 15 million to one in 150 million from n-toluene exposure. Because these ranges were well below the one in 1 million risk threshold that the FDA had defined to be significant, it decided to approve the listing of Green No. 6. Challenged in court, FDA's decision was upheld and implicit approval was given for using quantitative risk assessment to judge "reasonable harm."[33]

During the same period in which FDA struggled to interpret the application of the Delaney clause to feed additives and sweeteners, it also faced increased evidence of the carcinogenicity of widely used food-packaging materials. For example, by 1974 acrylonitrile had already been granted approval for use in cellophane, adhesives, paper, and various plastics, all of which were in contact with food. Increasing evidence of an association between cancer and occupational exposure to acrylonitrile was supported by then-incomplete animal bioassays. The issue drew even greater attention when Coca-Cola Company and Monsanto agreed to begin broadscale marketing of soft drinks in plastic bottles containing acrylonitrile. In retaliation, the commissioner of FDA reversed prior approvals of Monsanto's food additives.[34]

Monsanto won a court ruling requiring a hearing on safety, and the new FDA commissioner, Donald Kennedy, ruled that the compound was *not* safe because residues were detected migrating from the bottle to the drinks. Monsanto reformulated the bottle to reduce levels of migration, but Commissioner Kennedy still concluded that the bottle constituted a food additive and that migration would occur even if it remained undetected.[35]

Kennedy's narrow interpretation was overruled by the D.C. District Court, which seemed to expand FDA authority to interpret the meaning of key terms such as "safety" and *de minimis* risk. The court declared: "The Court is . . . concerned, that the Commissioner may have reached his determination in the belief that he was constrained to apply the strictly literal terms of the statute irrespective of the public health and safety considerations. . . . [T]here is latitude inherent in the statutory scheme to avoid literal application of the statutory definition of 'food additive' in those *de minimis* situations that, in the informed judgment of the Commissioner, clearly present no public health or safety concerns."[36] The ruling seemed to give FDA authority to judge when "no public health concern" existed, even when the substance in question would appear in the food. FDA finally approved Monsanto's acrylonitrile bottle, based on its conclusion that although the bottle was the "additive," the cancer-inducing agent was a "constituent."

The 1960 Color Amendments to FFDCA contain a prohibition against cancer-inducing additives similar to the Food Additive Amendments of 1958. FDA has broad authority to judge the safety of colors, even those demonstrating toxic effects other than cancer. The history of Orange No. 17 and Red No. 19 demonstrate the tension between a general safety standard and the zero-risk cancer standard. Both colors had been used in drugs and cosmetics at the time the color amendments were adopted, and therefore each became eligible for "provisional listing"—that is, their use would be allowed pending the outcome of toxicity testing. In August 1986, FDA concluded that each additive was "safe" for humans, despite the fact that both colors induced dose-related tumors in animal experiments. Previous policy within the administration would first have questioned if the chemical was legally a "food additive," and second, if it caused cancer in humans or animals. If the answer to both questions was yes, then FDA's policy was to ban the substance from food, drugs, and cosmetics.[37]

FDA employed quantitative risk assessment to judge the excess cancer risk, which it found to be one in 19 billion for Orange No. 17 and one in 9 million for Red No. 19.[38] FDA concluded that these risks are relatively insignificant—the average person's chance of being struck by lightning is more than a hundred times higher than the chance of developing cancer from normal exposure to Red No. 19. FDA still confronted the language of the Delaney clause, however, by invoking the concept of *de minimis* or "trivial" risks articulated in *Alabama Power Co. v. Costle.* In this case, the D.C. Circuit Court stated: "Unless Congress has been extraordinarily rigid, there is likely a basis for an implication of *de minimis* authority to provide exemption when the burdens of regulation yield a gain of trivial or no value."[39]

FDA concluded that the Delaney clause was not "extraordinarily rigid" and that it therefore provided the administration with the authority to classify some risks as "trivial." Using this logic, the significance of risk became inextricably tied to the magnitude of exposure. If one could determine that human exposure was minimal, risks would then become trivial and the compound could remain in the marketplace. With this reasoning, FDA approved Orange No. 17 and Red No. 19.[40]

But in 1987 Public Citizen sued FDA for their liberal interpretation of Delaney, and the same court that had opened the discretionary door for FDA slammed it shut. The D.C. circuit found that FDA's position was justified neither by court precedent nor by the statute's legislative history, even though the risk posed by the two colors in question seemed "altogether fair to characterize . . . as trivial."[41]

Pesticides as Food Additives: The Delaney Paradox

One of the most complex legal questions posed by the Delaney clause concerns its application to pesticides. As described above, pesticides that induce cancer are allowed to remain as residues on raw agricultural commodities and in processed foods as long as processed food contamination levels do not exceed those found in raw or "parent" foods. Among all classes of food additives, pesticides may therefore contribute the largest proportion of dietary risk, simply because so many different residues from pesticides classified as possible or probable carcinogens are permitted on a large number of foods.[42]

Only a small percentage of chemicals marketed in this country have been subjected to complete health and safety evaluations. In 1984 NAS estimated that among the 8,627 substances "regulated or classified by FDA . . . as direct food additives, indirect food additives, GRAS substances, colors and flavors," no toxicity information was available for 46 percent of them, and data then deemed adequate to conduct a full health-risk evaluation existed for only 5 percent.[43]

The National Cancer Institute and the National Toxicology Program have tested suspicious compounds for carcinogenicity over the past two decades. As more compounds are tested, the list of chemicals demonstrating either limited human evidence or sufficient evidence from animal bioassays has continually expanded. Among 540 chemicals tested by the National Cancer Institute in 1975, 86 demonstrated some carcinogenic activity. By 1984, Bruce Ames—a professor of biochemistry at the University of California who is well known for his work on "natural" carcinogens—reported that among two hundred chemicals tested by NCI over eight years, approximately 60 percent were carcinogenic. At that time, Ames concluded: "We have no idea of what the true percentage of carcinogens is among chemicals in general (including natural ones) when tested at the maximum tolerated dose in rodents. Even if it is 10 percent, our current regulatory policies, which assume that carcinogens are rare, are in trouble."[44]

Among pesticides, the universe has similarly expanded. In 1986, EPA reported that approximately forty compounds among 325 food use pesticides demonstrated some evidence of carcinogenicity. By 1987, EPA listed fifty-three of 289 tested compounds as demonstrating carcinogenicity. By 1993 this number had expanded to nearly seventy-five. Among all pesticides used in food, the current estimate by EPA toxicologists is that at least one-third of the 325 are likely to demonstrate some tumorigenic evidence when tested at the maximum tolerated dose in laboratory animals.[45]

Despite the steadily expanding evidence that some pesticides can cause cancer, EPA has moved at a glacial pace to revoke product registrations or reduce food tolerances. There has also been little effort by EPA to invoke §409 of FFDCA and issue tolerances for additives to processed foods in those cases where pesticide migration from raw to processed foods (for example, from tomatoes to tomato paste) is anticipated. Among all tolerances that EPA is responsible to monitor, 97 percent have been issued for raw foods only.[46]

In 1987, EPA provided NAS with evidence of carcinogenicity for fifty-three registered pesticides. These fifty-three compounds in the nation's food supply were legally permitted by their 2,525 separate §408 tolerances, which were in force in 1987.[47] For these same fifty-three pesticides, only thirty-one processed food tolerances had been set. This means that EPA had allowed residues of these compounds to exist on diverse raw foods and further had avoided establishing tolerances for their presence in processed food (required for cases in which residues concentrate). Overall, in 1987 there were 7,372 tolerances for raw foods, and 122 for processed foods, listed in the Code of Federal Regulations.[48]

Why are there so few processed food tolerances compared to those for raw foods, particularly when there is a significant potential for pesticide concentration in the production of oils and dried foods? Why has so little attention been directed toward processed foods when recent estimates show that the average American eats far more processed foods than raw foods? The answer lies in part in the enormity and expense of the task of tracing the fate of any pesticide residue through the nation's food supply. Rapidly changing food-processing technologies exacerbate this problem. Very few compounds have been thoroughly tested to identify how residues transfer from raw to processed foods. Nor has the fate of metabolites, degradation products, and impurities been tracked for most pesticides. In 1987, EPA estimated that it had complete residue chemistry data for only 25 percent of the compounds then registered.

The absence of residue evidence is often used to support claims that the pesticide is absent from the food supply. This bias clearly works to the manufacturer's advantage because without data, EPA has no need to set a processed food tolerance, and the pesticide continues to be judged under the risk-benefit balancing standards of FIFRA and §408 of FFDCA. If the pesticide is carcinogenic, the concentration of residues may have enormous financial implications for the producer because concentration invokes the zero-risk Delaney standard. The only way EPA can ensure the absence of processed food residues is to prohibit the pesticide's use on raw agricultural commodities.

The potential for errors in this logic, however, was realized in the case of Alar, a plant growth regulator applied to a variety of fruits, including apples. By 1989, EPA had known of the carcinogenic potential of Alar in test animals for over a decade. Although Alar—known also as daminozide—itself did not appear to cause cancer, one of the compound's metabolites, unsymmetrical dimethyl hydrazine (UDMH) (also a component of rocket fuel), was found to induce malignant tumors in young laboratory animals. Because Alar degenerates into UDMH as a result of heat processing—common in the preparation of apple juice and apple sauce—UDMH levels in these processed foods far exceeded the levels found on raw apples. The agency was forced to confront the Delaney clause directly due to the clear evidence of residue concentration of a carcinogenic substance. The product was banned following perhaps the most dramatic public controversy over a pesticide in this century.

Thus there are two primary ways that a pesticide may escape the need for a §409 tolerance. The first is the absence of convincing evidence regarding the carcinogenic potential of the final compound, its metabolites, or its degradation products. The second is the absence of evidence concerning the concentration of residues in processed foods. If both of these escape routes are closed, the Delaney clause applies, but only to the specific uses where both criteria are met. Uniroyal, for example, which manufactured Alar, could have decided to voluntarily refrain from using Alar on those crops processed in such a way that residues would concentrate. This strategy would have resulted in the risk-benefit balancing criterion being applied to all non-residue-concentrating uses of Alar, despite its evidence of carcinogenicity.

In 1985 NAS convened a committee to examine the scale of cancer risk to the public given that the overwhelming majority of tolerances had been set under a risk-benefit balancing standard, no data on processing effects had been compiled, and the universe of carcinogenic pesticides was expanding. The committee spent nearly a year trying to understand the logic that lay behind tolerances that EPA sets on both raw and processed foods. The committee at first wrongly assumed that tolerances had been set at a level that was health protective. This assumption was checked partly by testing the relationship between the toxicity of individual compounds and total allowable exposure. One would expect and hope that the allowable exposures would be lowest for pesticides exhibiting the highest toxic potency. The finding was unsettling: there was no such inverse relationship. Another source of confusion for the committee was that although the universe of carcinogenic pesticides was increasing steadily as systematic testing of older compounds was completed, there was no corresponding increase in the number of tolerances for processed foods.

The committee's attempt to understand these seeming contradictions led them to conduct their own analyses of pesticide residue, food intake, and toxicity data to try to estimate cancer risks associated with the fifty-three suspected carcinogens. The effort was made exceptionally difficult by the poor quality of residue data, particularly by outdated or nonexistent studies of processing effects. Benomyl, for example, is one of the world's most widely used fungicides, and EPA had issued nearly seventy-five separate tolerances for its use on foods such as apples, pears, peaches, and bananas. Yet these tolerances had been granted to its manufacturer, the du Pont Corporation, with very little evidence concerning potential transfer of residues to processed foods such as fruit juices, which make up a large proportion of children's diets. Often residue-test results relying on small samples were submitted to support original tolerances on raw foods or processed foods. In many cases, few if any studies on the effects of processing had been conducted. Given the speed of change in food-processing technologies, there is little hope that one or two studies could be representative of the residue levels likely to appear in billions of pieces of fruits or vegetables in the nation's food supply. Worse, EPA's scarce resources constrained it from conducting a complete reassessment of a chemical's health and safety data more than once a decade despite the Congressional mandate to review each compound every five years. Thus if the agency errs by underestimating risks, the tolerances were not likely to be reviewed for ten to fifteen years.[49]

The poor quality of national data on pesticide residues led the NAS committee to decide that they should not even use the data to judge human exposure to the carcinogenic pesticides in question. Instead, they decided to estimate exposure and cancer risk by making certain assumptions. First, they assumed that residues were on foods at the legally allowed tolerance level. Second, in the absence of either processing-effect studies or tolerances for a processed food, they assumed that residues would "flow-through" raw foods to processed foods at different transfer rates: 100 percent, 50 percent, or not at all. These are especially difficult assumptions to support given the variability in pesticides' metabolites and degeneration products in different foods. Also, there is the possibility of "flow-through" at levels above 100 percent—especially if processing involves drying, juicing, or oil extraction.

The cancer risk associated with any individual compound was believed to be dependent upon two variables. The first is the potency of the compound, or the strength of the dose-tumor response relationship. The second is the magnitude and duration of exposure.[50] More recently, as described in chapter 8, a third variable may be the age at the time of exposure, which may determine special periods of susceptibility. EPA has developed a system to classify compounds by the strength of evidence suggesting potential human carcinogenicity.[51] The strength of the relation is judged using a variety of criteria including corroboration among studies demonstrating the same effect; the strength of the dose-response relation; the diversity of tumor types produced; the proportion of benign to malignant tumors; the age of the animals at the time of tumor onset; the structural similarity of the compound to other known carcinogens; and the results of mutagenicity tests.[52]

The NAS committee reviewed exposure and dose-response data provided to it by EPA and found that among the fifty-three compounds with some carcinogenic evidence, only twenty-eight had evidence of sufficient quality to justify a quantitative estimate of risk.[53] Of these, ten were then classified as "B2" carcinogens, eleven as "C" carcinogens, one as "D," and six were not then classified.

Table 6–1 lists the pesticides that the NAS committee found posed the greatest carcinogenic risks from their use on all foods for which EPA has issued tolerances. A review of its data revealed several important insights into EPA's regulatory program. First, eighteen of twenty-eight pesticides posed a

Table 6–1. NAS Estimates of Cancer Risks from Selected Pesticides

Active ingredient	Number of crops	Estimated risk at tolerance	Type of pesticide	Class of carcinogen	Potency or q1*
Linuron	20	1.5×10^{-3}	Herbicide	C	3.28×10^{-1}
Zineb	83	7.2×10^{-4}	Fungicide	B2	1.76×10^{-2}
Captafol	34	5.9×10^{-4}	Fungicide	B2	2.50×10^{-2}
Captan	83	4.7×10^{-4}	Fungicide	B2	2.30×10^{-3}
Maneb	56	4.4×10^{-4}	Fungicide	B2	1.76×10^{-2}
Permethrin	43	4.2×10^{-4}	Insecticide	C	3.00×10^{-2}
Mancozeb	44	3.4×10^{-4}	Fungicide	B2	1.76×10^{-2}
Folpet	41	3.2×10^{-4}	Fungicide	B2	3.50×10^{-3}
Chlordimeform	24	3.2×10^{-4}	Insecticide	B2	9.40×10^{-1}
Chlorothalonil	47	2.4×10^{-4}	Fungicide	NA	2.40×10^{-2}
Metiram	11	1.2×10^{-4}	Fungicide	B2	1.76×10^{-2}
Benomyl	101	1.1×10^{-4}	Fungicide	C	2.07×10^{-3}
O-phenylphenol	22	1.0×10^{-4}	Fungicide	NA	1.57×10^{-3}

Source: NAS, *Regulating pesticides in food: The Delaney paradox* (Washington, D.C.: National Academy Press, 1987), table 3.9.

greater than one in 1 million excess risk, if one assumed that residues persisted on foods at the tolerance level. Second, the majority of these compounds were classified as "probable human carcinogens," predominantly due to animal evidence. Third, the majority of the compounds posing the greatest risk were fungicides. These fungicides are widely used on fruits and vegetables and some have the potential to migrate to fruit juices, which are heavily consumed by infants and children. Fourth, it was difficult to interpret the relative influence of a compound's potency (q1*) compared with its distribution in the food supply ("exposure") as contributors to carcinogenic risk. Still, the table does suggest a rough correlation between risk and the potency of the compound.

For all fifty-three presumed carcinogens studied, EPA had established nearly 2,500 raw food tolerances (§408), while it had set only thirty-one processed-food tolerances (§409). Among the twenty-eight compounds studied intensively by the NAS committee, only nineteen had processed-food tolerances. EPA appeared to have avoided setting processed-food tolerances by claiming the evidence was inadequate. This meant that EPA had not required adequate processing-effect studies to be submitted even though it permitted continued use of these pesticides. The committee anticipated that when fully tested, many of the compounds would be found to concentrate during processing and would therefore trigger application of the Delaney standard. If evidence of concentration emerged, EPA would need to either cancel the raw food tolerances underlying the acceptance of residues in processed foods, or construct a new interpretation of the Delaney clause.

The analyses also suggested that the majority of cancer risk from pesticides comes from raw foods, whereas the Delaney standard applied only to processed foods and their parent raw foods. This finding appeared to justify reform that would apply a uniform negligible-risk standard to both raw and processed foods. The committee also found that the severity of the Delaney clause could easily prevent an older, higher risk pesticide from being replaced by a new, lower risk pesticide, simply because the newer pesticide happened to concentrate during processing. The zero-risk standard therefore diminished the ability of EPA to encourage a pattern of pesticide substitution that would gradually reduce pesticide-related cancer risk in the food supply.

The academy released its study in 1987 to a storm of industry criticism. Pesticide manufacturers, growers, and grocery manufacturers complained bitterly about the way the academy had estimated carcinogenic risks by assuming residues would be on foods at legally allowable limits. They claimed that available residue tests demonstrated that contamination was far below tolerance levels. Cancer risks, they argued, were in reality far lower than those projected in the report.

The committee, however, had been very careful to explain that available residue data were inadequate to characterize national dietary risk from the pesticides examined and that their study was an exploration of how EPA set tolerances, particularly of the limited role that data on carcinogenic risk played in that process. What emerged from the study was an image of confusing and contradictory legal standards, a virtual nightmare of inadequate data for judging the health risks from pesticides and one that revealed the absence of any strategic plan for managing or reducing levels of risk allowed by current law. Although the NAS committee took great pains to claim that the food supply was safe, the data presented hardly supported their conclusion. Clearly, neither EPA nor the academy knew the extent of carcinogenic risks from pesticides in food.

EPA's Attempted Escape Through the De Minimis Window

EPA was put in a very difficult position by the NAS study. The domestic scientific community now understood that the health-effects data supporting nearly 8,500 tolerances were inadequate to judge the threat to health, whereas public exposure to pesticides in food—and increasingly in drinking water—was quite broad. George Bush was elected president in 1988, and one of his campaign promises was to enhance environmental protection. In part to legitimate his claim to be an environmentalist, he appointed William Reilly, then director of the World Wildlife Fund and the Conservation Foundation, to be the administrator of EPA. Reilly brought a high level of understanding of environmental problems to the agency, along with a personal commitment to improve the quality of scientific analysis within EPA. Partly due to his strong relationship with the president, he managed to restore funding for many of EPA's regulatory programs, which had been slashed during the Reagan administration.

One of the first major policy statements of Reilly's administration was a response to the NAS *Delaney Paradox* report.[54] First, EPA agreed with the NAS conclusion that the double standard for tolerance setting had no scientific rationale and that a single standard should be applied consistently, regardless of whether the foods are raw or processed or whether the pesticides are new or old. EPA proposed instead to apply the same risk-benefit balancing principle when setting all tolerances.

In addition, a "negligible risk" standard was proposed to apply to all carcinogenic pesticides. If a compound posed a less than one in 1 million excess risk, EPA would establish raw food tolerances. This standard would permit processed food tolerances to be issued even if residues concentrated, provided the risks were negligible. Under these circumstances, very little attention would be directed toward estimating benefits. Instead, the agency would assume that the manufacturers would not pursue registration unless the benefits outweighed negligible risks.

If a pesticide posed a greater than one in 1 million excess risk of causing cancer, then EPA would employ the risk-benefit balancing test, which entailed a more careful analysis of benefits. If benefits were deemed greater than risks, the tolerance would be granted. If the pesticides concentrate during processing, EPA would deny the tolerance and would be forced to prohibit raw "parent" food uses of the pesticide.

The EPA response to the Delaney Committee's report was hardly a change from past policy. The only significant distinction was the statement that the agency would set §409 tolerances for carcinogens that concentrated during processing, provided it found the risks to be "negligible" or beneath the one in 1 million excess risk threshold. The agency proposed a further escape from the Delaney clause if the pesticide in question was a Class C carcinogen and the evidence of toxic potency was of poor quality or ambiguous. Under these conditions, EPA proposed that carcinogens falling within the "high end" of the Class C category be subject to the standards described above, whereas those lying within the "low end" of the Class C category be judged according to a risk-benefit balancing standard. Under this policy, all compounds would be subjected to further analyses to see if other health risks are posed.

The "no observed adverse effect level" from the most sensitive toxic effect studies would then used to estimate an "allowable daily intake" of the pesticide, and used to calculate the "maximum permissible level of residues."

The "new" EPA policy was greeted warmly by industry, but it represented a bold and perhaps dangerous move given the D.C. Circuit Court's decision against FDA. In *Public Citizen v. Young,* issued only the year before, the court had demolished the FDA's argument that Delaney allows trivial risks associated with color additives. Enraged environmentalists saw the policy as an excuse to expand the influence of benefits in pesticide decisionmaking. In the murky world of risk and benefit assessment where the methods of analysis are highly technical and constantly changing, the clarity and consistency of the Delaney clause zero-risk standard was a beachhead that environmentalists would not yield without a fight.

In May 1989, the State of California, the Natural Resources Defense Council, Public Citizen, and the AFL-CIO (*Les et al.*) petitioned EPA to revoke several food additive tolerances as a way of challenging the agency's interpretation of the Delaney clause. They argued that food additive tolerances should be revoked for many pesticides and foods.[55] As a result, EPA decided to revoke those tolerances that were associated with pesticide uses no longer registered. But the agency refused to cancel tolerances for active uses, arguing that the risks were negligible, that there were insufficient data to demonstrate the magnitude of the risk, or that EPA was conducting an ongoing review of the chemicals in question that it should be allowed to finish before a decision was made.[56] The petitioners were not satisfied and filed objections, arguing that EPA had incorrectly interpreted §409 by allowing tolerances for residues posing negligible or *de minimis* risks. EPA responded by affirming its authority and decisions to issue tolerances for carcinogens on processed foods if residues concentrated.[57] Still unsatisfied, *Les et al.* challenged the EPA policy in the U.S. Court of Appeals, and in July 1992 the policy was reversed. The court found it to be inconsistent with the "plain language" of the statute, and the Supreme Court in 1993 refused to consider the case following a request by the National Agricultural Chemical Association.[58]

Thus in two separate cases, the escape routes from the Delaney prohibition were cut off because the courts found that the agencies' elaborate logic contradicted the plain language of the statute. Although the message being delivered by the courts to the executive branch was one of limited discretion, it has left EPA to face a very uncomfortable set of tolerance revocations. The same pressure that has been exerted on EPA and FDA to continue issuing tolerances under a negligible-risk rationale is now redirected toward implementing the statutory language of the Delaney clause, which has remained in its original form for nearly four decades.

By February 1993, EPA listed thirty-two pesticides and eighty different chemical-crop combinations that would potentially be affected by the *Les v. Reilly* decision. These chemicals and foods were selected because they were either Class B or Class C carcinogens, and because either processed-food tolerances had already been established or would be required given recent evidence of residue concentration.[59] Affected raw and processed foods included grapes and raisins, apples and apple juice, tomatoes and catsup, tea, citrus and pulp, wheat and bran, cotton and cottonseed oil, sugarcane and molasses, mints and oils, apples and pomace, and plums and prunes.

Since the court ruling, EPA has estimated that registrations and tolerances for as many as fifty separate pesticides may be revoked as the Delaney prohibition is strictly applied. In April 1994 the agency finally announced that if a pesticide is carcinogenic, it would no longer review requests for its tolerances in processed foods.

Continuing the Quest for Discretion: Change the Arena

Science lies at the heart of the EPA's mission by providing evidence for risk estimation, which leads to an analysis of the cost-effectiveness of various risk management options. But management choices by EPA may cost the public a good deal of money. The annual cost of private-sector compliance with all EPA regulations has been estimated to be over $100 billion—a figure commonly cited, but poorly supported. These expenses have also been estimated to constitute nearly 50 percent of the private sector's total costs resulting from federal regulations.[60]

These claims of cost have sharply focused attention on the quality of evidence that underlies EPA estimates of the environmental and health damages associated with specific technologies or behaviors. This attention was directed further by Executive Order 12299, signed by President Reagan, which gave the Office of Management and Budget responsibility to conduct cost and benefit analyses for significant regulatory actions. The Vice President's Council on Competitiveness, first implemented under Bush, was also created to further analyze the costs of regulation. Although the council has been disbanded, cost-effective regulation became a very high priority within the Clinton administration, which formalized a procedure for regulatory planning and review in a complex executive order (12866) signed by President Clinton in September 1993. This order requires the analysis of risks, costs, and benefits when making "significant" regulatory decisions.[61] It also requires risk-management planning by all federal agencies with regulatory responsibilities and has as a specific objective the comparative analysis of diverse types of health and safety risks.

The difficulty then is reconciling the need for "better science" (through more rigorous risk and cost-benefit analyses) with the delays and debates that this better science normally engenders. Few will argue that we should diminish the quality of analysis that precedes a potentially costly regulatory decision. But an acrimonious debate has erupted over where authority should reside to conduct the debate, in Congress or the executive branch, or at the federal or state level of government. Preferences seem less driven by the ability of different institutions to conduct quality science than about their relative vulnerability to manipulation by special interests.

Private-sector demands for better natural science to estimate risks and better social science to estimate the social costs of regulation naturally appeal to scientists, who make a profession of asking ever more questions and being very careful before reaching conclusions. This tentative approach is seconded by opponents of regulation, who have called for "comparative" or "relative" risk" analysis, which would permit the costs of risk reduction for individual regulatory decisions to be compared with the costs of reducing other types of risks. Comparative risk analysis appears to make sense, especially given the skyrocketing costs of restoring environments to "zero risk"—as demonstrated by the enormous expense of removing all detectable contamination from a Superfund site, all asbestos from schools, or all lead paint from homes. Opponents of relative-risk analysis, however, see it as a black hole for already scarce regulatory resources, an excuse for further delay, and most importantly, justification for avoiding regulation in a sector mandated by Congress such as pesticides, air quality, or water quality. It is important to realize that Congress never told EPA to avoid a new regulation in one arena because we could achieve more cost-effective risk reduction in another.

The *de minimis* interpretations of EPA and FDA have served as escape valves for pressures exerted by industry, which has been interested in avoiding loss of revenue from chemicals, even those posing carcinogenic risks. Given this background, and the relatively few processed-food tolerances issued for carcinogens, public interest groups were surprised that a "strict" interpretation of the Delaney clause would require cancellations of tolerances for raw foods. Foods affected turned out to be those with huge international markets such as wheat, apples, grapes, cottonseed oil, and citrus fruits. But following the *Les v. Reilly* ruling, industry sought relief from the zero-risk standard in Congress, whose members became far more sympathetic to their interests following the 1994 elections that swept Republican majorities into both the Senate and House.

Interest groups lobbying for environmental health were well aware that if EPA held the discretion to define acceptable carcinogenic risk and controlled the data and methods to estimate risk, the agency could maintain control over the decision process. Following *Les v. Reilly,* the agency had moved quickly and appropriately to revoke tolerances for offending pesticides. But if the Delaney clause were replaced by a negligible-risk standard, they would be forced to compete in the murky world of risk assessment, which demanded specialized expertise and quickly became bogged down in endless technical questions. Environmentalists have long been skeptical of the ability of EPA to openly express and control bias in their risk assessments.[62] The debate that followed, then, was not just about whether an administrative agency should have the authority or discretion to manage cancer risks. It also concerned EPA's right to control information and analytical methods in a way that rationalized its own registration and tolerance-setting decisions.

During the 1980s EPA was uncertain of its own methods of data management and risk assessment.[63] This uneasiness was nurtured by the agency's knowledge that its data sets were outdated and that as newly required data were submitted demonstrating higher risks than previously projected it would appear to have compromised the public trust. Risk estimates proved to be quite sensitive to alternative assumptions regarding residue level, food intake, or chemical toxicity; small shifts in assumptions concerning the magnitude of these variables could cause the risk estimate to breach an acceptable risk ceiling. The ceiling itself has been defined in different ways by EPA and FDA, and even within EPA it has been articulated differently among program areas. The agency was especially worried that if it was forced to consider risks to infants and children many pesticide uses and tolerances would be canceled.

The movement of the Delaney *de minimis* conflict from EPA to the Congressional arena forced the issue to be considered within a much broader debate over the effects and future of environmental law. The Delaney clause is only one of many risk-only standards that came under attack from a new and diverse coalition of interests during the early years of the Clinton administration. Chemical manufacturers, agribusinesses, and food processing companies have long been allied to promote broad EPA discretion to permit *de minimis* risks. By 1994, these groups had formed coalitions with advocates of private property rights, who have demanded compensation for any form of government regulation, and with state and local governments, which have resisted mandates by Congress to implement and enforce federal environmental and health standards without federal compensation.

The broadly based alliance had several common interests. First, the collective supported the use of risk assessment to demonstrate the significance of damages before government regulates. Second, they supported the replacement of risk-only standards—such as those contained in the Delaney clause or the Safe Drinking Water Act—with risk-benefit balancing standards, which require balancing the social costs of regulation with the environmental or health gains. Discretion to judge risks and benefits would lie with appropriate administrative agencies. Third, they promoted "comparative risk assessment"—an attempt to judge the risks of the technology or compound in question relative to other types of technological and normal risks.[64] Fourth, they supported compensation for the loss of private property rights, which may take the form of licenses, registrations, or permits previously granted by the government under the assumption that licensed products or behaviors did not injure health or the environment.

The retreat into the technical complexities of risk and cost-benefit assessment provided those bearing the economic burdens of regulation with room for argument. They may contest data quality, sampling designs, analytical methods, and the treatment of uncertainty in forecasts of risks and benefits. They can express risks as a range of possible outcomes, perhaps spanning a threshold of acceptability, such as a one in 1 million excess risk. They can also identify the near-term costs of reducing risks to zero, or to beneath some other health-based regulatory threshold. Because the marginal cost of risk reduction tends to increase dramatically as risks approach zero, these costs become a powerful incentive to relax standards.

Strategies for Maintaining Entitlements

The way in which USDA registered tens of thousands of pesticide products created an enormously complex set of rights. Despite the zero-risk cancer standard of the Delaney clause and the current classification of nearly one-third of food-use pesticides as carcinogens, nearly 97 percent of all tolerances were set under a risk-benefit balancing standard, and by EPA's own admission, the majority of these had little relationship to a current understanding of the health risks they pose. EPA may have inherited a risk management nightmare from USDA, but they too have done little to strategically reduce risk.

A new industrial strategy has emerged in the past two decades to protect entitlements, or use rights, formerly granted by USDA or, more recently, by EPA. This strategy appears to rest on four pillars. The first is to delay more prohibitive regulatory action by continually questioning the quality of evidence underlying claims of health or ecological risk. EPA has long suffered from an absence of criteria to decide when data are of sufficient quality to estimate risk, particularly risks of damages which might occur far into the future, such as cancer or multigenerational reproductive failure. In addition, EPA officials have been trapped between their wish to be recognized as high quality scientists and the agency's deadlines for making regulatory decisions on ambiguous or incomplete data. But when was the uncertainty too great? When should EPA delay regulation, particularly cancellation proceedings, to obtain results from another study or to secure another scientific advisory opinion? The pursuit of better scientific evidence always delays regulation. And because highly contentious decisions create a type of bureaucratic anxiety attack, EPA's response has been to control the rate of attack by stringing out their reviews. With tens of thousands of registrations and tolerances to review, virtual paralysis has resulted.

The second pillar rests on the FIFRA requirement that benefits be balanced with risks. Industry could count on USDA to estimate losses in jobs, crop productivity, and international trade with every proposed regulation. Moreover, the near-term costs of regulation are more politically potent that hypothetical long-term risks. Would EPA really trade jobs for risks extrapolated from rat and hamster studies, especially if damages are forecast to occur only after twenty to forty years? As long as the balancing standard, which provides EPA with infinite discretion, is applied, industry has felt confident that it could influence regulatory outcomes. The Delaney clause, by contrast, under tightly defined circumstances, gives EPA no administrative discretion, and thus the industry fight to abolish Delaney has been strengthened by EPA's pursuit of expanded administrative powers. Discretion really translates into control over the assumptions that underlie risk and benefit estimates. All of the advances in risk-assessment methods pioneered in the 1980s and early 1990s have done little to change the basic fact that risks vary dramatically depending upon the chosen set of assumptions.

The third industry strategy has become to compare the risks from a pesticide with other risks normally faced. For example, why worry about synthetic pesticides when we face a diverse set of natural toxins in food, some synthesized by plants to ward off insects, fungi, and viruses, and others formed while cooking on open flames? This argument is perhaps best articulated by Bruce Ames and Lois Gold, who believe that the public should care about natural rather than synthetic toxins. But it hardly seems prudent to avoid regulating synthetic toxins simply because we are commonly exposed to natural ones. After all, lead, mercury, and arsenic are all natural elements—and have been active ingredients in pesticides—that most of us avoid if we can. Government has a special obligation to prevent public exposure to toxins at levels that pose significant risks—especially if those toxins are invisible and unavoidable. The distinction between their natural and synthetic sources seems irrelevant to regulatory decision making.[65]

The fourth and final pillar of the industry strategy is the pursuit of the most cost-effective risk reductions across all areas of the environment. Given increasingly limited public resources, argue proponents, it seems logical to search for methods of reducing health risks—of different kinds—in the most cost-effective way. Using this logic, the costs of reducing the risks from pesticides in foods should be compared to the costs of lessening the dangers from pathogens in drinking water, airborne industrial toxins, or even automobile crashes. This approach, however, greatly expands the analytical complexity of risk analysis and economic valuation, often pushing comparative forecasts beyond thresholds of credibility. Sound public policy clearly demands comparative risk and valuation analyses, but they must rest on much firmer scientific foundations than those provided by current data.

Redefining the Problem

The field of environmental health risk assessment is still quite young, and perhaps it is most limited by a poor understanding of how we normally encounter complex mixtures of toxins and exactly how they may damage human health. We are just beginning to understand these problems as they pertain to some classes of pesticides, such as neurologically active insecticides (described in chapter 11), and this knowledge lays a rational basis for regulation. But we have poor data and primitive analytical strategies to understand the relative significance of different kinds of adverse health effects such as cancer, neurological damage, and reproductive failure. Some pesticides pose all of these types of risk, and given the sea of possible pesticide substitutes, prohibiting one compound simply trades one set of risks for another.

Moreover, our knowledge base has become fractured into highly specialized fields, and this process has in turn influenced the evolution of environmental health law and regulation. Given our limited understanding of the effects of being exposed to complex pesticide mixtures, how should we concentrate our scarce regulatory resources? How can we regulate to substitute low-risk pesticides for high-risk pesticides? These questions must be addressed if risks are to be reduced in the most cost-effective manner. The past three chapters have unfolded a history of regulation that is characterized by the granting of thousands of entitlements for using pesticides or for allowing their residues to contaminate food and water. I next explore how our understanding of pesticide risks has changed. Given the highly dynamic quality of this knowledge base, why should law have remained so stable?

PART 3

Evolving Knowledge

CHAPTER 7

The Human Ecology of Pesticide Residues

Each year, nearly 6 billion pounds of pesticides are sold in the global marketplace—2.5 to 3 billion pounds of which are purchased in the United States.[1] These compounds are applied to crops, forests, lawns, gardens, parks, highways, rail lines, power lines, lakes, ponds, swimming pools, office buildings, aircraft, ships, hospitals, schools, and day-care centers. Pesticides are also deliberate components of some clothing, shampoos, drugs, paints, wallpaper, shower curtains, rugs, blankets, and mattresses. Understanding the health and ecological risks incurred by these uses requires knowing where pesticides are released, how they move through the environment, and where they come to rest. This knowledge provides the basis for comprehending how humans are exposed to pesticides, which in turn is necessary for understanding the magnitudes of risks these contaminants impose.

Federal control of pesticides has primarily entailed granting permission for a pesticide's release and then attempting to prevent exposures believed to be unacceptably dangerous. But controlling the public's exposure to a potential toxin is much more difficult than other common risk-management strategies such as reducing the amount of contamination created, prohibiting its production, or securely containing it. Moreover, a legal strategy of exposure control has created an immense environmental monitoring problem, because it demands knowledge of contamination levels in food, water, and air. Not surprisingly, the simplest way to legally disarm this strategy has been to fail to produce reliable evidence about residues in the environment.

Misunderstanding Pesticide Use and Substitution

Knowledge about where and how pesticides are used is essential for allocating scarce chemical-monitoring resources. Little is known about specific patterns of pesticide use in the United States, and far less is understood about their use in foreign countries—which provide increasing proportions of the U.S. food supply. One reason for the absence of evidence is industry's interest in concealing sales and use records from competitors. Another difficulty is the enormous complexity and cost of tracking the fate of active and inert ingredients, which are mixed in tens of thousands of combinations at hundreds of formulation centers around the world.[2] This lack of understanding forces the government to take a comprehensive monitoring approach. But because broad monitoring is extraordinarily expensive—given the scale of the nation's food supply and the number of pesticides licensed to remain in the supermarket—this approach translates into most foods being tested only a few times. This sample-size problem will be addressed more fully later in a discussion of food-borne residues.

No organization, including the USDA, keeps a careful record of where and how pesticides are used in the United States.[3] Between 1979 and 1991, EPA estimated that more herbicides and fungicides, but less insecticides, were being used nationally.[4] The declining volume of insecticide use is deceptive, however, because it reflects the introduction and use of more potent insecticides, which are applied at lower volumes.[5] For example, new soybean herbicides are commonly applied at rates of 0.05 pounds per acre, whereas older compounds are applied at a rate of 0.5 to 2 pounds per acre.[6] A lower volume of application does not translate into a lower level of risk, though, because these lower volumes may be possible due to higher levels of toxicity or longer levels of persistence. Chlorsulfoton, for example, is an herbicide that has been used as a substitute for 2,4-D, which is widely used and suspected to be a human carcinogen. This compound is applied at a rate of 0.5 lb/acre, and chlorsulfoton is applied at 0.02 lb/acre. The soil persistence of 2,4-D, however, is only one to four weeks, whereas chlorsulfoton persists for three to four years. This level of persistence severely restricts the farmer's ability to rotate crops and may exert a selection pressure that encourages resistant species to multiply.[7]

Another way of understanding use and contamination potential is to consider the land area planted with specific crops along with the pesticides commonly used on these crops. Only 8 million acres in the United States are devoted to fruit, nut, and vegetable crops. For example, 500,000 acres are planted with apple trees, 250,000 acres grow lettuce, and one hundred thousand acres grow carrots. By contrast, 73 million acres are planted with wheat, 71 million acres with field corn, and 50 million acres with soybeans. To provide a sense of scale to these numbers, Connecticut is about 2.5 million acres in size. Regional differences in crop production, climate, and pest problems determine the mixture of pesticides used, and local ecological conditions will influence the scale and severity of contamination.

Between 1964 and 1982 the weight of active pesticide ingredients used in the United States nearly doubled, but total acreage in agricultural production remained nearly constant at approximately 340 million acres. Herbicide use rose from 210 million pounds in 1971 to 455 million pounds in 1982, and today herbicides constitute nearly 90 percent of the total weight of pesticides applied in the United States. This growth occurred principally because the application of herbicides replaced mechanical methods of weed control, such as field harrowing. Further, nearly 53 percent of all herbicides applied in 1991 were used on only three crops: corn, soybeans, and cotton. About 50 percent of all pesticides, measured by weight, are applied to corn in the United States, which in 1982 surpassed cotton as the crop demanding the greatest insecticide protection.[8] Because herbicides represent the largest class of pesticide use by volume, funding alternatives to herbicide use on these three crops is essential to any meaningful attempt at national pesticide use and risk reduction. Herbicidal uses in nonagricultural areas such as highways, power and rail corridors, fence lines, and on lawns have also grown dramatically during the past several decades.[9]

During the early 1990s, nearly 150 million pounds of only five herbicides were applied to corn and soybean fields each spring in the Midwestern United States. These compounds included atrazine, cyanazine, simazine, alachlor, and metolachlor. Spring rains wash some of these herbicides from the fields into streams, and from streams into rivers that provide drinking water for Midwestern communities. In 1995 nearly 14 million people drank water with detectable residues of these herbicides; 38 percent of samples from twenty-seven Midwestern drinking water reservoirs were contaminated by four or five of them. At Kansas City's municipal water-intake facility, for example, 61 percent of the samples contained two or more residues. Ironically, the USDA found that in 30 to 50 percent of the cases they studied, farmers lose money by applying these herbicides when they are not needed. Instead, farmers could employ integrated weed-control techniques such as weed-seed surveillance to determine the level of infestation, economic analysis to help decide whether herbicide application will cost more than crop losses, and application techniques that apply the herbicide to the crop rows instead of to the whole field.[10]

Local variation in pest problems and in the choice of insecticides, fungicides, and herbicides is enormous and extremely difficult to predict and manage. The general absence of high-quality use data about pesticides makes it expensive to track how and in what quantities residues are transported to the dinner table or drinking-water tap; these uncertainties in turn make it difficult to estimate human exposure and risk. Further, decisions by EPA to ban a pesticide due to excessive risk have traditionally caused farmers to simply substitute another compound. These choices are normally driven by market forces and the farmer's perception of a product's efficacy rather than by any comprehensive risk-reduction planning encouraged by EPA. For example, when aldrin, a chlorinated hydrocarbon insecticide, was banned in 1975, nearly 14 million pounds had been used annually on over 13 million acres of corn. Farmers first switched to several insecticides, including bufencarb, carbofuran, and phorate, and more recently to chlorpyrifos and fonofos. But these substitutions only exchanged one set of risks for another. Although human exposure to persistent chlorinated hydrocarbon insecticides diminished, exposure to organophosphorus insecticides posing acute neurological risks increased.

The fumigant dibromochloropropane (DBCP) provides another example. DBCP was banned in California in 1981, twenty years after it was demonstrated to cause severe atrophy and degeneration of the testes of mice, rats, and rabbits. The National Cancer Institute concluded that DBCP was an animal carcinogen, and by 1976 it was shown to cause decreased sperm counts among exposed farmworkers.[11] It also contaminated thousands of wells in California's Central Valley. After the 1981 ban, farmers commonly switched to ethylene dibromide (EDB) for use on tomatoes. EPA suffered an extreme loss of public confidence when EDB was then found to be a potent animal carcinogen and was banned in 1984. EDB was generally replaced by 1,3-D and metam sodium, a highly toxic herbicide that, as mentioned earlier, became notorious after its 1990 spill into the Sacramento River and Lake Shasta.[12]

Another compound banned in 1984 was the herbicide 2,4,5-T, which was well known as a defoliant used in the Vietnam War. It became controversial due to its contamination by 2,3,7,8-TCDD—a type of dioxin that is acutely toxic and cancer-inducing as well as a reproductive toxin in laboratory animals.[13] The compound is also suspected of playing some role in disrupting the human endocrine and immune system. In 1995, the predominant source of human exposure to 2,3,7,8-TCDD was believed to be through the food supply, especially dairy products, meats, and fish. Preceding its ban for agricultural uses, 2,4,5,-T was allowed for use on rice fields in Arkansas, and after 1984, it was anticipated that 2,4-D would be the predominant substitute. But 2,4-D commonly drifted from rice fields to adjacent cotton fields where it injured plants; it also became the subject of a contentious debate over its potential to cause lymphoma and soft-tissue sarcomas in farmers. A battery of other herbicides has thus taken its place, including acifluorfen, thiobencarb, and bromoxynil.[14]

Toxaphene provides another illustration of how incremental chemical regulation has commonly led to unmanaged substitution of pesticides and of their diverse risks. In 1976 nearly 26 million pounds of toxaphene, a chlorinated insecticide, were applied to 3.1 million acres of cotton. In 1979, however, the compound was found to be an animal carcinogen by the National Cancer Institute and the International Agency for Research on Cancer. Soon thereafter it was demonstrated to cause severe bone deformities in Great Lakes fish at the parts-per-billion level. It had volatilized from southern U.S.

cotton fields, traveled nearly a thousand miles in clouds, and had been deposited along with rain in the Great Lakes Basin.[15] Toxaphene was finally banned for further use on U.S. crops in 1982.

In general, then, patterns of pesticide use appear to vary according to a number of factors, including: (1) the availability of substitutes, (2) regulatory action by federal, state, and local agencies, (3) individual farmer's planting decisions, (4) variation in per-acre application rates among compounds, (5) regional variation in pest resistance to compounds, (6) the rate of innovation in the pesticide industry, (7) the availability of expertise regarding pesticide use reduction and integrated pest management practices, and (8) increasing consumer demand for greater control over pesticide residues in food and drinking water.

The absence of careful government monitoring of pesticide use inhibits our understanding of what happens to pesticides after they are applied. If we do not know what chemicals are being releasing to the environment or where they are being released, how do we know which pesticides to search for using expensive residue-scanning methods? In the absence of this knowledge, how can government officials accurately estimate what chemicals the public is exposed to, and therefore what risks they face?

But even if we knew which fields were sprayed with which compounds, estimating human exposure would still depend upon understanding patterns of finished crop distribution, the effects of food processing on residue concentrations, and the effects of food preparation or cooking on residue concentration. The overwhelming task of understanding these patterns of residue distribution and fate has caused government to instead focus monitoring efforts on raw foods. Yet, as discussed, the task and cost of searching for nearly 325 food use pesticides and additional metabolites in well over five thousand separate food products in the marketplace is also enormous.[16] These uncertainties have forced the government to rely on crude computer models that simulate patterns of chemical use, distribution, processing effects and finally, human intake of residues in contaminated food and water. Forecasts of exposure and associated risk are therefore highly uncertain. This regulatory tactic is similar to "end-of-the-pipe" pollution control efforts, which are far less efficient and effective than source reduction.

Global Transport and Fate

What happens to pesticides after they have been released? Their fate depends upon their chemical stability, level of solubility (either in water or lipids), and volatility. A large proportion of applied pesticides become airborne in the form of particulates or vapor because they may quickly evaporate from plant and soil surfaces or be vaporized as soil is tilled or eroded. Another factor governing fate is the weather at the time of application. Windy conditions normally cause less of the pesticide to reach its intended target.

Chlorinated hydrocarbons have been found in the air more commonly than have other classes of pesticides. Generally, concentrations of pesticides are measured by weight per unit area. The most common units are grams per cubic meter, or micrograms, nanograms, or picograms/meter^3. In early 1965, DDT was found in the air of six agricultural communities at levels ranging between 1 and 22 ng/m^3. Levels of DDT in the air above the Great Lakes region ranged between 0.1 and 10 ng/m^3 in the early 1970s, although these levels fell to 0.01 to 0.1 ng/m^3 by 1980.[17]

DDT was heavily used as a cotton insecticide before it was banned in the United States. The Mississippi River delta is one of the most productive cotton growing regions in North America, and a sequence of air measurements taken in the area between 1972 and 1974 demonstrate a significant decline in total DDT airborne concentrations. In 1972 the average monthly concentration was highest in August (515 ng/m^3) and lowest in December (6.3 ng/m^3). These figures dropped to 49 and 2.4 mg/m^{3,} respectively, by 1973; and to 37 and 2.1 ng/m^3 by 1974. Toxaphene levels were also detectable, with the highest monthly average concentrations reaching 1.0–1.5 µg/m^3 in 1973.[18]

Pesticides may become airborne indoors for a variety of reasons. In tropical countries, the primary cause of indoor residues is the use of compounds to control disease-carrying insects. As discussed in chapters 2 and 3, chlorinated compounds have been sprayed indoors for nearly half a century to prevent diseases such as malaria. They have also been used to protect against damage to buildings from termites. Chlordane, and before it, heptachlor, have been used extensively for this purpose, primarily next to building foundations. If heating or ventilating ducts are located beneath buildings, it is possible for vapors from these compounds to be blown into the building interiors. U.S. Air Force housing with this type of ventilation system, for example, was found to have between 1 and 263 µg/m^3 of airborne chlordane. In 1982, houses treated with chlordane contained airborne levels between 2 and 5 µg/m^3, and these levels persisted for at least twelve months.[19]

Other indoor uses of chemical compounds include attempts to control fleas, cockroaches, silverfish, and other pests. Pesticides are commonly sprayed on indoor plants; are components of paints, stains, and wood preservatives to prevent bacterial degradation; and are impregnated within rug fibers. In the 1950s, kitchen shelf paper was impregnated with DDT or dieldrin;[20] many people slept sandwiched between blankets mothproofed with dieldrin and mattresses impregnated with DDT; floors were waxed with insecticides that would kill bugs that walked over them;[21] and power mowers were outfitted with devices that spewed pesticide vapors.[22]

Pesticides may also bind to water droplets. They have been detected in fog both in Beltsville, Maryland, and in the Central Valley of California. The highest reported concentration in fog of any pesticide was for a parathion analog, detected at 184 μg/liter near a grape vineyard and dairy farm area near Lodi, California.[23] Pesticides have been found in twenty-three states in rainwater samples taken by the U.S. Geological Survey. Commonly detected compounds include those widely used herbicides that tend to volatilize within several months of application.[24] The herbicides methoxychlor and hexachlorobenzene, for example, were found in rainwater over the Great Lakes Basin,[25] and the herbicide atrazine was found in rainwater by 1977.[26]

Interest in the global dispersal of pesticides emerged in the 1980s.[27] Enewetak Atoll lies in the North Pacific Ocean far from agricultural and industrial sources of pesticides. In 1980 atmospheric sampling there detected BHC, chlordane, dieldrin, and DDE in the air and over ten times the detected vaporized concentrations in rainwater.[28] Airborne concentrations of these same chlorinated pesticides were also detected in the middle of the Arabian Sea, the Persian Gulf, the Red Sea, and in the middle of the North Atlantic near Newfoundland and Barbados. Significantly lower levels of DDT were found near Newfoundland than in the eastern seas. Supporting these findings, levels of DDT detected in the Gulf of Mexico were ten times higher than levels found near Newfoundland.[29] These findings may be explained by the continued use of DDT in surrounding nations both for agricultural purposes and disease control.[30]

Levels of airborne chlorinated hydrocarbons have been correlated with latitude over the open ocean.[31] For example, the levels reported at 69 degrees south in Antarctica are ten times the levels reported in the North Atlantic. These findings may be best explained by the higher continued use of chlorinated pesticides such as DDT in the southern hemisphere both for agriculture and control of insect-borne disease.[32]

The concentration of pesticides in the air is dependent upon a number of variables, including the chemical and physical properties of the mixture being sprayed, the distribution of particle sizes in the spray, the nozzle's height above the ground at the time of release, and atmospheric turbulence. Among these variables, the distribution of particle sizes appears to influence airborne concentrations most. For pest control, it is desirable to have a combination of particle sizes—some large enough to create a downward velocity of the mixture, and others small enough to encourage its dispersal. These smaller particles have a slower velocity and a lower weight; they are therefore more likely to be transported long distances.[33]

Airborne transport of pesticides is also affected by such factors as their density, viscosity, and the mixture's surface tension. Compounds may also be diluted with water or other chemicals that are designed to make them adhere to plant, water, or soil surfaces or to control their volatility. Once pesticides reach the ground, they may volatilize if they are in solution as water evaporates. This tendency is confirmed by the initial rapid loss of pesticides from plant surfaces. For example, 90 percent of heptachlor residues on leaf surfaces volatilizes within seven days, and 90 percent of dieldrin volatilizes within thirty days.[34]

The fate of pesticides is partly governed by how they behave in soils and sediments, which are important storage bins, or "sinks," for pesticides in the environment. There are several general principles that govern the attachment of pesticides to soil particles. First, fat-soluble compounds tend to bind quickly to organic matter in soil or sediment. Second, water-soluble compounds attach less easily to soil particles and are thus more easily detached. Third, pesticide attachment to soil or sediment increases as particle size decreases. These residues in soil may then become airborne depending on their rate of movement toward the soil surface, known as the wicking effect.[35] In short, as pesticides in solution evaporate from the soil surface, more of the solution is drawn to the surface.[36]

Some pesticides are formulated as water-soluble compounds in order to reach pests in the root zone of plants or to be absorbed by the plant and systemically distributed. These compounds will volatilize primarily through evaporation. Other pesticides applied to water surfaces for the purpose of killing disease-bearing insect larvae are often not water soluble and are formulated with oils such as kerosene to remain on the surface. Similarly, pesticides are sometimes mixed with oils to bind them to the plant surface, which decreases the amount washed away by rain. All of these factors influence the propensity of a pesticide to become airborne.[37]

Pesticide movement is also influenced by climatic conditions such as humidity and temperature. Organochlorine distribution, for example, has been found to be more highly concentrated in tropical air and water when compared with temperate regions.[38] Higher water concentration could be the cause; the tropics receive an enormous amount of rain, which would influence runoff from agricultural fields. One example of this phenomenon is provided by hexacholorocyclohexane (HCH). Residues of this compound were tested in south India, where 99.6 percent of HCH applied to rice paddies evaporated and only 0.4 percent drained into an estuary. Nearly 75 percent of the HCH in the estuary eventually volatilized as well; thus only 0.1 percent of the amount applied reached the sea. This rapid rate of volatilization appears to facilitate long-distance transport. Humidity at the time of application may also significantly affect the rate of evaporation. George Ware and his colleagues estimated in 1969 that less than 50 percent of pesticides applied by air in Arizona reached the ground due to low levels of humidity; the other half was volatilized.[39]

Persistent chlorinated pesticides such as HCH, DDT, and chlordane have been dispersed even more widely through atmospheric transport. Their movement from the site of application depends in part on their rate of transfer from the atmosphere to the oceans. The South China Sea and the Bay of Bengal, for example, appear to have become sinks for HCH, which is still heavily used in South Asia, especially China and India. Unexpectedly, the Arctic Ocean also appears to be a sink for HCH and its metabolites, as well as chlordane, which is distributed more uniformly in the oceans. By contrast, tropical waters, which are closer to the locus of initial release, appear to be a sink for DDT and its metabolites. This effect was predicted in 1975 as the use of DDT shifted from the developed, temperate parts of the world to less wealthy tropical and semi-tropical areas.[40] These patterns of residue distribution have important ecotoxicological and human health implications because chlorinated compounds tend to bioaccumulate within marine food webs, especially within marine mammals such as whales, dolphins, and seals, and within people who eat seafoods.

Bioaccumulation

Toxins that are fat soluble and resistant to degradation may accumulate as lower organisms are consumed by those higher on the food chain. This process is known as "bioaccumulation" or "biological magnification." These phrases describe the progressive accumulation or concentration of a chemical or compound along successive steps of a food chain. The storage potential of a compound such as DDT is extremely high due to this process; it bonds chemically to cells of any organism as well as to soil, silt, and even dead plants or animals. The deeper it lies in the soil, the less oxygen is available to promote its decomposition. DDT therefore has the potential to remain undisturbed for very long periods of time in ecological compartments such as lake sediments or agricultural soils.[41] Thus although its long, residual potency made DDT the compound of choice for insect-borne disease control and agricultural pest control, this characteristic also led to its eventual prohibition within high-income nations.

One can easily understand that planes spraying pesticides could overshoot their targets or that winds may blow the sprays beyond the bounds of fields. One can also imagine the potential for water-soluble compounds to be carried beyond the intended zone by streams or even into underground aquifers.

But the biological magnification of the effects of fat-soluble pesticides, especially chlorinated organic pesticides and dioxins, was a surprise. That bioaccumulation could result in global transport of contaminants was even less anticipated.

Knowledge of the residual life span of these chemicals emerged slowly within the fields of ecology and wildlife biology. Finally, however, it became clear that pesticides vary in their stability and persistence. Metals such as lead, arsenic, mercury, and cadmium are the most stable and persistent of pesticides. They do not degrade; they can be removed from an ecosystem or human tissue only through physical transport. The next most persistent class includes fat-soluble compounds. These can accumulate in plant oils and waxes, such as those that commonly exist on the skin of fruits, and they are transferred to wildlife and humans who eat the fruits. Those compounds that are most likely to survive the trip up the food chain are those that are fat soluble; these bonds make them generally resistant to degradation through microbial action or metabolism within each consumer. The chlorinated hydrocarbon insecticides have all of these characteristics, which explains their accumulation in the human food supply.

Persistence is a trait different than bioaccumulative potential. For example, triazine herbicides tend to be persistent within underground aquifers, which are oxygen deprived and have very limited levels of bacteria. Both of these conditions diminish the rate of triazine degradation to non-toxic products. Again, chlorinated hydrocarbon pesticides persist through a different biological mechanism: they chemically bond to fats. The ability of a pesticide to cross cellular membranes and to resist degradation and excretion all influence its potential to biologically accumulate from lower to higher organisms, moving up food chains. Because humans are at the top of both plant and animal food chains, our potential for accumulating fat-soluble pesticides such as DDT, aldrin, dieldrin, and chlordane is very high.

An example of biological concentration of contaminants was provided in 1967 by George Woodwell and his colleagues, who found DDT in sea water at a concentration of 50 parts per trillion. They then followed changes in the concentration of the insecticide as it moved along a food chain from detritus to plankton, from predatory fish to waterfowl. DDT was found in plankton at 40 parts per billion, at 1 part per million in pickerel, and at 26 parts per million in cormorants—520,000 times higher than water concentrations.[42]

One case that demonstrates the process of biological accumulation occurred at Clear Lake, which lies about ninety miles north of San Francisco. The lake is shallow, turbid, and provides an ideal habitat for a small gnat, *Chaoborus astictopus,* which annoyed local residents and anglers. By 1949, the effectiveness of chlorinated compounds to control mosquitoes was well known, and the gnat was found also to be susceptible. The volume of the lake was calculated, and DDD, a relative of DDT that poses a lesser threat to fish than DDT, was chosen. It was applied to achieve a concentration of 1 part DDD to 70 million parts water and resulted in a rapid decline in the gnat population. But by 1954 the gnats had returned and the treatment had to be repeated, this time at a higher concentration of 1 part DDD to 50 million parts water. This application was followed by yet another treatment in 1957.

The lake was also the nesting site of western grebes, a bird commonly found in shallow lakes of the western United States and Canada.[43] By the winter of 1955, the grebes started dying, and no one knew why. Infectious disease was suspected, but no evidence was found. Then someone thought to analyze the birds' fatty tissue for evidence of DDD. What they found was baffling: it contained DDD at a level of 1,600 parts per million. Next they tested the fish, which were found to have DDD levels between 40 and 300 parts per million, with one bullhead carrying a level of 2500 parts per million. The fish were ingesting plankton that contained 5 parts per million. Moreover, if one measured the concentration of the chemical in the water, it was immeasurable shortly after application. Yet it had not disappeared; "it had merely gone into the fabric of the life the lake supports." Every fish, frog, bird, and plankton examined contained DDD for years following its application. The California Department of Health prohibited further use of DDD in the lake, although they found no human health hazard.[44]

Another route for biological magnification of residues was discovered during efforts to control Dutch elm disease. This disease is caused by a fungus that was first recognized in the United States in the 1920s. It is found in about half of the United States and attacks a prized shade tree, *Ulmus americana*. The fungus moves from tree to tree transported by several species of bark beetles. Heavy applications of DDT were used to control the disease (0.68–1.36 kg of DDT per tree). This intensity of application resulted in the immediate death of many birds, and indirect damage was suggested by 1958.[45] The leaves themselves contained residues of DDT in the range of 200 ppm, and residues that contaminated the soil were taken up by earthworms, found with 33 and 164 ppm of DDT, roughly nine times the residue levels found in the soils. Because earthworms are a primary food of robins, those robins living in the vicinity of sprayed elms died in great numbers. As described later in this chapter, nearly eighty other species of birds are also affected by DDT.[46]

DDT also became the cornerstone of a gypsy-moth eradication campaign waged in the northeastern United States during the 1940s and 1950s.[47] The moth, native to Europe, was introduced to the United States in 1869 by Leopold Trouvelot. Several moths escaped from Trouvelot while he was attempting to crossbreed them with silkworms in Medford, Massachusetts. Slowly but steadily, the moth spread from Medford. It occasionally leaped broad distances when carried by storm winds, or when egg masses hitched rides on plants shipped by rail or truck. The moth defoliates trees in roughly seven-year cycles. At one point in the cycle, its population explodes, devouring tens of thousands of acres of foliage.

USDA became fascinated with the potential of DDT to control the moths' expanding habitat and declared chemical war on the insect. In 1946, it proposed spraying a 30 million acre area—then the entire zone infested—with DDT for a period of five years. DDT was mixed with fuel oil, tanker planes were hired by both USDA and the State of New York, and in 1957 1 million acres were blanketed with the mixture. By 1957, 3 million acres were showered. On Long Island, the planes misted suburban lots, commuters at train stations, dairy farms, truck gardens, and suburban tree-lined streets. On hot summer afternoons, children would run behind the spray trucks, cooling themselves and frolicking in the toxic mists.[48]

One of the key lessons learned from the 1957 spraying program was the susceptibility of dairy products to the accumulation of DDT. Only forty-eight hours after one farm in Westchester, New York, was sprayed (despite the farm family's pleas to avoid the pastures), milk from cattle that fed on the grasslands contained 14 parts per million of DDT. Although even then FDA permitted no residues of DDT in milk, some milk from this farm made it to the marketplace and was consumed in American homes. Several well-publicized lawsuits finally drew considerable public attention to the program, and by 1958 only one-half million acres were sprayed—a number that dropped to one hundred thousand acres between 1959 and 1961 in response to vehement public protest.[49]

Understanding biological magnification became a crucial basis for managing diminishing populations of fish-eating birds and falcons as well.[50] Understanding the cause of declining reproductive success of these larger birds of prey took two decades. Some evidence of egg breakage was discovered in 1951, and the problem was believed to be caused by agricultural chemicals.[51] By 1963, peregrine falcon populations were declining in the United States, France, Finland, Germany, Sweden, and England. Ospreys, large fish-eating birds which nest in coastal marshlands, were also in trouble, as was the brown pelican along the U.S. Pacific coastline.

Hickey and Anderson believed that the declines in population were somehow linked to a change in the composition of the birds' eggshells, which are composed largely of calcium carbonate. By examining eggs in museum collections, they demonstrated that the shells of peregrine falcons collected between 1947 and 1952 weighed almost 19 percent less than those collected between 1891 and 1939. (DDT was first heavily used in the United States during the late 1940s.) These same researchers found that twenty-two other species' eggshells had thinned after 1945.[52]

The cause of the shell thinning was believed to be an agricultural chemical, and gradually, through the late 1960s, correlations between levels of DDT and shell thickness became clear. As levels of DDE, a DDT metabolite, increased, shell thickness decreased—not only in peregrines, but also in bald eagles, American kestrels, Japanese quails, prairie falcons, brown pelicans, white pelicans, California condors, cormorants, great blue herons, and gannets. Although many other potentially toxic compounds were also detected in the eggs—including DDE, DDD, dieldrin, PCB's, and mercury—DDE was consistently identified as the primary contributor to reproductive failure.[53]

The bans on broadscale application of chlorinated pesticides that EPA adopted in the early 1970s were followed by a clear improvement in the reproductive success of many of the affected species. The peregrine falcon has returned to several urban centers and now nests in skyscrapers; formations of brown pelicans once again patrol the California coastline; and dead treetops and telephone poles in the coastal marshlands of New England are once again favored nesting sites for the fish-eating osprey.

Some have argued that biomagnification is an ecological rarity because it requires chemically stable toxins, a successful transfer of the toxin among species along a food chain, and that each species be unable to degrade or excrete the compound.[54] Still, the intense use of chlorinated hydrocarbon pesticides during the later half of the twentieth century has caused measurable contamination even in the most remote parts of the world. DDT residues, for example, have been detected in Adélie penguins and crabeater seals of Antarctica. Although airborne transport of residues is suggested as a cause by a finding of 0.04 ppb of DDT in water melted from Antarctic ice, the most likely source would be from the ocean currents that in general facilitate worldwide distribution and accumulation of fat-soluble pesticides within marine food webs.[55] On Signy Island in Antarctica, penguins and skuas (a scavenger bird that migrates to the tropics) were found to have residues of DDT, DDE, heptachlor, and dieldrin. In this same area, residues of these compounds were found in krill, the most common food for blue whales.[56]

Marine organisms have the ability to amplify the concentration of fat-soluble contaminants along food chains despite relatively low levels of contamination in sea water (in the parts per trillion range). For example, residue levels in the striped dolphin, the top predator in its food chain, were found to be 10 million times higher than surrounding waters.[57] This magnification occurs because cetaceans have a thick layer of subcutaneous fat that acts as a storehouse for chlorinated contaminants such as DDT and its metabolites. Although males appear to maintain their level of contamination, residue levels in females decline during maturity, perhaps because they transfer residues to their young during lactation. In 1981, 60 percent of the PCB residues in striped dolphin mothers were discovered to transfer to their young during lactation. This extremely high level of transfer is likely due to the correspondingly high level of fat in milk.[58] The toxicological significance of these levels of contamination are not yet well understood for marine mammals, nor is it clear that sampling designs were well controlled to account for variation in migratory patterns and demographic characteristics within species. Small sample sizes also make it difficult to reach generalized conclusions concerning levels of contamination of marine mammals, oceans, and the atmosphere. Still, there is clear evidence of global pesticide dispersal, and convincing evidence that lipophilic pesticides have bioaccumulated along food chains. Each of these findings surprised experts and raised very difficult questions concerning pesticides' possible toxic effects on fish, marine mammals, and the humans who consume these species.

Sources of Human Exposure

Estimating human exposure to pesticides is complicated by the number of pesticides on the market, their patterns of transport and fate, and the diverse routes by which humans may be exposed. The dominant routes of human exposure include air inhalation, oral ingestion, and exposure through the skin. The toxicity of any single compound may vary by route of exposure, but oral ingestion generally results in the highest level of toxicity.[59]

Pesticides that are inhaled do not necessarily reach the lung in the same concentration that exists in the ambient air (that is, air outside the body). Instead, some particles are likely to be captured by mucous membranes in the upper respiratory tract, which will drain into the gastrointestinal tract. The percentage of material that reaches the lung is also partly dependent upon particle size: the smaller the particle, the deeper it may penetrate the lung. Particle size is determined both by the formulation (that is, powder versus solution) and by the method of application (fine mist versus droplets). The choice of nozzle design is also important to the distributional pattern of pesticides sprayed from aircraft and may govern the degree to which the spray drifts.

Estimating inhalation exposure is thus extremely difficult because many factors are involved. It is common for airborne pesticides to concentrate in clouds for very short durations following application, so that a worker holding his breath or moving out of a visible cloud may be exposed dermally but have inhaled very little. Greenhouse workers, who work in an enclosed, humid environment; those who mix active ingredients of pesticides; and those who load spraying equipment generally experience the highest levels of inhalation exposure. Field workers are also potentially exposed from drifting sprays applied either by aircraft or by ground equipment. Use of respirators is a common exposure-management strategy under these conditions, but in tropical and many arid climates, or under normal greenhouse conditions, use is rare because workers wearing them find them intolerably hot and claustrophobic.[60]

In addition, many rugs are still impregnated with compounds to kill insects and mildew, and these can become airborne from vacuuming or rub off on young children who crawl or roll over them. Paints normally include fungicides to attack bacteria, which otherwise may break down the paint and cause it to chip and peel. Until the early 1990s, paints were formulated with mercury, which is an effective bactericide. The mercuric compounds volatilize and while airborne can be inhaled. Once settled, they can become airborne again from floor vacuuming. It is important to realize that children are likely to be exposed at a level higher than adults for several reasons: they spend more time indoors, live closer to the ground, put more non-food material in their mouths, and have higher respiration rates than adults. All of these variables are likely to yield a higher dose of contamination per unit of body weight or surface area than would be expected for an adult. Again, quantifying the amount of exposure is a formidable analytical task.

Homes are commonly fumigated to kill insects such as fleas, silverfish, and cockroaches that either carry diseases or are simply nuisances. Fumigation will leave a residual concentration of pesticides in the air that may be inhaled for days and may also persist on walls, floors, rugs, clothing, bedding, cookware, uncovered food, toys, and any other exposed surfaces. Perhaps the parents love houseplants and meticulously spray them with common over-the-counter insecticides. Only a small portion of the pesticide leaving the spray container will reach and remain on the leaf and flower surfaces of the plant; the remainder will drift from the plant to any uncovered surface, and it can be transported throughout the home via forced air heating, cooling, and ventilation systems. Once the particles have landed, many will volatilize and again become airborne—perhaps as the result of vacuum cleaners, which are unable to trap the small residue particles. Estimating human exposure to the contents of a can of home insecticides requires quantifying amounts sprayed into the air, recirculation potential, and rates of volatilization, degradation, and inhalation.

Few think that skin contact with toxins such as pesticides poses any significant risk, but the majority of documented farmworker poisonings have come from absorption through the skin. Among all pesticides, parathion appears to be responsible for the most documented poisonings—tens of thousands of poisonings around the world, with most occurring in poorer tropical or semi-tropical countries. Mexico, for example, reported one thousand to 1,500 cases each year during the middle 1960s with a mortality rate between 0.8 and 2.9 percent. Japan reported 1,500 cases and hundreds of deaths per year in the early 1950s. In the United States, parathion was a dominant source of farmworker poisoning during the 1950s, even when workers did not enter the fields for several weeks.[61] The persistence of parathion's toxicity is related to its conversion to paraoxon, which has nearly twice the half-life of the parent compound (5.5 versus 2.9 days) and has been estimated to be ten to fifty times as toxic.[62] Nearly twenty incidents of intoxication from parathion as long as one month following field application were documented between 1949 and 1970. In response to these studies, California enacted field reentry regulations to protect farmworkers from these and other toxic chemicals.

Those who live on the Connecticut shore are familiar with pinhead-sized deer ticks that have become infamous for carrying the spirochete that causes Lyme disease—an illness with flu-like symptoms that can progress to chronic arthritis and severe neurological degeneration. As one precautionary measure, pets wear flea collars and are dipped, sprayed, and powdered on a regular basis; this way, they carry the insecticides wherever they go. Children hugging animals may absorb the pesticide through their skin, through the eyes, or by breathing airborne molecules.

Outside the home, garden and landscape use of pesticides, herbicides, and fungicides has increased dramatically during the past two decades.[63] Many lawn care products are sold as combinations of fertilizers, herbicides, and insecticides that can constitute a significant hazard if not carefully applied. Lawn care companies now market and spray mixtures of fertilizers, insecticides, herbicides, and fungicides to promote a deep green, weed-free, insect-free, and disease-free lawn. Golf courses commonly receive more pesticides per acre than any other type of landscape, including specialty cropland. EPA and many states have become concerned about harmful exposure to the poorly regulated complex mixture of pesticides. Connecticut, for example, now requires that sprayed lawns be posted with warnings for a twenty-four-hour period. Unfortunately, however, this strategy is not based on scientific evidence that rolling on the lawn after the twenty-four-hour period is safe, and it does nothing to prevent possible contamination of shallow drinking-water wells.[64]

Pesticide Residues in Water

In 1969, the U.S. Department of Health, Education, and Welfare prepared a landmark report on the environmental health consequences of pesticides.[65] The Mrak Commission, as it came to be known, directly considered the seriousness of water quality contamination. At this point in the regulatory history of pesticides, attention was focused primarily on chlorinated compounds, none of which had yet been banned. Regulatory authority still lay with USDA, because EPA was not created until 1970.

Evidence of water supply contamination by pesticides was sparse and generally restricted to chlorinated compounds. For example, the intake pipes to the Chicago water-supply filtration plant were sampled in 1969, and Lake Michigan's water was found to contain lindane, heptachlor epoxide, aldrin, and DDT, although generally in the less-than-part-per-billion level. At that time it was well known that lake sediment acted as storage sinks for chlorinated contaminants, which could become resuspended when disturbed. Suspension would then expose microorganisms to the pesticides, and contaminants would continue to accumulate in the food chain.

Transport of chlorinated pesticides within estuaries was also of concern because coastal agriculture commonly drained into them and because by 1969 accumulation of chlorinated pesticides in shellfish and fish was scientifically established. Algal exposure to DDT caused a dose-related decrease in rates of photosynthesis. There was also some speculation that DDT could result in reduced phytoplankton populations and that continued use of persistent pesticides "might then have long-term impact on total photosynthetic activity, perhaps inducing a change in oxygen and carbon dioxide partial pressures in the atmosphere."[66] Many of the commonly used chlorinated hydrocarbon insecticides were found in California estuaries sampled in 1966 and 1967. High levels of DDT, DDD, and DDE were found in king crab, cape salmon, oysters, mussels, and clams. Shellfish levels exceeded 10 ppm in estuaries receiving runoff from agricultural and urban areas, whereas more isolated areas tended to have levels one hundred times lower.[67]

The detection of pesticides in surface water may depend on the season of study. Sampling during the fall and winter is unlikely to reveal residues, but spring and summer samples, particularly if taken soon after heavy rains, were more often contaminated. Insecticides, including parathion, aldrin, and endrin, were detected migrating in surface water from field to drainage ditches, streams, and rivers. Contamination levels were found to be related to application times and subsequent weather; heavy rains washed the pesticides downstream.

Given limited evidence that 2,4-D was detected in drainage ditches, the

Mrak Commission suspected herbicide use was to blame for the contamination of surface water supplies. This compound was sprayed directly into drinking-water reservoirs by the Tennessee Valley Authority in the 1960s to control nuisance vegetation, especially water milfoil. Nearly nine hundred tons of 2,4-D were applied in a granular formulation to nearly 8,000 acres in seven reservoirs. It was then detected in the part-per-billion range at the water treatment plants, and, ten months following treatment, at a level of 58 parts per million in bottom mud.[68] Paris green, or copper acetoarsenite, was also still being sprayed on surface water as a mosquito larvicide during the 1960s, along with light oils and organophosphate compounds such as malathion and chlorpyrifos.

Citing a rainwater study of the British Isles reported in 1968 that found DDT, DDT metabolites, alpha-BHC, gamma-BHC, and dieldrin in the parts-per-billion and trillion range, the Mrak Commission concluded that these concentrations were insignificant and that the primary pathway by which pesticides contaminate water was through the direct application to surface waters and from water that runs off treated lands. The impression given by the commission was that groundwater is generally protected from pesticide contamination. It referred to studies that claimed that organochlorines moved only a short distance by leaching from soil; that parathion was "very unlikely" to contaminate groundwater; that normal agricultural uses of lindane and heptachlor posed little threat; and that dieldrin was highly unlikely to move through the top twelve inches of soil.[69] The commission also cited state claims that pesticide contamination of public wells was rare, with the exception of accidental spills or the application of termiticides such as chlordane around buildings. Michigan at that time required only a 150-foot separation between any pesticide mixing or storage area and a well.[70]

The Mrak Commission was perceived to be precautionary in its interpretation of the dangers posed by pesticides. And because few publications other than *Silent Spring* had attempted such a broad interpretive sweep across evidence on the transport, fate, and environmental health effects of pesticides, it set an important tone for important decisions on individual pesticides and particularly influenced EPA's ban of DDT, aldrin, and dieldrin in the 1970s. It also increased the visibility of pesticide contamination as an important national and international environmental problem comparable to the issues of air and water pollution. It may, however, have diverted attention from exploring the groundwater contamination potential of insecticides.

The generation of pesticides to follow the chlorinated compounds such as DDT was designed especially to avoid the problems of persistence and bioaccumulation. Water solubility was part of this solution; water was used to dilute and disperse compounds. Water-soluble pesticides have surprised chemists, geologists, and hydrologists, however, in their speed and scale of movement—they may contaminate water supplies miles from the application site. Atrazine is so widely used on Midwestern corn and soybean farms, for example, that it is now possible to predict with considerable accuracy the increase in atrazine in surface-water drinking supplies following its application each spring. Residues are washed by rains from the field into drainage ditches and streams, which empty into reservoirs.

Underground aquifers may also be contaminated by water-soluble pesticides as water leaches through soils to underlying storage areas. Whereas simple gravity explains much of the compound's movement, underlying bedrock geology (particularly the structure of rock fissures) and layers of impervious soil materials may facilitate the rapid long-distance movement of contaminants. Ethylene dibromide (EDB) is one of the most potent animal carcinogens among pesticides, and it was widely used to protect shade tobacco in Connecticut fields. EDB contaminated large areas of ground and drinking water beneath the meandering floodplain of the Connecticut River, which became more valuable for suburban home and shopping center development than for tobacco production. Owners of new homes were shocked to find their drinking water had been contaminated by tobacco companies that had long since left town.

Exposure to pesticide-polluted drinking water poses a particularly serious threat to young children because such a large proportion of their diets is made up of water. The diets of newborns and very young infants are predominantly liquid and commonly consist exclusively of infant formula and fruit juice, both of which may be supplied as concentrates to be diluted with tap water. By comparison, adults are more likely to drink from a diversity of sources, such as milk, soda, juices, or other commercially processed beverages. Also, less than 30 percent of average adult food intake by weight is liquid. These trends are described more fully in chapter 9.

Although the persistence and bioaccumulation of chlorinated compounds was one of the great surprises of the earlier generation of pesticides, the scale of movement of water-soluble compounds through soil, from field to stream, and through subsurface groundwater supplies has been the dominant surprise of the later generation. Contamination of underground water aquifers on Long Island by aldicarb, a carbamate insecticide, became well known during the early 1980s. Widescale contamination of Midwestern aquifers by herbicides beneath corn and soybean fields was recognized by the middle of that decade.

Whereas bioaccumulative insecticides achieved global distribution primarily through contamination of the food chain, water polluted by pesticides is normally a local or regional problem. Contamination potential depends on numerous factors, including the type of pesticide (its solubility, formulation and method of application); soil type (particularly its amount of clay and organic matter); superficial geology, including slope and characteristics of the water table; climate; season; and cropping patterns.

Today nearly 53 percent of the U.S. population derives drinking water from groundwater sources, and 47 percent drink water from surface supplies. Many Midwestern cities lie next to rivers that are the dominant sources of their drinking water. Given our scientific understanding that pesticides could easily migrate from field to surface waters or to underground aquifers, it is remarkable that the quality of data on contamination levels is so limited. One reason for the fragmented image we have is that the patterns vary locally and seasonally, as do patterns of pesticide use. Clearly, another reason is the expense of sampling and analysis.

As was the case with food, the history of pesticide detection in water supplies has demonstrated that when pesticides are sought, they are commonly found. Various studies have detected between thirty-nine and forty-six different pesticides or their metabolites in groundwater, primarily groundwater in agricultural areas. The most commonly detected compounds have been water-soluble compounds, soil fumigants, and herbicides. Aldicarb, a highly toxic carbamate insecticide, has been found in twenty-four states and became notorious as a contaminant of water supplies within Long Island's sandy aquifers. EDB was found in twelve states and dibromochloropropane (DBCP) in five states. Herbicides have been found in groundwater throughout the country and include atrazine, alachlor, cyanazine, dicamba, dinoseb, metolachlor, metribuzin, simazine, trifluralin, and 2,4-D.

The Monsanto Company produces alachlor, in 1995 the world's most sold herbicide. In 1987 Monsanto sampled 1,430 private, rural, domestic wells from agricultural areas where alachlor was most heavily used. These wells were selected from nearly 6 million located in forty-five states that serve nearly 20 million people. The study concluded that one hundred thousand people in the sampled areas may be exposed to detectable levels of alachlor from their drinking water. Nearly 13 percent of the wells tested contained some herbicide residue; atrazine was most commonly detected, followed by alachlor, metolachlor, cyanazine, and simazine.[71]

In 1990, EPA conducted its own survey of drinking-water wells that included both community water supplies and private rural wells. They found that 10.4 percent of the community water supplies and 4.2 percent of the rural domestic wells contained more than one pesticide. The herbicide dichloranchloranphenicol-peptone agar (DCPA) was most commonly found, followed by atrazine, simazine, prometon, and DBCP.[72]

Pesticides, particularly herbicides, are far more commonly detected in surface water samples than in groundwater tests. Surface reservoirs, lakes, and rivers provide nearly 60 percent of the water in public supply systems and serve approximately 47 percent of the U.S. population.[73] In a 1989 analysis of "raw" (pre-treatment) and "finished" (post-treatment) drinking water from surface supplies in Illinois, Iowa, Kansas, and Ohio, the herbicides alachlor, atrazine, cyanazine, metolachlor, and 2,4-D were found in 67 percent of the raw and finished samples tested.[74] EPA has classified all of these compounds as possible or probable human carcinogens. Atrazine in particular was found in raw water in 77 percent of the Illinois samples, 93 percent of Iowa samples, and 100 percent of Kansas samples.

There are significant seasonal fluctuations in levels of surface water residues; the lowest levels occur between August and April, and peak levels appear in May and June. These detections correspond well with seasonal patterns of herbicide application, which commonly occur early in the spring.[75] Herbicide levels in Midwestern surface waters, for example, may be ten times higher during early spring than after the fall harvest.[76] The contamination problem appears to be primarily related to the water-soluble herbicides, although the insecticides chlorpyrifos and carbofuran, both cholinesterase inhibitors now under scrutiny for their neurological effects, have also been found (albeit less frequently). Similar seasonality has also been found for levels of herbicides applied to California rice fields in the Sacramento River valley.[77]

The responsibility for defining acceptable levels of contamination of drinking water lies with EPA, which derives its authority from the Safe Drinking Water Act.[78] By May of 1993, EPA had established maximum contamination levels (MCL's) for only twenty-three of the six hundred active ingredients registered for use in the United States. MCL's are set primarily to protect the public health, but the technological feasibility and cost of attaining standards may also be considered by EPA. If pesticides are classified as drinking water contaminants and are suspected of causing cancer, then MCL's have historically been set to prevent exposures that will increase cancer risk levels by one in 1 million or more. If chemicals are not believed to cause cancer, then the highest exposure level that caused no observed adverse effects in laboratory animals (NOAEL) is reduced by a safety factor, normally of one hundred, but occasionally as high as ten thousand. Uncertain scientific evidence of pesticide toxicity and the general failure to test drinking-water supplies for pesticides all help to explain the lethargic pace of EPA in setting MCL's. Further, municipalities strongly resist more rigid testing requirements. Every time EPA sets a new MCL, municipalities around the country must routinely demonstrate that their water is not contaminated beyond the legal limit. Despite this bureaucratic resistance, these are perhaps the most important limits to set and monitor, given the dominance of water in the human diet and that contaminated water may easily become the primary pathway of human pesticide exposure.

Pesticide Residues in Food

As explained in earlier chapters, in the nearly thirty thousand separate pesticide products registered by EPA, six hundred active pesticide ingredients are combined with nearly 1,600 inert ingredients. Roughly 325 of these active ingredients are allowed for use on food crops, and EPA has set nearly 9,300 separate tolerances, which limit maximum allowable residue levels in various foods. This general discussion of pesticide residues in food provides background necessary for later critiques of the ways that government has estimated human exposure and health hazards.

The worst cases of pesticide poisoning from foods have come from accidental contamination rather than from low-level residues that remained following crop protection efforts. The overwhelming majority of cases have involved either mercury compounds or those insecticides containing organophosphates, carbamate insecticides, or both.[79] Perhaps the worst recent U.S. case of pesticide poisoning occurred during 1984 from California watermelons. Aldicarb was illegally used on the watermelons, resulting in eight deaths and 1,350 illnesses. Aldicarb sulfoxide, a primary metabolite of the insecticide aldicarb, is suspected as the primary toxin, and the fact that many watermelons were eaten during the Fourth of July holiday contributed to the high numbers of injured.[80]

One of the fundamental principles of pesticide residue chemistry is that if a compound is used on a food crop there will be a residue in the food—whether or not it is detectable by the chosen analytical method or even by state-of-the-art techniques. EPA, and USDA before it, however, have granted registrations for food uses of pesticides even when no analytical method of residue detection had been available. This means that effective regulation of human exposure to pesticides presents a colossal analytical task—one that requires that the government establish statistically reliable sampling and testing protocols to ensure that residues from each compound do not exceed a safe level. How big is the problem? EPA regulates pesticides on nearly 675 different foods, and as one example, this year over 1 billion bananas will be shipped from numerous tropical countries into Connecticut alone. During 1988 and 1989, the FDA, responsible for testing pesticide residues in food throughout the United States, tested only seventy-two banana samples for benomyl, a fungicide commonly used on bananas and classified as an animal and possibly a human carcinogen by EPA.

Moreover, by 1995 FDA still did not keep track of cases in which multiple pesticide residues occurred on single food samples, so that very little data are available to forecast probable exposures to mixtures of pesticide residues in the diet. These data are necessary to predict and control possible additive, synergistic, or antagonistic effects from exposure to these mixtures. Because many fruits and vegetables are allowed by law to have over fifty different pesticide residues, the analytical burden of estimating human dietary exposure and associated health risks is staggering.

Sampling and Detection

Any crop treated with a pesticide will never be completely free of residues, although residue concentrations will eventually degrade beneath a detectable level.[81] Most residue levels will degrade and volatilize after application is stopped. How then should the environment be sampled to detect pesticide residues in food and water? Many mistakes in sampling design are possible and common and may lead to significant errors in estimates of contamination, human exposures, and risks. Most errors are attributable to a misunderstanding of how residues are distributed in space and time. This knowledge is essential to produce statistics that can then be used to predict the distribution of the residues in the population, based upon sample results. There is an obvious circularity in logic here, in that some understanding of residue distribution is necessary to create an experimental design to understand it better. The design of sampling strategies is best done iteratively, in a knowledge-dependent manner. As new knowledge of the behavior and distribution of residues in the environment is derived, it should be used to strategically restructure later sampling designs. For example, early USDA food-intake surveys sampled few children beneath age one, thus the results were of little use for predicting intake patterns for U.S. infants. Our gradual understanding that children's diets vary considerably from those of adults, however—and that these variations are tied to probable differences in pesticide exposure—has encouraged USDA to modify their sampling design to include more young children.

Understanding the magnitude and distribution of pesticide contamination in the environment is complicated by the large number of pesticides allowed to be released and the diverse ways in which they move through the environment. The scale of these problems calls for strategic and iterative sampling designs, rather than a comprehensive approach. Sampling all pesticides in all foods, for example, would quickly consume limited monitoring budgets of regulatory agencies.

Current government policy has done little to discourage sampling bias. For example, although peppers constitute a relatively small proportion of the average diet in the United States in 1988 and 1989, FDA sampled eighteen thousand California peppers following several detected violations. During this same period, FDA tested fewer than one hundred samples of apple juice for pesticide residues, even though apple juice is one of the most commonly consumed foods by young children. Sampling bias also results from the type of analytical method used. FDA's sampling program is highly biased to search for pesticides that are detectable using multi-residue screens capable of finding as many as seventy different compounds. Some compounds, however, may require their own individualized test. Thus, one would hope that federal pesticide sampling would be biased toward detection of pesticides that are the most toxic, not those which are least expensive to detect, and toward environments and foods where the public is most likely to encounter the most toxic residues. These criteria have not been employed by FDA, in part due to budgetary constraints—a problem likely to become more severe.

There are several important spatial types of sampling bias as well. FDA, for example, may sample peaches nationally for a fungicide used by peach growers only in the northeastern United States. If summary statistics are produced, the average national residue levels will be lower than those experienced in the northeastern United States (if the peaches are regionally rather than nationally distributed). Other foods, such as beef, tend to be nationally distributed, whereas milk tends to be produced even closer to the point of consumption.

Spatial bias may also occur because of differences among sampling sites. Farmers who neglect to turn sprayers off at the end of rows while tractors slow to turn around will dose the ends of fields more heavily than the middle. Spray nozzles may be adjusted or clog during application and cause higher pesticide concentrations in some parts of the field. If there is overlap among passes of airplanes or spray bars, some rows may be more heavily contaminated than others. In addition, wind turbulence can produce extremely uneven distribution of pesticides applied by aircraft.

There is also an important temporal dimension to sampling bias. Sampling to detect pesticide residues should consider the time lapse between application and testing, which can significantly reduce residue levels through processes such as photodegradation and volatilization. In fact, the process of residue degradation is purposely reduced for some compounds to extend product life. Fungicides, for example, may be sealed by waxes sprayed on apples, tomatoes, cucumbers, eggplants, apricots, and other fruits and vegetables.

Sampling error may result as well from the selection and combination of pieces to be analyzed. Bunches of grapes, apples, or heads of lettuce, for example, are taken from the field or the crate, and "composited"—combined with other pieces of similar food—to form a sample slurry for analysis. Food such as oranges are then peeled to test residue levels on the part that is consumed and regulated. This composite sampling method is controversial because residues from one contaminated portion, say a single orange, may not be detected if the sample has been diluted beneath the detection limit by other uncontaminated pieces of food.

To improve estimates of human exposure to pesticides, residue sampling should also be conducted as close to the time of consumption as possible because some pesticides degrade to less toxic substances over time and others tend to concentrate during food processing. In practice, samples are most commonly taken from raw commodities at distribution warehouses. This practice posed a significant problem with Alar, a plant-growth regulator commonly used on apples during the 1980s. As explained earlier, Alar degrades to UDMH when it is heated during the pasteurization of apple sauce or apple juice. UDMH was believed by some to induce tumors in laboratory animals at a young age and at low doses. If raw apples are tested for UDMH residues, normally none are found. If Alar is used on fresh apples and these are used to make apple juice, however, UDMH is normally detectable, unless it has been diluted by juices from apples not treated with Alar.

Apples and their juice provide a simple example of a far more complex problem: sampling residues in processed foods. Anyone who has visited a supermarket recently can appreciate that many foods are now extremely complex mixtures of more basic foods and a huge and ever-growing collection of artificial colors, flavors, preservatives, plasticizers, antibiotics, growth hormones, dispersants, waxes, and other food additives. Understanding residue fate during processing is the least understood area of pesticide chemistry but one of the most relevant for estimating human exposure to toxic pesticides. Many current registrations and tolerances are unsupported by any evidence of residue fate during processing. Under these circumstances, EPA simply assumes that 100 percent of residues will transfer from the raw to the processed food. If the processed food is high in fat, such as a vegetable or seed oil, and the pesticide is known to be fat soluble, an additional concentration-multiplier is used.

Samples taken from a field known to have been treated with a pesticide will normally contain detectable residues. As soon as produce is crated and shipped to a warehouse for distribution, however, residues often dissipate. Moreover, sampling of foods once they reach the marketplace can be misleading because produce is commonly mixed with food from many different farmers, parts of the country, and increasingly from different nations. Marketplace residue analyses are the most difficult to conduct because it is rarely known what pesticides were applied, particularly to foreign-grown produce. The influence of these varied sampling sources on exposure estimates is explored in chapter 10.

The Detection Limit Problem

Over a period of only two decades, detection limits have been reduced by nearly six orders of magnitude. By 1952, a method known as paper chromatography[82] was used to detect a variety of chlorinated hydrocarbon insecticides, including DDT and its metabolites, in the part-per-million range.[83] Gas chromatography, the next advance, operates by volatilizing and separating compounds in a column. By 1962, gas chromatography was capable of identifying seventy-one separate pesticides, with some detectable in the 0.02 to 0.2 ppm range. The design of an electron-capture GC detector in 1963 permitted the detection of chlorinated compounds at levels between 1 ppb and 0.1 ppm.[84]

Refinements of the mass spectrometer in the 1970s have revolutionized environmental chemistry and our understanding of how pesticides move through the environment. It breaks chemicals into ions and records each separately, producing a "mass spectrum" that uniquely and definitively identifies chemicals. Detection beneath the part-per-billion level became possible through the introduction of selected-ion monitoring, which is the product of the marriage between gas chromatography and mass spectrometry. This method, for example, can detect the difference between an aflatoxin B1 ion with a mass of 312.0633 and a peanut-oil molecule with a mass of 312.2183. This distinction permits the identification of aflatoxin B1 in the 100 parts-per-trillion range. Two mass spectrometers can also be linked together. The first can thus act as a filter for the second and permit the detection of dioxin in low part-per-trillion concentrations.[85]

Most residue testing by regulatory agencies is conducted to find out whether contamination levels exceed the legal maximum, or tolerance. Thus the detection limit is commonly set at 1/10th the legally allowed residue level by laboratories in regulatory agencies. FDA, responsible for monitoring residue levels in the nation's food supply, most commonly sets the detection limit at 0.1 to 0.01 ppm, regardless of the legal limit. These tolerances were commonly established by USDA and after 1970 by EPA, and are based on the amount necessary to control pests in the field. These limits, however, may have little if any relationship to a toxicologically safe level of human exposure. Moreover, as I've suggested, "tolerance," a legally allowable residue limit, is a confusing term because it suggests human tolerance without adverse health effects.[86]

The FDA employs two types of sampling designs. One type monitors residue levels in raw commodities (surveillance sampling); the other is focused to detect violative residue levels (compliance sampling). Other surveys are designed to test food in the market basket, rather than the shipping dock, in order to more closely approximate human exposure. Both single-residue and multiple-residue detection methods are available to monitor residue levels in foods and drinking water.[87] The U.S. food supply has predominantly been monitored using multi-residue screens that can detect most of the 325 pesticides allowed to exist as residues in the diet. These screens use liquid or gas chromatography, and sometimes both.

In 1989, the NAS Committee on Pesticides in the Diets of Infants and Children decided to collect and interpret pesticide residue data in order to estimate the risks facing children from contaminated food and water. Regulatory responsibility for pesticide residues in food is fragmented between EPA, which establishes allowable residue limits for each pesticide-food combination, and FDA, which is responsible for monitoring and enforcing the EPA limits for all foods except meats, poultry, and eggs, which are separately monitored by USDA.[88] The committee found that there is significant variation in sampling designs and residue testing methods within and among these authorities—differences that make it extremely difficult to interpret their findings. Considerable variation, or "improvisation," in multi-residue methods and the choice of detection limits has occurred among laboratories, differences that are not always subject to peer review and are rarely published. Single-residue methods, which are more precise but more expensive, have normally been reserved for cases in which a pesticide that was not discernible by a multi-residue method becomes a public health concern. Additional sources of error could also include poor training of analysts, equipment that is less than state-of-the-art, and the use of inappropriate reference standards to quantify the results.[89]

Detected Pesticide Residues in Food

What pesticide residues have contaminated the U.S. food supply? A study of residues in food between 1969 and 1976 found traces of 124 different pesticides in 56 percent of the FDA samples, and residues of more than one pesticide were found in 30 percent of the samples. Chlorinated organic pesticides such as DDT, DDE, dieldrin, heptachlor, BHC, endrin, and lindane were commonly detected in dairy, eggs, beef, and poultry.[90] The source of these residues appears to have been contaminated animal feed. Many of these pesticides were also routinely found in beans, vine and root vegetables, peanuts, and soybeans. With the exception of malathion, they were also commonly detected in grains. In a separate analysis of infant and junior foods, DDT and metabolites were found in 18.4 percent of the samples tested.

FDA also conducted eight "market basket" surveys between 1982 and 1984, more than a decade following the domestic ban of DDT and seven years following the ban of aldrin and dieldrin. Over one hundred pesticides and metabolites were detected, with an average of sixty-four pesticides and other industrial chemicals detected per basket. The most commonly found pesticides included DDE (25 percent incidence), malathion (22 percent), dieldrin (15 percent), pentachlorphenol (15 percent), BHC-alpha (14 percent), diazinon (13 percent), HCB (9 percent) and heptachlor epoxide (8 percent).[91]

In 1993, the FDA published another in a series of studies on pesticide residues in infant foods and in adult foods eaten by infants or children.[92] The 1993 paper was remarkable in that it summarized FDA's sampling of foods consumed by infants and children over a seven-year period. It also provides an opportunity to compare patterns of residue incidence with those of an earlier study that covered the period 1969–1976. The 1970s were a period of active regulation and prohibition of additional use of chlorinated hydrocarbon insecticides, so one would expect their relative incidence and levels in food to decline.

FDA used three different sampling designs to monitor pesticide residue levels in the domestic food supply. The first was implemented to test pesticide residue levels in both domestically produced and imported foods. Between fifteen thousand[93] and nineteen thousand[94] samples were analyzed annually. FDA analyzed a subset of ten thousand samples collected over the seven-year period 1984–1991. The total number of samples of the following foods are presented in table 7–1.

Perhaps the most important conclusion of these analyses has little to do with detected residues; instead it pertains to the way in which sampling is conducted. For example, nearly ten billion bananas are shipped into the United States every year. Sampling an average of 167 bananas per year has little hope of offering any protection to consumers from illegal residue levels. Hundreds of millions of apples are produced and sold in the United States every year. Hundreds of millions of gallons of orange juice are sold yearly, whereas FDA tested only an average of ten samples, including only seven samples of grape juice.

In this same study, FDA presents the residue results for domestic samples. Residues of forty-eight different compounds were found in the 2,464 apple samples tested, with captan, a fungicide, and chlorpyrifos, an insecticide, detected most frequently. Chlorpyrifos was also the most frequently detected pesticide in bananas and oranges. For apple juice, five different compounds

Table 7–1. Number of Samples Tested in FDA Pesticide Samplings of Foods Heavily Consumed by Children, 1984–1991

	Domestic samples	Samples/ year	Import samples	Samples/ year
Apple	2,464	352	735	105
Apple juice	114	16	351	50
Banana	72	10	1,097	156
Grape juice	32	5	24	3
Milk	2,919	417	—	—
Orange	862	123	474	68
Orange juice	13	2	64	9
Pear	571	81	816	117

Source: Yess et al. (1993); see note 92.

were detected in 54 of 114 samples (47 percent), with daminozide (Alar) detected most often before its use was completely phased out in 1993.

Continued findings of chlorinated hydrocarbon residues banned during the 1970s is difficult to interpret. DDT was banned in 1971, yet between 1984 and 1991 it was the most commonly detected residue found in milk (11 percent of samples). Dieldrin was found in 9.5 percent of the milk samples tested and heptachlor was found in 8.6 percent, even though both were banned before 1980. When FDA tested infant foods in a market basket survey, DDT was the most commonly found pesticide in infant meat and poultry dinners, as well as in milk samples. In beef and dairy products, these findings imply that the residues continue to contaminate animal feeds. In other crops, they may have indicated the persistence of residues in soils and uptake by plants.

Among all ten thousand food samples tested, only fifty violated federal tolerances, and most of the violative samples were cases where no tolerance for the pesticide food combination had been established. Does this mean that the U.S. food supply at that time was safe? FDA interprets the evidence this way. But any conclusion about the safety of the nation's food supply must account for the appropriateness of the sampling design, particularly the small sample sizes. Also, the dilution effect of blending or compositing numerous pieces of food, before analysis, leaves little confidence that residues are uniformly as low as those reported. FDA's consistent failure to report cases for which multiple residues appear on single foods also leaves us with a poor understanding of the potential for numerous residues to contribute to a single effect, such as neurological damage or an unacceptable risk of cancer.

Using the *de minimis* argument to justify low levels of bioaccummulating residues seems poorly grounded for two reasons. The first is that the compounds may accumulate in human tissue, unlike organophosphorus, carbamate, and pyrethroid insecticides. The second problem is that the residues continue to appear in foods that children consume in high quantities, such as milk, beef, poultry, and egg products. Although the levels detected are generally low, a milk residue was found at nearly 1 ppm (the FDA "action level" is 1.25 ppm). These conditions create a pattern of tissue contamination that persists for a lifetime given the stability of metabolites such as DDE in human tissue.

A more recent review of residue data collected by FDA from 1990 through 1992 found significant differences in residue detections among FDA laboratories. Most FDA labs use the Luke extraction method, which is capable of detecting more than three hundred different pesticides if all screens and methods are employed. Those labs that used three or more detection screens found pesticide residues in a higher percentage of the food tested than those that used two or fewer detection screens. Those labs using the most rigorous methods found nearly twice the level of pesticides in apples, pears, bananas, tomatoes, and green beans.[95]

If one wanted to portray the food supply as being free of residues, the "nondetection" results from labs using insensitive methods would be included in data summaries. Restricting the summary to the results from labs using the most sensitive methods will demonstrate a more significant level of contamination. The labs with the least sensitive methods have tended to lie in the northeastern and southern parts of the country, whereas those with the most sensitive methods were in the Midwest and the West, with the exception of Atlanta.

A significant number of pesticides that pose health risks are not detectable using the Luke or other multi-residue screening methods and require single residue screens. These compounds include benomyl, ethylene-bis-dithiocarbamates (EBDC's) fungicides, and O-phenylphenol. FDA tested only 1 percent of the apple samples and 2 percent of peach samples for benomyl, despite its known common use on these crops. Similarly, only 1 percent of potatoes and 2 percent of apples were tested for EBDC fungicides. No detection method exists for some pesticides that may appear as residues in specific types of food. By 1995, the maximum "safe" level of atrazine, the most commonly used herbicide in the United States, was not detectable in milk. Instead, EPA is forced to rely on a model of residue fate in cattle based upon levels of atrazine in feed. And although milk is the food most consumed by children between the ages of six months and five years, the absence of a reliable detection method is normally used to claim the absence of evidence of risk. It could also be argued that allowing a residue to persist in the food supply without a reliable method to test for its presence constitutes uncontrolled experimentation.

The unreliability of FDA residue inspection programs has caused grocery manufacturers to increasingly rely on private testing laboratories to ensure that fresh fruits and vegetables are free from detectable pesticide residues. Third-party certification companies now test produce in more than 450 supermarkets in the United States. In some areas, the certification company will contract directly with growers to ensure that dangerous pesticides are not used and that any other pesticides are used in a manner that causes them to dissipate below any detectable level by the time they reach the supermarket. The contracts with growers allow the private residue labs to strategically test for compounds they know were used.

This type of focused analysis has found more residues more often and at higher levels than have FDA tests. Apple juice provides an example. Between 1990 and the end of 1992, FDA tested only forty-one samples of apple juice for pesticide residues. Of these, seven samples contained residues of one pesticide, and two samples had residues of two pesticides. No sample was tested using single-residue tests necessary to detect the fungicides described above, commonly used on apples.

By contrast, a private supermarket certification lab analyzed twenty-five samples of apple juice during the same period. It found that five of eleven samples tested for benomyl were positive, and four of seven samples tested for Alar were positive. More than half of the twenty-five samples had residues of one pesticide, and four samples had two different residues. Of the positive samples, the majority were from foreign sources, including Argentina and Canada. Generally, the levels detected by the private certification labs were close to that found by FDA labs when FDA used screens capable of detecting all possible compounds, including single residue scans.[96]

The Environmental Working Group, a recently formed environmental research and lobbying organization, constrained their analyses of residues in the diet to results from labs using the most sensitive methods, which reduced the sample size from 17,000 to 14,629 and allowed it to cover twenty-two crops most consumed by infants and children. Nearly 50 percent of these samples had detectable pesticide residues; some crops, however, were more contaminated than others. For example, 76 percent of peaches and 73 percent of strawberries had detectable residues, whereas only 18 percent of broccoli and 7 percent of cauliflower samples contained residues. FDA found a total of 108 different compounds on twenty-two foods. On five foods, thirty-eight different pesticides were detected. Endosulfan was found on twenty-one different crops, chlorpyrifos on twenty, carbaryl on nineteen, dimethoate on nineteen, and methamidophos on nineteen. DDT and its metabolites, DDE and DDD, were found on eleven crops. This is not a claim that the levels detected were toxicologically significant, merely a suggestion that if sensitive detection methods are focused more pesticides are detected than are normally reported by FDA.

These results are largely confirmed by those of the private supermarket testing labs, which tested nineteen of the same foods chosen for the FDA analysis. The private supermarket labs found 59 percent of the samples contained residues, with strawberries (82 percent) and oranges (80 percent) most often contaminated by at least one compound. A total of 81 different pesticides were found on these nineteen foods, with twenty-nine pesticides found on five or more foods. The list of most commonly found pesticides by the private labs is similarly dominated by cholinesterase inhibiting insecticides. In addition to the FDA findings, the private labs commonly detected fungicides: benomyl was detected on eleven of nineteen crops, and EBDC residues were found on twelve crops.

One important limitation of FDA residue analyses is their failure to report instances where multiple residues were found on single food samples. Apples provide an example of the problem. In 1991, at least forty-eight different insecticides and fungicides were applied to apples in the United States. Although no grower would use all forty-eight, USDA estimated that 25 percent of the nation's apple crop was treated with thirteen different compounds, and 50 percent of the crop was treated with four compounds: captan, guthion, chlorpyriphos, and petroleum distillates. Apples should not be singled out for criticism; grapes were treated with as many as twenty-six different pesticides and oranges with as many as twenty.[97]

Reviewing private lab results compiled by EPA, the Environmental Working Group study found that two or more pesticides were present in 62 percent of orange samples tested, 44 percent of apples tested, and between 33 percent and 25 percent of all cherries, peaches, strawberries, celery, pears, grapes, and leaf lettuce tested. Some individual apples had traces of as many as eight different pesticides, and several oranges had seven different residues. These findings were generally confirmed by an independent analysis of FDA data.[98]

Pesticide Residues in Imported Food

The United States currently imports nearly six hundred different types of food from nearly 150 different nations.[99] Over 6 billion pounds of fruit and 8 billion pounds of vegetables are imported into the country annually.[100] Yet FDA tests only eight thousand imported fruit and vegetable samples yearly, which means that on average roughly one residue test is performed for each 2 million pounds of imported food.

Pesticide contamination of food consumed in the United States is obviously influenced by agricultural practices in other countries. Furthermore, rapid growth of multinational corporations and the explosion in worldwide trade in pesticides, agricultural technology, and food have created a pattern of exchange that is extraordinarily difficult to monitor and lies largely uncontrolled. For example, one-half of all winter fruits and vegetables imported to the United States come from Mexico, and this percentage may rise appreciably given the recent North American Free Trade Agreement. Mexico produces 97 percent of our imported tomatoes, 93 percent of imported cucumbers, 95 percent of imported squash, 99 percent of imported eggplant, 68 percent of imported melons, and 85 percent of imported strawberries. Although Mexico is creating a national laboratory system to monitor residues in exported produce, it has relied on its private sector to monitor these contaminants.[101]

This problem has been well recognized since at least 1979, when the General Accounting Office reported to Congress on the need for better control of both U.S. exports of pesticides and residues on imported foods.[102] At that time, U.S. imports of food exceeded $13 billion per year; they now exceed $25 billion per year.

It will surprise many to learn that it is still legal for a U.S. firm to produce and export a pesticide banned for use in this country. This practice permits what environmental groups have called the "boomerang effect" or the "circle of poison," whereby pesticides banned in the United States are exported for use on crops abroad and then imported, illegally, as food residues. Although FIFRA requires that manufacturers submit annual reports to EPA about exports, the makers of these compounds have successfully prevented EPA from disclosing these data publicly by arguing that the information is confidential. The only other source of data is from U.S. Customs records of individual shipments, which are allowed to be described in the vaguest of terms, such as "weed killing compound."[103]

U.S. manufacturers exported over 465 million pounds of pesticides in 1990, and U.S. Customs records demonstrate that over 52 million pounds were banned, restricted,[104] or unregistered for use in the United States. Of those pesticides identified by active ingredient, nearly 42 percent were either banned, restricted, or unregistered. More than 140 companies shipped pesticides from the United States to 119 different nations, a list dominated by the Netherlands, Belgium, Japan, Switzerland, and Colombia. Banned, restricted, or unregistered pesticides were most often shipped to Argentina, Belgium, Colombia, Ecuador, Japan, the Netherlands, and the Philippines.

It is difficult to place U.S. exports into a broader picture of international exchange, but some statistics are available. The United States purchases approximately one-third of all pesticides produced in the world each year, whereas the European Economic Community and Japan together account for another one-third. Developing nations consume approximately 25 percent of world production, with Latin American nations purchasing approximately half of this amount.[105] The world's largest exporter of pesticides is Germany, followed by the United States and the United Kingdom.[106] The largest importers in 1989 included the former Soviet Union, France, and the United States. Global production of pesticides is concentrated. The top ten companies account for 50 percent of total production, and the top thirty-six firms produce 90 percent of the pesticides manufactured in the world. The top fifteen firms are located in the United States, western Europe, or Japan.[107]

Despite these summary statistics, knowledge of patterns of international pesticide exchange is quite limited. It is common for a nation to import an active ingredient, combine it with other active and inert ingredients, and export the commercial product. Thus it is extremely difficult to track patterns of production, mixture, and distribution. It appears that the Netherlands and Belgium play an important role in importing active ingredients, formulating commercial products, and then exporting them abroad. Industrialized nations that produce large volumes of staple crops—such as wheat, corn, soybeans, and cotton—are heavy consumers of pesticides. Tropical nations producing crops for export also rely heavily on pesticides, due to the diversity of pests.

In the absence of consistently applied international standards for pesticide use, the only real hope any nation has of keeping exposure beneath safe levels is to monitor residues in both domestically produced and imported foods. Yet the scale and costs of comprehensive protection are enormous. By 1979, FDA normally used only two multi-residue screens to detect pesticides on imported foods. These two screens were capable of detecting only ninety of approximately 325 pesticides that had food-use tolerances. GAO found that 130 different pesticides were allowed, recommended for, or used on only ten different imported foods, and that none of these could be detected by FDA's most commonly used multi-residue tests. Compounds that escape the multi-residue tests require the more expensive and time-consuming single-residue scans, which generally necessitate specific knowledge of the pesticides used on the imported food. We have just seen how poor our understanding is of U.S. patterns of pesticide use and food distribution. Patterns of foreign pesticide use and international distribution of food are far more complex and therefore virtually impossible to monitor.

Between 1979 and 1991, Mexican produce violated FDA tolerances two to six times as often as did domestic produce according to findings derived from both the surveillance and the compliance sampling methods. But foreign use of compounds that are neither registered in the United States nor tested for their toxicity pose an even more serious problem. In 1979 six countries provided the bulk of U.S. coffee imports. Of ninety-four pesticides allowed on coffee by these governments, seventy-six had no U.S. tolerances. Foreign governments allowed aldrin, dieldrin, chlordane, and DDT to be used on coffee after they were banned in the United States. And in the two nations that provided nearly 40 percent of all tea, twenty of twenty-four commonly used pesticides had no U.S. tolerances. Residue levels on imported fruits and vegetables are poorly understood. In the 1980s, imports of fruits and vegetables increased substantially as knowledge of their nutritional value improved and an understanding of the link between fat intake and heart disease and cancer grew. Between 1980 and 1986, fruit imports tripled and vegetable imports doubled. In 1990, the United States imported $100 million more fruit than in 1989.[108]

A 1993 study by the General Accounting Office found that U.S. firms exported twenty-seven different pesticides that are unregistered in this country. Of these, eight have been either completely or partially canceled and seventeen have never been registered. FDA did not even know how to test for all of these compounds, and USDA tested for only three of them. Moreover, U.S. firms have been under no legal obligation to report export destinations, foreign registration requirements, or residue testing methods, a policy that has left domestic regulatory agencies in a sea of uncertainty.[109] Since 1988, FIFRA has required that U.S. firms notify foreign governments of shipments of banned, canceled, suspended, or restricted-use pesticides. Yet in 1989 Congress found that EPA had no consistent procedure for preparing these notices and ensuring they were received by appropriate agents of foreign governments.

In 1990, the General Accounting Office reviewed the pesticide control laws and administrative procedures adopted within five Latin American nations, including Mexico, Chile, Costa Rica, Guatemala, and the Dominican Republic. It found that although all five nations had pesticide control legislation that prohibited the use of compounds that had been canceled or suspended in the United States, there were, in general, not enough resources to keep track of patterns use, residues, and changes in U.S. law. Given similar problems in our own government agencies, this situation is understandable.[110]

Nearly 7 percent of all U.S. agricultural imports in 1979 originated from Nicaragua, Honduras, Guatemala, and El Salvador. At that time these nations permitted the use of DDT, dieldrin, toxaphene, endrin, ethyl parathion, and other compounds that were either banned or strictly regulated in the United States. In 1988, forty-two thousand pounds of beef contaminated with chlordane and heptachlor, both prohibited for use on foods in the United States, were imported from Honduras. By the time FDA had detected the problem, thirty-nine thousand pounds of the beef had been shipped to a processing plant in Minneapolis, distributed around the country, sold, and presumably consumed.[111]

Even when pesticide residues are detected in imported food by federal inspectors, the food is still likely to reach the marketplace. In 1986, the General Accounting Office found that FDA often had not prevented the marketing of foods found to contain illegal pesticide residue levels. In 1986, seventy-three of 164 shipments containing illegal pesticide residues were allowed to reach the marketplace. Of these, only in eight cases were the importers assessed damages.[112] One reason that contaminated food reaches the consumer is that importers, rather than FDA or USDA, retain possession of suspect shipments. FDA lacks the authority to impose fines on importers who distribute adulterated foods, and even when imposed and collected the penalty is not large enough to deter this practice. Importers in 1995 were not required to export contaminated shipments; instead they were allowed to pay a penalty.[113]

Residues in imported beef are monitored by USDA's Food Safety Inspection Service (FSIS). In 1986, the General Accounting Office found that FSIS fulfilled its annual testing quotas by May 1, and therefore the service stopped testing imported beef for the normal full range of possible residues. Further, FSIS has not always removed the remainder of lots of beef from the marketplace once contamination has been detected. Mexican beef, for example, has presented a considerable problem for inspection. Since 1984, Mexico has been prohibited from exporting processed beef to the United States due to chemical residues. Yet in 1986, 60 percent of imported live beef came from Mexico, despite the absence of FSIS knowledge of chemicals used or methods for their detection.[114]

Efforts to regulate pesticides vary tremendously among nations. Although the United States is the acknowledged world leader in attempting to protect public and environmental health from significant pesticide risks, other high-income nations have statutes that express the same mission. Consistency in mission, however, hardly translates into consistency in effectiveness. Requirements for testing toxicity, environmental fate, and transport vary significantly, as do procedures for interpreting uncertainty in scientific evidence. Nations that are a part of the Organization for Economic Cooperation and Development (OECD), for example, tend to use data on product effectiveness to reduce the quantity of pesticides allowed for use, whereas in the United States these choices are left to the marketplace if health and ecological threats are deemed to be manageable. The OECD nations similarly apply residue-monitoring funds to test imported foods rather than domestically produced food. Resources available for scientific review, monitoring, and enforcement also vary tremendously among nations, with the developing nations applying the fewest resources to these efforts.[115]

Mexico, for example, has registered seventeen compounds for use on food that have never been registered in the United States. Because Mexico has applied few resources to monitor exported foods, the responsibility for detecting these compounds falls to FDA and USDA. But Mexico is not alone. Chile, Guatemala, Costa Rica, and the Dominican Republic have all shipped food to the United States containing pesticides that have no U.S. registration. All of these nations complained that their resources were inadequate to disseminate U.S. regulations to their growers who exported crops to the United States.[116] Moreover, information about these regulations may have been difficult to obtain. EPA was criticized in 1989 by Congress for failing to keep other nations informed of current pesticide guidelines, especially concerning canceled, suspended, or restricted products.

Many developing nations, especially those in tropical parts of the world, have continued to permit the use of persistent chlorinated insecticides such as DDT after developed nations banned their domestic use. High-income nations basically traded the risks associated with the persistent and bioaccumulating insecticides for those attributed to more acutely toxic insecticides such as the organophosphates and carbamates. These high-income nations believed that they could control these acute toxins through worker education and by setting maximum contamination limits for various environmental media such as food and water. Low-income nations have found that they are less able to control acute risks, as demonstrated by the high number of accidental deaths of farmworkers, animals, and fish. These poisonings are associated with a number of factors, including illiteracy, the failure of manufacturers to print pesticide labels in a language understood by users, the absence of protective clothing and worker training, and very limited regulatory and monitoring efforts. Thus continued use of chlorinated hydrocarbon insecticides in the developing world offers greater protection to applicators but increases the risks from exposure to contaminated crops, meats, dairy products, and fish—thereby posing an overwhelming and global monitoring problem.[117]

The proportion of the U.S. food supplied from abroad is increasing, but during the last decade of the century we have had less and less knowledge of and control over the pesticides used. The recent passage of the North American Free Trade Agreement and recent amendments to the General Agreement on Tarriffs and Trade will facilitate trade in agricultural technologies. Rapidly improving international communication technology and the trend toward multinational, fractured corporate ownership will also promote these exchanges. But these developments also threaten to make detecting pesticide residues more difficult. This problem will be especially acute for foods produced in tropical nations where pest control problems are most severe, where tremendous pressure exists to expand agricultural exports, and where environmental health regulations are generally weakest. Tomato growers in Florida, for example, recently announced their intent to move their farming operations to Mexico if hazardous pesticides on which they rely are banned by EPA. Together these conditions demand more vigilant and strategic screening of imported foods, with particular attention to compounds used abroad but not registered in the United States.

Residues in Human Tissue

DDT residues were first found in human tissues—including blood and the liver, kidney, heart, and central nervous system—in 1944. Higher concentrations of DDT in fat tissues were also found.[118] Technical-grade DDT includes approximately 4 percent DDE, and most species transform a portion of the DDT they consume into DDE. Levels of DDE in the human body therefore tend to continue to increase even after exposure to DDT has stopped. In addition, the metabolite binds more tightly to fat than does DDT.

The average level of total DDT residues (including the metabolites DDE, DDD, and DDA) in the U.S. population rose from 5.3 ppm in 1950 to 15.6 ppm in 1955. Since then, they have declined from 8 ppm in 1970 to 3 ppm in 1980. If exposure were stopped, it would take the average person twenty years to eliminate all DDT residues, and it appears that DDE residues can never be fully eliminated.

Residue values have been higher for older people. Children under the age of fourteen have levels roughly one-third the level of those over forty-five, and African-Americans experience levels roughly three times those of whites for corresponding age classes. It remains unclear whether the higher levels in African-Americans result from variation in environmental exposure (for example, dietary patterns) or from genetic differences. There is a significant correlation, however, between DDT levels and a deficiency of the enzyme glucose-6-phosphate dehydrogenase, which is more common in African-Americans than in whites.[119]

Humans bioaccumulate, or concentrate, organochlorine residues just as other mammals do. Once stored in adipose tissues, these residues are stable unless they are mobilized either through lactation or significant weight loss, which burns fat. Also, the organochlorines appear to transfer freely across the placenta from mother to fetus. Whereas earlier research focused on the detection and movement of DDT and its metabolites in the human body, more recent research has concentrated on cyclodiene insecticides, including chlordane, oxychlordane, trans-nonachlor, heptachlor, heptachlor epoxicide, aldrin, and dieldrin.[120]

DDT and Human Milk

A finding that has had greater emotional effect than the discovery of residues in human fat has been evidence that milk from nursing mothers was contaminated at an average level of 0.13 ppm DDT. First discovered in 1951 by Edward Laug,[121] this phenomenon has now been more fully explained. Because the human breast is largely adipose tissue, and blood flow to the breast is high relative to other tissues, fat-soluble compounds such as DDT or DDE tend to concentrate there. The high lipid content of human breast milk (3–5 percent) thus explains the relatively high levels of DDT found in it. Excretion in breast milk was first demonstrated in dogs, and soon after in rats, goats, and cows.[122] Rachel Carson summarized available evidence by 1962 in the following paragraph:

> Insecticide residues have been recovered from human milk in samples tested by Food and Drug Administration scientists. . . . In experimental animals the chlorinated hydrocarbon insecticides freely cross the barrier of the placenta, the traditional protective shield between the embryo and harmful substances in the mother's body. While the quantities so received by human infants would normally be small, they are not unimportant because children are more susceptible to poisoning than adults. This situation also means that today the average individual almost certainly starts life with the first deposit of a growing load of chemicals his body will be required to carry thenceforth.[123]

Since Laug's study was published, many analyses have been conducted on DDT levels in breast milk, and in general they indicate a gradual decline in concentration that may be partially explained by the U.S. phaseout of the compound in 1972. Its continued detection, however, may demonstrate DDT's high level of persistence or our continued exposure through the food supply. Several studies have shown that DDT residues or their metabolites are found in nearly every sample tested.[124] Detected residues of DDT and DDE reported prior to 1986 ranged between 0.2 to 4.3 ppm in total milk and between 1.2 to 14.7 ppm in milk fat.[125] Other chlorinated pesticides have also been detected in human milk, including dieldrin, which was found in all of the samples tested at concentrations between 0.05 and 0.24 ppm.

In addition, heptachlor, chlordane, and their chief metabolites have been consistently detected in human milk at between 0.035 and 0.13 ppm. Hexachlorobenzene, which was formerly used as a pesticide, contaminated human milk and is highly toxic to infants. This compound is no longer registered for agricultural uses; few studies, however, have been conducted to analyze its concentration in fat and breast milk, or its source. Of the few reports produced, earlier concentrations in breast milk were higher than more recently conducted studies.[126]

In many tropical nations, DDT is still used to spray crops and the interior of homes. The highest national average levels of DDT detected in breast-milk fat were recently found in China (4.4 ppm DDE, and 1.8 ppm DDT) and India (4.8 ppm DDE, and 4.4 ppm DDT).[127] KwaZulu mothers in northern South Africa were exposed to relatively high levels of DDT from interior spraying; this incident resulted in 24.8 ppm total DDT in breast-milk fat. This area, near the shore of Lake Sibaya, is malaria-endemic, and DDT has routinely been applied in this part of the world to the walls of dwellings, which are normally constructed of mud, branches, and thatch. DDT has been banned for agricultural purposes since 1986 in South Africa; therefore little exposure is presumed to come from food crops. But women could be exposed to DDT from food contaminated by indoor spraying, from eating fish or wildlife contaminated from malaria-control efforts, or from crops grown in soil still contaminated by the persistent compound.[128]

Infant exposure to DDT is especially significant there because children are often breastfed for several years. Breastfeeding reduced the DDT concentration levels in breastmilk as the KwaZulu mothers had more children. DDE levels also were reduced, perhaps because DDT was eliminated before it could be metabolized to DDE. In a subsequent study of KwaZulu mothers, an attempt was made to determine the source of infant exposure to total DDT residues. If the infant blood levels of DDT were highly correlated to breast-milk levels, then breast milk could be presumed to be the dominant source of exposure, rather than contaminated food, air, or dermal exposure in the sprayed household. H. Bouwman and his South African colleagues found that breast milk indeed proved to be the dominant source of exposure, and that mean total DDT residues in milk fat were 15 ppm, with a maximum of 57 ppm. While examining blood levels of DDT, they found that children between the ages of three and nine had significantly higher levels of total DDT than adults between the ages of twenty and twenty-nine. They attributed these differences to breastfeeding.[129] What effect is DDT having on children in South Africa? In one of very few studies that have been conducted to identify the health effects of DDT and its metabolites on infants, it was found that exposures greater than 4 ppm in milk fat were correlated with hyporeflexia—either delayed or non-elicited reflexive responses.[130]

With this discussion, I by no means intend to undermine the advantages of breastfeeding one's child. It is important to realize that the definitive immunological, nutritional, and psychological benefits of breastfeeding appear to far outweigh any risks to an infant of exposure to chlorinated organic contaminants.[131] In general, the toxicological implications of exposure to former and current DDT residue levels in human tissue remain unclear. Some good news is that the residue levels of chlorinated compounds in human tissues and milk are declining in the United States.

The Risk Connection

Understanding possible sources, routes, magnitudes, frequencies, and durations of human exposure to toxic substances is essential for understanding the nature of the risks we face from toxins. In this study, I have chosen to explore the significance of childhood exposure to a complex mixture of compounds permitted to exist as contaminants of human and animal foods and drinking water. We have been regularly exposed to an ever-changing mixture of these compounds in our diets, homes, schools, playgrounds and athletic fields, workplaces, and hospitals. To understand the dangers of these mixtures we must be able to analyze patterns of exposure and to determine how these patterns vary within the population, perhaps by age, ethnicity, and region.

Attaching a meaning to exposure requires knowing a compound's toxic potency. But because there are at least several thousand compounds that are combined to make pesticides, testing each compound individually and in combination with others for every type of possible toxic effect is an enormous task. As a result, exposure analyses are not normally conducted unless a compound is detected in air, water, or food.

Advances in detection technology have helped scientists to discover pesticides in many media where they had been presumed not to exist. Finding pesticide residues, of course, immediately raises a concern about possible toxic reactions. DDT has been a wonderful teacher of human ecology and has offered invaluable lessons about toxicity. Tracing DDT's use, persistence, movement through the environment, biological magnification along food chains, and accumulation in human tissues has taught us much about the connectedness of humans to their environments. DDT is joined by other chlorinated organic pesticides, PCB's, and various forms of dioxins that are still widely distributed, persistent, and bioaccumulative. Although there is enormous uncertainty concerning the ways that these residues influence the health of humans, their ubiquity and longevity in human tissues ensure that we are conducting a global-scale experiment on ourselves—an experiment to which we continue to subject our children. How cautious should we be with these kinds of chemicals? The answer seems to depend upon how we respond to two additional questions. First, how toxic is the compound likely to be? Second, how certain are we of our ability to restrict human exposure to a safe level?

CHAPTER 8

The Susceptibility of Children

All substances are poisons; there is none which is not a poison. The right dose differentiates a poison from a remedy. —**Paracelsus**

Paracelsus, a physician-alchemist living in Germany during the sixteenth century, wrote a defense of his use of chemicals to treat illness, including the use of mercury to treat syphilis. Paracelsus claimed simply that all substances are potentially poisonous and that dosage alone differentiates a poison from a harmless compound and even a remedy.[1] This principle—that the dose makes the poison—has since become the foundation of modern toxicology and provides the basis for government management of drugs, pesticides, and other contaminants in food, drinking water, and air.[2]

But at what dose does a substance change from harmless, or even health-promoting, to damaging? This deceptively simple question has become the Achilles' heel of modern U.S. environmental law. If toxicity is related to dose, then the central management problem is to identify that dose of a toxin that threatens health, so that human exposure may be restricted beneath this level. In the case of pesticides, two types of dose are important for making this distinction. The first is the effective dose necessary to kill a pest, which determines the desired application rate. The second is the dose that humans could receive from pesticide residues in the field, on food, in drinking water, in the air, or through the skin from the tens of thousands of products that deliberately contain them. Unfortunately, most legal pesticide tolerances have been set based upon the amount necessary to control pests, rather than on that required to protect against dangerous levels of human exposure.

Even if our governmental agencies were to change their priorities, however, the threshold of exposure that distinguishes a harmless from an adverse effect is often difficult to identify. The maximum safe level for any pesticide often depends on which type of adverse effect we explore, such as cancer or neurological disease. It also can change according to the susceptibility of the group studied or even to changes in an individual's vulnerability over time.[3] In this chapter, I explore children's special susceptibility to pesticides, which provides a basis for later claims that current law and regulation fail to protect children's health.

Children and Pesticide Toxicity

Children differ from adults in a variety of physical and functional characteristics, many of which influence their susceptibility to toxic substances, including pesticides.[4] The phrase "heightened susceptibility" implies that some subset of individuals experiences a higher incidence of an adverse effect than others exposed to the same dose of a toxin. Susceptible groups might be defined by their youth, ethnicity, genetic traits, illness, or other characteristics.

Physical growth is a function of the birth, growth, and death of cells. Importantly, the growth of human cells, tissues, organs, and body systems occur at different rates. Overall body growth proceeds from the head downward over time. For example, at age two a toddler's brain size is nearly that of a mature adult, whereas his or her total body weight is 20 percent, and height is 50 percent, of an adult's. And although the midpoint in the height of an infant is at mid-abdomen, the midpoint of an adult is normally where the legs and trunk join.[5] Another way to look at this differential rate of growth is to realize that although the brain is 50 percent of its adult weight by six months, it takes two years for a child's height to be 50 percent of its mature size and nine years for the weight of the liver, heart, and kidneys to reach 50 percent of a mature weight. The total body weight of an eleven-year-old is still only 50 percent of adult weight. Thus from conception to adulthood, growth is characterized by constantly changing proportional relations among tissues and organs.

The most rapid periods of growth occur in utero, during infancy, and during puberty. At these times, the body is adding new tissue faster than at any other period in life. Changes in height provide a striking example. During the first year of life, a child grows on average 50 centimeters each year. By age four, that growth is reduced to 10 cm/year, although it rises to 12 cm/year during puberty.[6] Even though the maximum neuronal cell population is reached by age two, full development of neuronal sheaths, or myelination of nerve tracts of the peripheral nerves and the spinal cord, is not completed until age eighteen.

These periods of rapid growth may also be periods of heightened susceptibility to toxic substances. The exposure of an immature organ or system such as the central nervous system to a toxic compound could prevent its normal maturation.[7] Moreover, growth is regulated at puberty by the hypothalamus, pituitary glands, and gonads, among other tissues, and it is correlated with the presence or absence of growth hormone. Both androgens and estrogens control this development, and in 1993 the NAS recognized that high levels of estrogens, or molecules that are recognized as estrogen-like, may be responsible for reduced growth through their effects on bone development.

Although there are numerous and sometimes conflicting theories of carcinogenesis, there is some consensus that the speed at which cells proliferate or reproduce and the rapidity with which DNA is synthesized are somehow related to the risk of tumor formation. We may therefore be most susceptible to carcinogens during periods of most rapid growth. There is in fact some evidence that direct carcinogens are more potent in rapidly growing animals.[8]

Growth from conception through infancy—roughly until a child reaches age two—results from a rapid increase in the number of cells in the body, termed hyperplasia. After infancy, however, most growth is accounted for by increases in cell size, termed hypertrophy. This generalization is complicated by differential rates of hyperplasia among different organs and an increase in the number of cells produced during puberty in response to hormone secretion. Breast-tissue cells, for example, proliferate most rapidly near the time of the first menstrual period, or menarche. Risk of breast cancer from radiation is known to be highest if exposure occurs either just before or after menarche.[9]

Although little is known about children's susceptibility, a variety of factors appear to play a role in the greater incidence of tumors in the elderly. The latency period between exposure to a carcinogenic agent and the development of cancer, for example, is often between twenty and forty years. Reduced efficiency and accuracy of DNA repair also seem to be important. Changes in the rate of cell proliferation in target tissues, along with changes in hormones and in other growth factors that regulate proliferation, together may also contribute to the increased incidence of cancer among the elderly.

Although epidemiological evidence has been important for understanding the cancer-causing effects of a number of compounds, most of our understanding of potential age-related susceptibility to carcinogens is derived from animal test results. This is due to the extraordinary difficulty of knowing the causal effect of a single toxic agent that acts over a very long period of time, during which the population of interest is likely exposed to a complex mixture of other carcinogenic compounds. In one analysis of fifty-two separate animal studies of this issue, V. N. Anisimov found that in nineteen of the studies, aging decreased tumor incidence—indicating heightened susceptibility among the young—whereas in twenty-eight studies tumor incidence increased with age. In the remaining five studies, no effect was observed. Anisimov also studied the age-related sensitivity of rats to N-nitrosomethylurea and found that as the age at exposure increased, mammary gland and kidney cancer decreased (implying higher susceptibility among the young); and cervical, uterine, and vaginal cancers increased. No effect on the hematopoietic system was found.[10]

Figure 8–1 shows the rate of maturation of different human organs. The

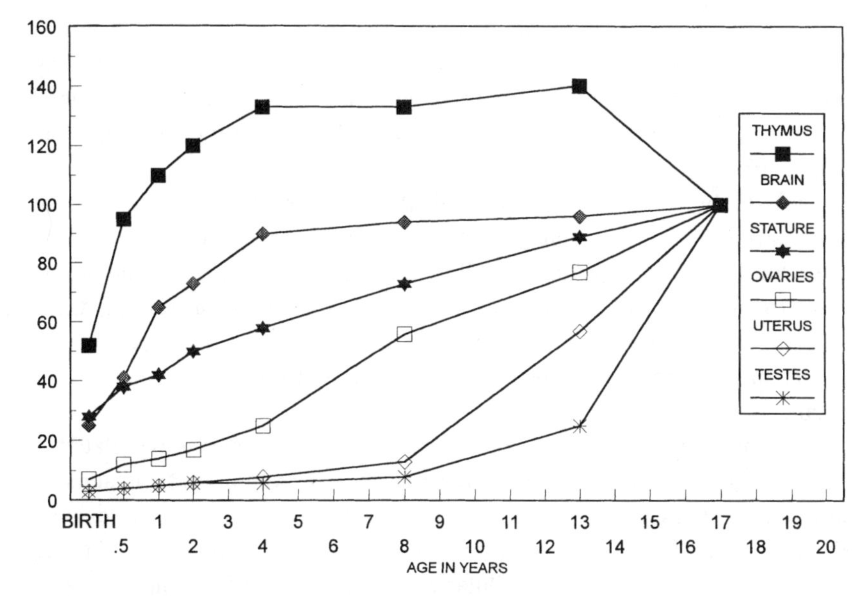

Figure 8–1. Maturation rates for various human organs by percentage of adult organ weight. Growth rates differ among types of human organs. The brain is almost fully mature by age five, whereas growth rates of the uterus and testes are most rapid during adolescence. Stature is a poor indicator of most organ growth rates. The size of the thymus decreases after age thirteen until it reaches adult weight by age seventeen. (Adapted from P. L. Altman and D. S. Dittmer, eds., *Growth* [Federation of American Societies for Experimental Biology, 1962])

testes and uterus develop most slowly, with the most rapid rates of growth occurring during puberty. The rate of growth of the ovaries, kidneys, and spleen is more stable during childhood, and the most rapid growth of the kidney and spleen occurs between birth and age two. The brain rapidly approaches 63 percent of adult weight by age one, and reaches nearly 90 percent of maturity by age four. The thymus similarly grows with greatest speed during the first four years and is larger than adult size during most of childhood.

If rates of cell proliferation are positively correlated with a heightened susceptibility to carcinogens, then the variation in the rates of organ development described in figure 8–1 may indicate different "windows of vulnerability" for infants and children. Interpreting epidemiological and animal evidence is difficult because the specific mechanism by which carcinogens act is often unknown. Some toxic agents, for example, are believed to be "initiators" that damage DNA. But initiators are not believed to be capable of inducing tumors alone; instead they seem to require later exposure to chemical "promoters," which further alter the genetic code governing cell reproduction. It is extremely difficult to predict human exposure to the substances now presumed to be a complex mixture of initiators or promoters. Similarly we know little about how the process of enzyme repair is activated following DNA damage. The efficiency of repair may be higher during childhood than in old age.

Despite these cautionary statements, there is an emerging consensus that if a cell requires progression through several stages prior to tumor induction, then "initiation" of cells during childhood statistically increases the probability that the cell will be "promoted" through all additional necessary stages of tumor development. A child simply has a longer period of time, compared to an adult, during which exposure to "promoters" from diverse sources may occur.[11]

The susceptibility of children to pesticide toxicity may also be related to body composition. The relative proportions of body water, fat, skeletal mass, and other tissues change quickly during childhood. Stature increases only threefold from birth to age seventeen. But during the same period, surface area increases eightfold; extracellular water increases tenfold; and total body water fifteenfold. Total body fat increases seventeenfold in males and thirty-six-fold in females.[12] Age-related variance in body composition may influence how dangerous pesticides are for children. Water-soluble pesticides are likely to become more diluted during ages when water volume constitutes a higher percentage of body weight. Extracellular water as a percentage of total weight is greatest in infancy (42 percent), whereas intracellular water peaks between ages eight and thirteen for females (35–36 percent), and during adolescence for males (38 percent). Age-related susceptibility to water-soluble pesticides may also depend upon where and how water is stored in the cells. The cell size of infants, for example, is smaller than that of adults. The proportion of cell surface area to total cell mass is therefore higher in infants than in adults. This could mean that infants are more susceptible to pesticides that act on the cell membrane.

The proportion of body weight made up of fat tissue and the distribution of fat may also have an important influence on childhood risks from pesticides. Young children have a lower level of fat per kilogram of body weight than adults. At birth, for example, 14 percent of body weight is fat—on average—whereas at six months, 25 percent of body weight is fat. Lipid-soluble pesticides such as chlorinated hydrocarbons bind to body fat. Storage of pesticides in fat tissues may be one mechanism of controlling toxicity because it prevents the circulation of toxins to organs, where they may cause greater damage. If an infant and an adult were exposed to pesticides in equivalent doses per unit of body weight, pesticides may become more concentrated in the fats of the infant, because the infant's fat as a percentage of total body weight is lowest. The implications of these findings are uncertain but unsettling given that infant exposure to chlorinated, lipophilic compounds normally occurs through breast milk. Once these compounds enter the body and bind to fat, some may persist for a lifetime. For example, the half-life of DDE, a DDT metabolite, is nearly the length of the average human life span.

To this point I have considered how variance in the *physical growth* of cells, tissues, organs, and organ systems may influence age-related susceptibility to pesticides. Development, for the purposes of this analysis, will instead refer to *functional maturation* of cells, tissues, organs, organ systems, and the whole organism. Metabolic rate, for example, is a function measured as an organisms' total energy expenditure. As humans mature, the total energy expenditure increases. But if measured per unit of body weight, infant metabolism is far higher than that of adults. This effect may have two significant, and somewhat contradictory, implications for judging the toxicity of pesticides. First, infants may be better able than adults to break down toxic substances into harmless compounds. Some breakdown products, or "metabolites," however, are more toxic than the parent compounds. Thus, for some compounds, increased metabolic rates may increase risks to children.

Food and water intake are also closely related to metabolic rates. If consumption is expressed per unit of body weight, a child's intake will be far higher than an adult's for many foods, especially water, milk, fruits, juices, cereals, and certain vegetables. This higher intake fuels the growth of children. These differences in food intake mean that for any level of contamination in foods preferred by children, they will be more highly exposed than adults.

In addition, the functional development of kidneys and the liver may also have important implications for the toxicity of pesticides in children. Kidneys are immature at birth, with a relatively low filtration potential—which for some compounds may result in a slower rate of drug excretion and an increased potential for drug toxicity. Similarly the slow evolution in the liver's ability to detoxify some drugs such as chloramphenicol make these compounds toxic to infants.

It is easy to see that the issue of pesticide toxicity in children is complex and confusing. But even more troubling, there appears to be no systematic method or model appropriate to predict variation in age-related susceptibility to toxins. Variance in susceptibility may be related to a number of independent and potentially related factors, including the precise mechanisms of toxicity, metabolic rates, development of enzymes, and rates of organ system growth and functional development.[13] Further, most studies of chronic toxicity that have supported pesticide registration were structured to test for adverse effects in sexually mature animals and therefore provide a poor basis for estimating effects in human infants, children, adolescents, and young adults.[14] This is not to say that conducting these tests would be easy. There are very real difficulties in dosing infant animals during lactation. The mother must be exposed at a level that produces the desired concentration of residue in milk, and this is difficult to achieve given the potential for the contaminant's biotransformation and metabolism. Thus even those studies that claim to include the effects on young animals are begun between weeks six and eight in the lives of laboratory animals, a life cycle stage more comparable to human adolescence than early childhood. Results of these studies therefore provide a poor reflection of possible human exposure *in utero,* during infancy, and during childhood, and may therefore underestimate the incidence of tumors or the rate of tumor development.[15]

In 1993, NAS recommended that testing protocols be redesigned to require exposure in utero during the first trimester of pregnancy, in the mother's milk, and through oral, dietary exposure. In 1992, the National Toxicology Program tried this protocol in the study of ethylene thiourea, which is a metabolite of the fungicide EBDC, and found that in utero exposure produced an increase in the number of malignant thyroid tumors in both mice and rats.[16]

Limitations of Human Evidence

A variety of toxicological conventions have evolved to help in setting standards for acceptable levels of human exposure to hazardous contaminants. The results from carefully structured tests on humans obviously provide the best possible evidence on the toxicity of the compound to humans. Data on human response to toxins is available from a variety of sources: (1) records of human poisonings, (2) exposure in a controlled environment, such as the workplace or home, (3) the pharmaceutical use of the compound, and (4) testing of compounds on volunteers. A substantial body of human evidence has been produced for pesticides in all of these ways.

Between 1 and 3 million people, many of them children, are poisoned by pesticides in the world each year, and at least 168 different pesticide compounds are known to have caused significant human illness or death. Although 80 percent of pesticides are used in the industrialized nations, 99 percent or more of the deaths from pesticides occur in low-income nations, where safety precautions such as label instructions, protective clothing, and field reentry intervals are less likely to be employed.[17] Most acute poisonings reported to the U.S. Poison Control Centers are from insecticides, followed by rodenticides, herbicides, and fungicides.[18] But data from poisoning episodes are of limited use for establishing safe levels of exposure. Although human poisonings provide some data on the effects of exposure beyond a safe level, these events normally yield only crude estimates of a victim's exposure. And whereas mortality from pesticide ingestion is unfortunately easy to diagnose, illness associated with less-than-lethal doses may produce symptoms difficult to distinguish from other diseases. Physicians normally treat symptoms of poisoning rather than attempt to accurately identify the magnitude of dosage.[17]

Occupational exposure has been studied for 142 different pesticides, with tremendous variance in the purpose and quality of research. The results of epidemiological studies for chronic effects such as delayed neuropathy or cancer are generally limited by a low level of confidence in exposure estimates and are designed to demonstrate the difference in the relative risk, or "odds ratio" (OR), between an exposed and unexposed population. Again, these data rarely provide an opportunity to identify a sensitive threshold between a safe and unsafe level of exposure because it is so difficult and expensive to identify and monitor exposed individuals and because it is so hard to be certain of the magnitude and duration of exposures.

Nearly sixty-five pesticides, including arsenic and mercury compounds, have at some time been used as drugs. In 1995, twenty-six pesticides were still administered therapeutically, which provided an unusual opportunity to explore age-related susceptibility to intoxication. One example has been the use of chlorinated insecticides to control head lice. One recent study suggests that this use may increase the risk of brain cancer in children.[18]

Clinical trials of human toxicity are normally conducted only for new drugs. The FFDCA requires that the effectiveness of drugs, but not necessarily their safety, be tested in humans before they may be licensed and marketed.[19] In 1977, A. K. Done and his colleagues reported that drug safety and efficacy for children had not been demonstrated for 78 percent of drugs marketed in the United States in that year.[20] FDA has allowed these drugs to be marketed if the labels included no instructions for use in pediatric populations, as well as disclaimers. Unfortunately, physicians may therefore assume that the drugs are safe for children if the dosage is reduced to reflect differences in body weight or surface area. A survey conducted by the American Academy of Pediatrics in 1990 found that no information on pediatric use was included for 78 percent of new drugs marketed in the United States between 1984 and 1989.[21]

One body of evidence that provides important insight into age-related susceptibility includes research results from drug experiments to fight cancer in children. Clinical trials in ninety-day animal studies attempt to establish the acute toxicity of these compounds by finding the maximum tolerated dose (MTD) for anticancer drugs. Human drug experimentation has also been common among both adults and children who are being treated for cancer. Although the dosing regimen normally lasts for days or weeks rather than a single exposure, these experiments provide valuable information about age-related response to short-term exposures of xenobiotic compounds. In one comparative study published in 1982, Glaubiger and his colleagues found that among thirteen of sixteen compounds tested, the MTD was higher for children than adults.[22] In other words, children's short-term tolerance was higher than adults'. The ratio of child to adult MTD was never greater than 2.2 nor less than 0.83, demonstrating a rather narrow range of variation.[23]

Vincristine, an anticancer drug, provides an example of the difficulty in predicting age-related response. Neonates and infants are at higher risk of neurotoxic and hepatotoxic effects than are older children receiving the compound. Vincristine is metabolized in the livers of the very young at a lower level of efficiency than in older children.[24] Other anticancer drugs such as N-oxide and cyclophosphamide are metabolized into more toxic compounds. Although metabolic activation of these drugs may be most rapid in children, detoxification and excretion may act together to reduce their toxic effects in the young. One final example of higher childhood tolerance is azidothymidine (AZT), which has been widely tested for its effectiveness in fighting the human immunodeficiency virus (HIV). Although the side effects were similar for adults and children, children appear to tolerate a longer period of therapy than do adults.[25]

Neurotoxins

Just as various organ systems have different rates of growth, so do various structures within the brain. The brain is not structurally mature until four to six years of age. During periods of rapid growth, the human nervous system may be particularly susceptible to either structural or functional damage from xenobiotic compounds such as neurotoxic pesticides. Evidence for this increased vulnerability is derived from several sources. Perhaps the best demonstration is the case of lead, which has toxic effects on prenatal and young children at levels far below those which induce similar effects in adults. Lead is persistent; it accumulates both in the environment and in the body; it produces toxicity in numerous organs; it produces higher toxicity among immature humans; it has either no threshold, or one that is extremely low; some of its effects are irreversible; and there is widespread exposure because it has commonly contaminated indoor and outdoor environments.[26]

The effects of lead on the maturing brain have been examined more than any other neurotoxicant, and it is also known that the developing brain is generally more sensitive than the mature brain to irradiation. Fetal exposure to addictive drugs and alcohol suggest a heightened vulnerability of the fetus; this exposure often results in irreversible changes in brain capacity.[27] As tests of nervous system dysfunction have become more sensitive, our understanding of this process has grown. Susceptibility appears to result from a variety of factors. There is no placental-fetal barrier to the transport of lead; maternal and fetal blood-lead levels are nearly identical. Lead has been measured in the fetal brain as early as the thirteenth week of pregnancy. It appears to be sequestered in the brain only during the later stages of fetal development, which means that before this time it can freely interact with vital subcellular organelles, especially mitochondria. Lead toxicity in the nervous system results in the swelling of endothelial cells, which appears to result from lead-induced changes in cell permeability. Immature endothelial cells are less resistant to these effects, and they thus allow both fluid and lead to reach newly formed neuronal structures in the brain. In addition, lead within cells may affect the binding of calcium, which in turn could influence the release of neurotransmitters.[28]

Measurement of childhood behavioral and cognitive function in response to lead exposure demonstrated significant impairment at blood lead levels (20 to 40 μg/dl) well below those previously found to cause clinical disease in adults.[29] Studies of lower lead levels in the human fetus during the 1980s found neurotoxic effects and behavioral dysfunction at blood lead levels less than 20 μg/dl.[30] It is important to note that these researchers did not presume a threshold effect for lead. When they found biochemical changes at levels of exposure lower than those expected to induce clinical effects, they designed more sensitive tests and methods capable of detecting subtle but important changes in behavioral and cognitive function in children.[31]

In 1992 Ellen Silbergeld suggested that lead may be toxic to the central nervous system in different ways. One effect may be the "miswiring of the central nervous system," which can result in permanent dysfunction. A second effect may be the disruption of signal transmissions at nerve synapses, which possibly results from the interactions between lead and calcium, and between lead and zinc.[32]

Pesticides as Neurotoxins

Many pesticides are neurotoxicants that can affect organisms in the short term, longer term, or both. These effects may include mild cholinesterase depression, which in turn can cause the abnormal nerve-conduction velocities associated with low-level chronic exposures; unusual neurobehavioral and psychiatric effects; the "intermediate syndrome," which involves paralytic effects following acute exposure; dysfunction of the visual system; severe systemic disease; and if exposures are high enough, death.

Nearly fifty different pesticides known to inhibit human cholinesterase are permitted to persist as contaminants of the U.S. food supply. Cholinesterases of various types regulate the rate of electronic signal transfer among nerve synapses, and its depression allows more rapid signal firing. People who have this syndrome will experience twitching and convulsions, and if cholinesterase levels are low enough, they will die. There is evidence that levels of at least three types of esterases increase rapidly with age from gestation through the first year of life, and thereafter, more gradually through childhood until they reach adult levels.[33] Thus lower levels, especially in the very young, could make them more susceptible to adverse effects from further pesticide-induced depression.

Because the neurotoxic effects of organophosphate insecticides is considered more fully in chapter 11, only a brief summary of the literature is presented here. First, there is a substantial body of epidemiological evidence that high-level exposure to organophosphorus compounds causes direct damage to the central and peripheral nervous system; and in the most extreme cases, death results. Second, low-level chronic exposure to some organophosphate compounds may cause subtle but significant and measurable effects on neurologic function and behavior. Third, there is some evidence of synergism among anti-cholinesterase compounds, but additive-dose effects appear to be more common. The precise biochemical mechanisms that cause toxic reactions to complex mixtures are often unknown and rarely studied. But this is an extremely important area for future research because most people are exposed to pesticide residue mixtures in the diet and in other contaminated environments. Fourth, as noted above, there is emerging evidence that the developing central and peripheral nervous systems of children may be particularly vulnerable to neurotoxic agents such as anticholinesterase compounds. This is due primarily to the delayed maturation of certain nervous system components such as the brain until the age of four to six years. One exception to this conclusion appears to be the increasing susceptibility with age to delayed neuropathy.

Fifth, there is a small but growing body of evidence that neurotoxic agents known to produce cognitive, motor, and sensory deficits in humans generally result in corresponding deficits in laboratory animals. Sixth, there is some evidence of significant variance—both among and within individuals—in normal ChE levels and ChE recovery rates. This variance may be due to genetic factors, age, sex, pregnancy, drugs, or enzyme repair efficiency, and may be an extremely important factor for governments to consider in setting acceptable exposure levels. Seventh, compared with other types of toxic effects, the quality of data on acute effects resulting from high level exposures to cholinesterase-inhibiting compounds is quite good. Data on subchronic effects such as delayed neuropathy from high-level exposures is also quite good due to epidemiological and clinical reports. There is emerging though inconclusive evidence suggesting the mutagenic effects of specific compounds and their possible role in the etiology of cancer. Finally, there is clear evidence that pesticides are only one class of neurotoxins that commonly contaminate the U.S. environment. Among the twenty-five most released compounds registered in the National Toxics Release Inventory, seventeen are believed to have neurotoxic potential.[34]

Age-Related Susceptibility in Animals

Evidence of the acute toxicity of pesticides is derived from both animal studies and cases of human poisoning, accidental and intentional. Although animal evidence often has limited value for estimating age-related susceptibility due to the small number of dosing categories used and the general absence of tests performed on young animals, several papers explore the question. In 1964, for example, A. K. Done found that immature animals were more susceptible to thirty-four compounds, whereas adults were more susceptible to twenty-four compounds.[35] In a study conducted by E. L. Goldenthal in 1971, LD50s (the dose at which 50 percent of animals exposed were killed), were compared among newborn and adult animals for pharmaceutical chemicals. The results were intriguing: 225 general pharmaceutical compounds were more acutely toxic to neonates than to adults, and forty-five were more toxic to adults. Nearly all of the differences were less than tenfold, and most were less than threefold.[36] Several fungicides, herbicides, and the insecticide heptachlor, were more toxic to newborn than adult rats. These differences in susceptibility provide important guidance for the choice among possible "safety factors" designed to offer protection to the most vulnerable.

There is also variance in susceptibility among species for any given age. The level of maturity of nonprimates at birth is less than humans, and lowest in rats and mice among test animals. Further, these rates of maturation also vary by organ, raising the probability that organ-specific and function-specific "windows of vulnerability" may exist during crucial phases of physical and functional development. Functional maturation, such as metabolic rate or kidney filtration efficiency, does not necessarily change at the same rate as physical growth, and it appears to be species- and age-related. For rodents, metabolic rates and kidney filtration potential of newborns may exceed those of adults, reducing the toxic potential of xenobiotic compounds in these young animals.

Although older rats have been found to be more sensitive to dieldrin than young rats,[37] the opposite seems to be true for malathion, an organophosphate insecticide. In this case older rats were less susceptible than the young because they were somehow better able to break down this compound into nontoxic components.[38] G. M. Benke and S. D. Murphy also found that parathion and methyl parathion were less toxic to adults than young animals because the adults could more efficiently detoxify the insecticides' oxygen analogues.[39] The lower metabolic and excretory capacity of newborns have also been associated with higher-than-adult toxicity in the cases of benzyl alcohol,[40] hexachlorophene,[41] and diazepam given to mothers and has resulted in the "floppy infant syndrome."[42] Rapid maturation in the functional processes that detoxify drugs in humans again suggests that the youngest are the most vulnerable. For example, glomerular filtration in the kidneys increases quickly during the first week of life, exceeding adult levels between three and five months of age. Anticonvulsant drugs such as phenobarbital, carbamazepine, and diazepam are excreted very slowly in newborns, but more rapidly in infants and young children than even adults. The very young are therefore more susceptible to toxic effects than adults, whereas older children are less susceptible.

Pesticides and Cancer

Cancer is among the leading causes of death in the United States. The disease was responsible for 23.7 percent of all deaths in 1991 (530,000 individuals), up from 17.7 percent in 1973, according to the National Cancer Institute (NCI). During the same period, the percentage of all deaths from heart disease declined from 38.4 percent to 33.2 percent. Among those less than age sixty-five, mortality from cancer rose from 21 percent of all deaths in 1973 to 26 percent. NCI estimated that 1.2 million new cases of cancer occurred in 1994.[43]

Interpreting changes in cancer mortality statistics, or rates, is complicated by improvements in diagnosis, treatment, and recordkeeping. Increased incidence rates may simply reflect improved screening, diagnosis, and recordkeeping efforts. Declines in mortality may mask increasing incidence of types of tumors treated successfully, or a lengthening of the average survival period following diagnosis. For these reasons, NCI established a special program to collect data on a routine basis from several cancer registries in nine different regions of the country, known as the Surveillance, Epidemiology, and End Results (SEER) Program.[44] These areas were selected for several reasons, including the existence of careful cancer reporting programs that covered a population that well represented the different demographic and epidemiologic characteristics of the U.S. population.

Results from the SEER regions paint a grim picture of increasing cancer incidence combined with decreasing mortality for some forms of cancers, including those that affect the colon, rectum, ovaries, testes, bladder, and thyroid. Increasing incidence combined with increasing mortality was registered for cancers of the breast, prostate, brain, liver, esophagus, and lung. Both incidence and mortality declined for cancers of the stomach, pancreas, larynx, cervix, as well as for Hodgkin's disease and leukemias. Increases in cancers of reproductive organs within SEER areas were notable: prostate (126.3 percent), testes (42.7 percent), female breast (23.9 percent), and ovaries (4.4 percent).[45]

Childhood cancer incidence—for ages 0 to 14—increased for most tumor types within the SEER areas when incidence is compared between 1973–74 and 1990–91: the incidence of tumors increased in bones and joints by 13.1 percent; in the brain and other parts of the nervous system by 38.4 percent; and in the kidney and renal pelvis by 11.9 percent. In addition, the incidence of acute lymphocytic leukemias increased by 20 percent; of non-Hodgkin's lymphomas by 6.8 percent; and of soft tissue tumors by 26.7 percent. The incidence of Hodgkin's disease declined by 10.5 percent, although if children from birth to age nineteen are included, incidence increased by 5.9 percent.[46] The good news over this period is that mortality rates have declined sharply—at an average of 42 percent across all sites—for the zero to fourteen age group. The most dramatic improvement has been for Hodgkin's disease (−76.2 percent), soft tissue sarcomas (−70.5 percent) and non-Hodgkin's lymphomas (−70.5 percent). Mortality rates for brain and other nervous system cancers showed the least improvement; they declined by 16.3 percent.[47]

Epidemiologists try to understand the causes that underlie these trends and to suggest methods for cancer prevention. Some influences are clearly natural, such as the effects of aging or genetic susceptibility, whereas others are environmental or "exogenous." In many cases the interactions of endogenous and exogenous variables are poorly understood. There is a special interest in understanding and controlling exposure to exogenous carcinogens—such as tobacco, sunlight, alcohol, heavy metals, and some foods—especially animal fats—as well as many food and water contaminants. Thus the diet is of special concern because it is a primary route of human exposure to a broad group of suspected carcinogenic agents—both natural and synthetic.

Diet is believed to be responsible for roughly one-third of all cancers in the United States, and it may contribute to as much as 50 percent of breast cancer incidence.[48] Alcoholic beverage intake may be responsible for 3 to 5 percent of domestic cancers, and there is recent evidence of its association with breast and colon cancer. Intake of red meat has been associated with incidence of colon and prostate cancer. Drinking-water contaminants such as arsenic and radon increase cancer risks regionally. Smoking is believed to cause another one-third of cancers and one-quarter of all heart disease. Hormones may be responsible for as much as one-third of all cancers, because of their role in regulating cell division. Estrogens and progestogens, for example, appear to cause breast cancer cells to proliferate. Risk of breast cancer increases along with early menarche, late menopause, and extended estrogen therapy, all of which increase the cumulative exposure to estrogens.[49] Inflammation associated with chronic infections may be responsible for nearly one-third of the world's cancer. Viruses may also play a role in cancer formation. Hepatitis B and C viruses—which induce inflammation and cancer of the liver—are known to infect nearly 500 million people, especially in Africa and Asia. Aflatoxin, a natural mold found in peanuts and corn, appears to promote liver cancer among those with chronic hepatitis. Epstein-Barr virus is associated with Burkitt's lymphoma under conditions of immunodeficiency. Shistosomiasis, a parasitic disease caused by a blood fluke that is common in Africa, Asia, and parts of South America, is a cause of inflammation and cancer of the colon or bladder, depending upon the species of Shistosoma.[50]

Heredity also appears to play an important role in some forms of cancer, especially those that afflict children or young adults. Skin color influences susceptibility to melanoma, a deadly form of cancer related to sun exposure. Genetic factors appear to play a role in approximately 10 percent of breast cancer cases.[51] Variables other than heredity appear to be more important influences in the evolution of most forms of cancer.

Cancer, or uncontrolled cell division, may result from the coincidence of several conditions. These may include mutations in a gene known to suppress tumors—gene p53—which guards a cell cycle checkpoint. Checkpoints act to inhibit tumor formation by preventing a cell from dividing if it contains too many DNA lesions, which are damaged bases or chromosome breaks. Some damage occurs naturally and some results from exogenous influences such as environmental contamination.[52] More rapid rates of lesion formation appear to be well correlated with increased probability of a mutation. Also, more rapid division of stem cells tends to increase mutations and therefore cancer incidence. The rate of cell division is affected by hormones, inflammation, high caloric intake, and exposure to some chemicals. If rapidly dividing cells are exposed to a mutagenic agent, such as some pesticides, the chance of cancer formation increases. The effectiveness of a lesion in promoting a tumor is controlled by the rate of DNA repair accomplished by various enzymes. These defenses appear to be strengthened by the presence of micronutrients in the diet such as the antioxidants ascorbate, tocopherols, and carotenoids, which are found in some fresh fruits and vegetables. DNA lesions caused by oxidation and mutations appear to increase with age.[53] Special precaution seems warranted to protect against childhood exposure to known and suspected mutagens and carcinogens because the most rapid rate of cell reproduction occurs before age nineteen.

If the diet plays such an important role in cancer formation, it seems important to understand the mutagenic and carcinogenic potential of chemicals—both synthetic and natural—present in food. Bruce Ames and Lois Swersy Gold have demonstrated that many natural chemicals found in plants act in some protective capacity against predators.[54] They have tested these "natural pesticides" and found that nearly half induce cancer in rodents. Gold and her colleagues also found that nearly half of all chemicals tested—both synthetic and natural—are carcinogenic in rats or mice.[55] Ames and Gold have interpreted these findings to suggest that natural chemicals contribute more to cancer risk from the diet than do synthetic chemicals and that the public is overly preoccupied with regulating pollutants such as pesticides, which they argue contribute a relatively low level of risk. When they make their comparisons, however, they consistently rely upon contamination data collected using sampling designs that tend to underestimate the level of residues in the food supply; in addition, they have not considered how variance in the distribution of pesticide residues can cause one group—such as children or pregnant women—to be more exposed to carcinogens than others.[56]

The high proportion of chemicals tested that have proven to be carcinogenic in rodent studies is either cause for significant public alarm and careful regulation or proof that there is something drastically wrong with our methods of chemical testing on animals and the way we infer human cancer risk from the results. I now turn to examine this question.

Extrapolating Human Risk from Animal Studies

Formal testing of chemical carcinogenicity using animal studies began in the 1960s, as a way of judging these chemicals' potential to induce cancer in humans. By 1991, the National Toxicology Program had studied 382 chemicals in laboratory animals to estimate their human carcinogenic potential. Among these, 51 percent (195) induced a carcinogenic response in at least one species-sex grouping. Among those tested, 67 percent (255) were selected based on a suspicion of carcinogenicity, and among these, 66 percent (169) were positive. When the 127 remaining compounds were tested because of known human exposure but with no suspicion of carcinogenicity, 20 percent were positive. This testing protocol appears to have credibly identified those compounds that are either noncarcinogenic in animals, or have such low potential for carcinogenicity that statistical detection is impossible given normal sample sizes. Chemicals were classified as noncarcinogens if no statistically significant increase in tumors was observed in a standard animal bioassay involving two species—each with both sexes—exposed at the maximum tolerated dose.

Although these are serious impediments to understanding what causes disease and at what potencies, the animal carcinogenicity studies described in this chapter provide important assurance that our testing strategy has not been irrational. When we compare the results of human epidemiological studies to the results of animal studies, we find a strong correlation. All known human carcinogens are also animal carcinogens; this overlap, however, exists for only thirty-nine compounds. By contrast, there are literally hundreds of compounds that have demonstrated carcinogenic effects in animals but for which no reliable human evidence exists. In the vast majority of cases, there is simply no evidence of an adverse human effect, rather than evidence that there is no effect.

There is also considerable replication of findings among animal studies. If an animal bioassay demonstrates the carcinogenicity of a toxic agent, then how reproducible are the results? Gold and her colleagues have compiled a database on the results of animal bioassays used to estimate carcinogenic potency, and it now includes nearly 3,700 experiments on 975 different chemicals conducted by the National Cancer Institute and National Toxicology Program (NCI/NTP) and by others.[57] The authors searched their database for studies conducted on the same chemical, the same sex and type of rodent (hamster, mouse, or rat), and the same route of administration. They found 161 experiments conducted on thirty-eight different chemicals, which allowed them to make seventy comparisons.[58] They also discovered concordant results for sixty-one of seventy cases, and in all but two of thirty-five positive comparisons the tumor occurred in the same organ. Studies that demonstrated higher carcinogenic potency tended to be reproduced more often, and experiments on mice that produced conflicting results tended to have been conducted over shorter periods. These findings increase confidence in estimates of relative carcinogenic potency and target organ from animal bioassays.[59]

Understanding the potency of neurological toxins provides a different sort of problem. As I explain more fully in chapter 11, neurotoxins such as the organophosphate and carbamate pesticides may cause a variety of adverse acute and chronic effects—each of which may be dose-dependent. When tested for a chronic effect such as cancer, animals are separated into groups and each group is administered a specific dose for eighteen to twenty-four months. One group normally receives the maximum tolerated dose, another one-half this dose, and a third group remains unexposed as "controls." The incidence of tumors is recorded for each dose and a mathematical relation between dose and tumor response is derived. Many computer models have been designed to then extrapolate the dose response curve to below the lowest dose tested in animals, which is still normally far higher than likely levels of human exposure. In contrast to acutely toxic effects, such as cholinesterase inhibition, no threshold effect is presumed for carcinogens—it is simply assumed that any exposure results in an increased risk of cancer.

Humans are normally exposed to contaminants at doses far lower than those administered to test animals. Could high-dose testing produce carcinogenic responses in animals through secondary biological mechanisms that would never occur in humans? This possibility has produced one of the most acrimonious controversies in the history of regulating environmental contaminants.[60] EPA has assumed for several decades that carcinogens have no "threshold effect"—that is, even the smallest doses of a carcinogen pose some cancer risk—and this view has been supported by consumer and environmental interest groups. If a threshold effect is believed to exist, it becomes a rationale for permitting low levels of environmental contamination. The dosing regimen for animal tests thus becomes crucial to the arguments of both sides.

Precise definition of a threshold, however, is difficult given the constraints normally surrounding long-duration animal tests. One practical limitation is the cost of each study, which has reduced the number of animals tested to fifty for each dose-sex-species group. Testing so few animals makes the tests incapable of detecting an increase in tumor incidence greater than 1 percent.[61] In response to early concern over the possibility of falsely negative results, two significant changes were made to these studies' design. First, the period of exposure was extended to include the entire life of the animal, with the exception of the nursing period. Second, test animals were separated into at least three dosing regimens: the highest dose tolerated with no apparent toxicological effect; some proportion of the highest dose tested; and a control group with no exposure.[62]

The choice of the highest experimental dose is important for several reasons—not the least of which is that it could cause toxic effects that kill the test animals before tumors form. A practice therefore evolved that prescribed setting the highest dose at the maximum tolerated dose (MTD), "the highest dose of the test agent during the chronic study that can be predicted not to alter the animals' longevity [through] effects other than carcinogenicity . . . and [it should cause] . . . no more than a 10 percent weight decrement, as compared to the appropriate control groups, and . . . not produce mortality, clinical signs of toxicity, or pathologic lesions (other than those that may be related to a neoplastic response) that would be predicted to shorten an animal's natural life span."[63]

The MTD is first established through a preliminary short-term study during which as many as six separate doses are administered.[64] Use of the MTD in cancer bioassays, however, has come under increasing scrutiny. The first criticism is that the majority of compounds tested by the National Toxicology Program have induced cancer in animals using the MTD protocol and have thereby focused public concern and resources on risks opponents believe to be relatively small and hypothetical. The Office of Technology Assessment, however, has a dramatically different concern. It has found that a high proportion of compounds identified as carcinogenic by NTP have not been regulated.[65] A second and even more serious criticism is that some compounds found to cause cancer only at high doses may do so through mechanisms that do not occur at the lower doses commonly experienced by humans. Saccharin and butylated hydroxyanisole (BHA) are nongenotoxic, but they do induce toxic cellular responses. Saccharin also produces increased rates of cell proliferation at high doses, which indirectly increases the probability that tumors will develop. BHA produces cellular erosion and ulceration at high doses. This line of reasoning is commonly used by pesticide manufacturers to support the claim of a threshold effect, and therefore to justify registration of the pesticide under the supposition that human exposure may be managed to ensure that the threshold is not crossed. Yet the National Toxicology Program has found that "90 percent of chemicals that were carcinogenic at high doses were also producing tumors at low doses."[66] This appears to support the conclusion that the MTD, although extreme, is an important anchor for estimating carcinogenic potential.[67] A third argument against the use of MTD is that results may do little to inform us about the shape of the dose-response relationship, knowledge that is vital to the accuracy of risk estimates. Opponents of the MTD argue that if lower doses were tested, tumor incidence would be lower than predicted by extrapolating from effects at the MTD.

A final criticism is that estimates of carcinogenic potency and other measures of toxicity (including MTD) are highly correlated. This raises a suggestion that toxic effects of a compound (other than its carcinogenicity) could play a role in carcinogenicity.[68] In 1993, the majority of an NAS panel recommended the continued use of the MTD in animal bioassays: "The assay identifies substances that do or do not increase the incidence of cancer under the conditions of the assay and provides an operational definition of *animal noncarcinogens*. The assay also identifies target organs, demonstrates tumor types associated with exposure, provides a consistent basis for interspecies comparisons, and can serve as a guide in designing followup studies."[69] They also recommended testing at lower doses, ranging between MTD/2 and MTD/10, to provide a better basis for extrapolating responses to the lower doses more likely to be experienced by humans. The shape of the dose-response curve at lower doses is crucial to estimates of cancer risk, and it requires more refined testing. Yet six of the eighteen committee members disagreed, concluding: "The HDT should be selected as the highest dose that can be expected to yield results relevant to humans, not the highest dose that can be administered to animals without causing early mortality from causes other than cancer."[70]

Models of Carcinogenesis

Our understanding of how most carcinogens cause cancer is primitive. Because we rarely understand the precise biochemical mechanisms at work, we are left with the difficult task of mathematically modeling biological responses to contaminants. The most widely accepted models of carcinogenesis are multistaged and assume that healthy cells are transformed to a malignant state through one or more intermediate stages. Some evidence supports the hypothesis that cancer evolves in at least several steps. The mutations are assumed to occur in a specific sequence, to be irreversible, and to happen at dose-related rates.[71]

S. Moolgavkar, D. Venzon, and A. Knudson accepted the multistaged theory of P. Armitage and R. Doll, but modified it to account for how quickly exposed cells originated, mutated, and died. Their model, called the MVK model, attempts to account for rates of cell mutation or "initiation"; the clonal expansion of a population of initiated cells via "promotion"; and their "progression" from initiated cells to malignancy through further mutation.[72]

Neither the Armitage-Doll nor the MVK model directly accounts for the possibility that mutation itself may be a multistaged process. M. Anderson, for example, suggested that mutation begins with DNA damage, but that cells are capable of repair before the damage is "fixed" and transmitted to future generations through replication. C. Portier and A. Kopp-Schneider then redesigned the multistaged models to account for damage and repair.[73] Although this model appears to reflect biological processes more accurately than earlier models, there is rarely data available to validate rates of damage, repair, and mutation at different periods during the multistaged process. The authors themselves acknowledge validation problems.

Another complication emerges from the recognition that rates of cell division vary by age and by organ type. The most rapid rates of division generally occur in utero and among infants. Organ weight is sometimes used as a rough surrogate for estimating the number of cells, but this can be misleading because growth occurs by means of hyperplasia (increases in cell numbers) and hypertrophy (increases in cell size). Without data, only rough approximations of these variables are possible.

Animal bioassays that test the carcinogenic potency of chemicals are normally designed to begin only after weaning, and doses are given at levels far higher than humans are expected to experience. There are at least three different types of extrapolation that are therefore necessary to interpret the relevance of bioassay results to the regulation of human exposure. The first is extrapolation from animals to humans; the second is from high experimental doses to low human doses; and the third is from constant animal dosing to the intermittent exposures more likely faced by humans. The models considered above do not account for the possible effects of various patterns of intermittent exposure. Occupational exposures, for example, occur only in the workplace and often only while employed at a specific firm, while performing a certain task within a firm, or until a specific manufacturing process is changed. And dietary exposure to pesticides and other food contaminants will vary depending upon patterns of food intake, which appear to vary by at least age, affluence, and ethnicity.

Thus, two critical variables appear to work together to increase the cancer risk of pesticides to young children. The first is that rates of growth or cell proliferation in the young are far higher than adults, and as the birth-death-mutation (BDM) model suggests, higher rates of replication of initiated, or intermediate, cells increase the probability of progression to malignancy. The second is that children appear to be exposed to certain carcinogenic pesticides at levels higher than adults.

In an attempt to account for both of these factors, Duncan Murdoch, Daniel Krewski, and I revised the BDM model to assume that the net birth rate of initiated cells (those mutated during the first stage) was proportional to organ weight, which is a rough indicator of normal cell proliferation rates.[74] This choice was made based upon the assumption that initiated cells will divide at roughly the same rate as normal cells. By assuming variable rates of cell proliferation by organ, the model now accounts for the more rapid growth rate of children as compared with adults.

Higher childhood exposure to carcinogens was also considered, and the study was based upon food intake data that show higher intake of some foods by children than adults. Any given level of pesticide residue in apple juice will therefore result in a higher level of exposure in children.[75] In some models of cancer development, childhood exposure is considered a bigger contributor to malignancy than are exposures during adulthood. This idea has been explored since 1985 in a rapidly expanding body of scientific literature.[76] The effectiveness of any dose in producing malignancy varies with the birth rate of initiated cells (which is related to the amount of exposure to the toxin). This rate, in turn, varies with the birth rate of normal cells—and as discussed, children's cells proliferate faster than do adults'. When higher birth rates for initiated cells are assumed in these models, the incidence of malignancy is projected to increase by a factor of 3.74.

Age-Related Sensitivity in Animal Studies

Age-related susceptibility to chemical carcinogens, particularly among newborn rodents, was suspected by 1961.[77] S. Vesselinovitch at the University of Chicago studied age-related effectiveness of radiation and toxic chemicals in tumor induction. By 1971 he had found additional age-related susceptibility for urethane, ethylnitrosourea (ENU), and radiation.[78] This work was supported by a 1981 study by M. Naito and his colleagues. They examined the effect of administering a single dose of ENU to rats at various stages of the neonatal period and found that the brain was the most susceptible to tumor formation during the perinatal period—from birth through age four weeks. The spinal cord was most susceptible in neonatal and one-week-old rats.[79]

As early as 1968, B. Toth reported that young experimental animals may be more susceptible to cancer in some organs, such as the lung and liver, whereas they may be less susceptible to tumors in others—such as the breast and skin.[80] In 1966, a research team headed by Berenblum found a sixfold higher rate of leukemia among mice exposed to urethane shortly after birth compared with others whose exposure began at forty-five days. Perinatal exposure to ethylene thiourea, a metabolite of EBDC fungicides, followed by two years of dietary exposure, produced somewhat higher incidence of thyroid tumors than were experienced by those whose exposure began after birth.

In an effort to clarify some of these conflicting results, Anisimov reviewed a cross-section of contradictory evidence in 1983. His survey shows that age may be an important factor in governing the susceptibility to cancer, but not always. In cases where heightened susceptibility among the young is observed, both higher rates of cell proliferation and differing metabolic capacities appear to be important factors. For example, aflatoxin B1 and polycyclic aromatic hydrocarbons each cause liver tumors in newborn rodents but not in older animals. This phenomenon appears to be a function of the more rapid rate of liver growth immediately following birth. ENU is also a more potent carcinogen in newborn test animals than in adults, and it does not require metabolic activation. Diethylnitrosamine, by contrast, does require metabolic activation, which is less efficient in young animals—so in this case adults have the higher incidence of cancer. The young more accurately replicate DNA during cell reproduction and may also be more effective than the old in repairing damaged or altered DNA.[81]

In their 1975 study of benzo(a)pyrene, commonly produced from burning meats and fats, Vesselinovitch and his colleagues found that newborn and infant mice developed both liver and lung tumors more readily than did young adults. They concluded that age at the time of exposure was the most important mediator of rates of tumor development in the liver, lung, stomach, and lymphoreticular system.[82] By 1976 Vesselinovitch's group had expanded their studies to include DDT, dieldrin, aflatoxin B1, benzidine, and diethylnitrosomine (DEN), the results of which further supported their conclusions regarding the heightened susceptibility of rodents to carcinogenic effects later in life—and, in the case of some compounds, during infancy or youth. Reflecting on the implications of their findings for human risks, the scientists suggested: "The greater sensitivity of certain tissues to carcinogenesis at perinatal age periods in conjunction with dose-response data might prove useful for the establishment of tolerance levels for human exposure during similar physiologic age periods."[83] Nearly seventeen years later, NAS echoed their suggestion, while many at the time wondered if EPA would ever pay attention.

In a 1983 study of eight hundred rats, eight hundred hamsters, and twelve hundred mice, scientists led by R. T. Drew examined the effects of the age at the time of first exposure and the duration of exposure on tumor incidence. The compound chosen for analysis was vinyl chloride, a well-known animal carcinogen. Nine major tumor types were found, and the incidence was higher among animals exposed earlier in their lives. The incidence of mammary tumors in rats was higher when exposure began earlier, but the increase was not related to the duration of exposure. In hamsters, the highest incidence of stomach adenomas and hemangiosarcomas occurred when the animals were exposed early in life, for only six months. Mammary tumors were highest when six-month exposures were delivered earliest in life. Later exposure did not increase tumor incidence. A similar trend in mammary tumors was found in both strains of mice tested, B6C3F1 and Swiss. In mice and hamsters, a six-month exposure induced eight of the nine major tumor types found in the study. The authors concluded that a higher incidence of tumors occurred when exposure occurred early in life, regardless of the duration of exposure.[84]

Evidence of heightened cancer susceptibility among young, rapidly growing animals led many scientists to suspect that risk was related to more rapid rates of cell proliferation. Yet the mechanism of tumor formation appeared to be specific both to the biochemical properties of individual agents and to the target organ's reaction to it.[85] Examining DEN, M. C. Dryoff and his colleagues found that the liver cells of younger rats were fifteen times more susceptible to being initiated through the mutagenic effects of DEN than were liver cells of older rats. The effect appeared to be growth-rate dependent.[86]

In their explorations of another concept of susceptibility, W. Lijinsky and R. Kovatch examined both tumor incidence and the rate of death. Nitrosomorpholine administered to both young and old rats induced liver tumors in nearly all exposed, but the neoplasms killed the younger rats much more quickly. The researchers suggest that the time it takes for a compound to cause death is an indicator of susceptibility, and they demonstrated that young rats given a 2.5 times smaller dose as the old rats died in the same amount of time. Examining a second nitrosomine, nitrosobis (2-oxopropyl) amine (BOP), they found that 35 percent of the older rats had liver tumors at the lower dose, and 50 percent at the higher dose. This compares with no tumors in the young male rats treated at similar doses. The contradictory nature of these findings demonstrates, once again, that age-related susceptibility may be compound and organ specific.[87]

Finally, differences in tumor susceptibility based upon dietary fat intake were tested in 1983 by P. Chan and T. Dao using N-methyl-nitrosourea (NMU). Mammary tumor incidence was clearly age-related; it decreased in incidence as age increased. Similarly, for every age studied tumor incidence was higher in rats fed a high-fat diet than in those that consumed a low-fat diet. Also, young rats fed a high-fat diet had nearly a fourfold increase in risk over older rats fed a low-fat diet.[88]

Epidemiological Studies of Pesticides and Cancer in Children

Environmentally induced disease has reached epidemic proportions in the United States. Diseases contracted in the workplace are now blamed for between fifty thousand and seventy thousand deaths per year. Deaths from asbestos exposure are forecast to exceed 300,000 lives. Between 3 and 4 million preschool children currently have levels of lead in their blood that are believed to cause long-term neurological and cognitive impairment.[89] In the cases of asbestos and lead, there are clear biological markers of exposure and a reasonable understanding of dose-related human health effects. These landmarks also exist for studying numerous contaminants of workplaces, where the magnitude and duration of exposure are reasonably well known. Under these circumstances and given this knowledge base, environmental epidemiology has been relatively successful in guiding well-informed policymaking.[90]

In general, confidence about what causes disease comes from knowledge of the differences in disease incidence between exposed and unexposed populations. This understanding follows more easily if the disease is rare or clearly identifiable; if there is a clear distinction between exposed and unexposed populations; if the magnitude and duration of exposure is known; if the time elapsed between exposure and disease recognition is not extremely long; and if there are few if any other possible causes of the disease. For example, the occupational exposure to PVC among workers who manufactured the compound was eventually blamed for three cases of angiosarcoma, a rare liver tumor. Similar effects have been found among animals exposed to this toxin.[91]

When epidemiologists attempt to identify the environmental causes of cancer, they commonly face many obstacles. One difficulty is that onset of the disease may be delayed twenty to forty years following exposure to a carcinogenic agent. A second problem is that during the exposure period an individual may simultaneously have been exposed to numerous other carcinogenic agents, some of which may be capable of inducing the same effect as the suspect compound. To differentiate among probable causes, then, scientists depend on estimates of relative exposure. But accurate estimates of exposure, especially of exposures that vary in magnitude and are of intermittent duration, are notoriously difficult to obtain. Retrospective studies, particularly those looking back over long periods of time, often rely on recollections of former conditions of exposure or crude estimates. In some cases, this strategy works marginally well. The number of cigarette packs smoked per day, for example, is a relatively good indicator of exposure to the toxic compounds in cigarette smoke, and most smokers are aware of the number of cigarettes they smoke daily. By contrast, estimates of human exposure to pesticides in food, drinking water, or the household depend on both estimates of food and water intake and their levels of contamination.[92] In addition, because most epidemiological studies are conducted after the chemical under suspicion is already in use, and exposures have already occurred, they often have little effect on regulatory judgments concerning the licensing of thousands of new compounds introduced yearly.

In 1988 the EPA found that 7 percent of all pesticides used in the United States were applied in the home, garden, or yard.[93] This discovery led to an interesting series of epidemiological studies that have explored how pesticides are used in the home and the relationships of these patterns of use and exposure to disease incidence. In 1992, for example, a team lead by J. R. Davis found that among 238 families studied in Missouri, 97.8 percent used pesticides in the home, garden, or yard at least once per year, and 66 percent used pesticides more than five times each year. The most common site of application was in the home. Further, 80 percent of those pregnant reported using pesticides during their pregnancies, and 70 percent of those with children reported applying compounds during the first six months of the child's life. Nearly 57 percent of the families used herbicides, 50 percent used insecticides to control fleas and ticks on domestic animals, and 10 percent used the chlorinated hydrocarbon lindane to control head lice in children.[94]

In 1981 three pediatricians in California reported in *Lancet* the incidence of acute leukemia in seven cases shortly following in-home exposure to organophosphate insecticides. The interval between exposure and diagnosis ranged between one and twenty-eight weeks. Six of the children had been exposed to a combination of DDVP and propoxur, and the duration of their exposures ranged from two minutes in a shed to several days in a recently fumigated home. The authors were careful to determine that the parents had previously used these compounds in the home, and although they recognized that their anecdotal evidence hardly proved a causal relationship, they recommended extreme caution in using these compounds around children.

Even though their study was not conclusive, this type of finding motivated more careful ecologic-epidemiological analyses. By 1987, a case-control study of childhood leukemia in Los Angeles county explored causes of the disease. For children under age ten, a 3.8-fold increase risk of leukemia was found if parents used pesticides in the home, and a 6.5-fold increased risk was revealed among families using pesticides in the garden.[95] A study of 309 leukemia cases in Shanghai that used 618 healthy controls demonstrated a 3.5-fold increase in risk associated with maternal occupational exposure to pesticides.[96]

In 1989, the Children's Cancer Study Group reported that among families of 204 children with acute nonlymphoblastic leukemia, the most consistent association found in their analysis of potential causes was pesticide exposure. If the fathers held jobs involving pesticide exposure for more than a thousand days, the child had a 2.7 times higher chance of contracting the disease when compared to controls. As the parental exposure declined, the odds also declined. Children regularly exposed to pesticides in the household had a 3.5 times higher incidence of leukemia than those not exposed there. The frequency of household exposure also played an important role.

To determine if the risks varied by age, the population was restricted to children under age five. If either parent was occupationally exposed for more than a thousand days, the risk was 11.4 times higher than controls ($p=0.003$), and triple the risk if all cases were included. Eight mothers exposed to pesticides in the household during pregnancy had children with leukemia. The odds of having a child with leukemia doubled if the mother was exposed during pregnancy. These findings suggest strong age-related susceptibility to pesticide toxicity.[97] Risk also appears to vary with the intensity of exposure.

The relationship between chemical exposure and Ewing's bone sarcoma was evaluated in a 1992 study conducted in the San Francisco Bay area by F. A. Holly and his colleagues. They found that children of fathers in agricultural occupations for the six months prior to conception until the time of diagnosis had 8.8 times the risk of controls. For children of parents whose occupational exposure was specifically to pesticides, herbicides, or fertilizers, risk was 6.1 times higher than controls.[98]

In 1993, a team led by J. R. Davis explored possible connections between pesticide use in Missouri and childhood brain cancers.[99] Previous reports of an association were made in 1979 by E. Gold and others, who found that children exposed to pesticides from in-home exterminators faced a 2.3 times higher risk of brain cancer.[100] In 1985, T. H. Sinks found an elevated risk of brain cancer among children born to mothers who used aerosol pesticides during pregnancy and after birth.[101]

Davis and his colleagues found that the odds ratios for brain cancer varied significantly by "use situation" and by duration of exposure. The use of lindane to control head lice among children between the age of seven months and diagnosis increased the odds of brain cancer 4.6-fold over controls. The use of pesticide bombs in the home to control nuisance insects during pregnancy increased the odds of childhood cancer 6.2 times. The use of no-pest strips for nuisance insects also caused risk to increase by 3.7 times. Use of flea collars and flea shampoos on pets also caused elevated risk, although the compounds used as the active ingredients were not reported. Garden use of diazinon elevated risk by 4.6 times. Finally, use of herbicides in the yard elevated risk by a factor of 3.4, although again the specific formulations were unreported.

A group of Danish researchers studied a cohort of 4,015 gardeners between 1975 and 1984 to observe cancer incidence, which was then compared with national incidence rates. For all cancer sites combined there was little distinction—for soft-tissue sarcoma, however, a 5.2-fold increase in incidence was found; chronic lymphatic leukemia was 2.75-fold higher than national background rates; and non-Hodgkin's lymphoma incidence was two times higher.[102]

In a 1991 review of the literature on parental occupational exposure to toxic chemicals and childhood cancer, L. O'Leary and his colleagues evaluated thirty-two studies published between 1974 and 1991. They began with the hypothesis that parental occupational exposure to toxic substances could increase childhood cancer risk and concluded that occupational exposure to chemicals (paints, petroleum products, hydrocarbons, solvents, and pesticides) and metals increased the risk of childhood malignancy. Although they were aware of the overlap among categories (pesticides also contain petroleum distillates, and both paints and pesticides commonly contain solvents), and commented that this overlap made it extremely difficult to disassociate the effects of different classes of agents (let alone individual compounds), they still argued that the exposure, en masse, contributes to childhood cancer incidence. Exposures may have come from the parents' clothing, or in the case of chlorinated compounds, from dietary sources. Moreover, these authors suggested that chromosomes damaged from some occupational exposures may pass from parent to child. There is some recent evidence, for example, that mutations that give rise to retinoblastoma, rhabdomyosarcoma, and Wilms's tumor may originate paternally.[103]

Interpretations

During the past twenty-five years, U.S. environmental law has grown exponentially. If there is any common purpose to these laws, it is to manage potential damages to human health and the environment. Risk estimation has become the dominant analytical method used to judge the relative seriousness of various threats, and it has rapidly become the foundation for the management of environmental contamination. During this same period our understanding of the toxic properties of pesticides has also grown, but in a piecemeal and incremental way. For most pesticides we still do not understand the biological mechanisms by which they cause toxic effects. This uncertainty forces experts and others to rely on computer models to simulate biological processes and patterns of their disruption by toxins.

This chapter opened by reviewing a fundamental principle of toxicology: that the dose makes the poison. But to make policy based on this principle requires that environmental managers understand the level of environmental contamination and probable ranges of human exposure (via air, water, food, and so forth); be able to estimate the toxic potency of the contaminant as it varies by dosage; and comprehend how susceptibility may vary with the exposed population. A sense of urgency should propel their efforts. While these environmental managers struggle to understand all of these factors for complex mixtures of environmental contaminants, human exposure continues.

In this chapter, I chose to review briefly only two types of toxic effects—cancer and neurological damage—to illustrate how chemical toxicity can vary among groups. Experts have traditionally relied on animal evidence to estimate the risks contaminants pose to humans, but this has been an inadequate strategy for several reasons. The first is that it forces extrapolation of results from test species to humans. Second, test animals are normally exposed at doses far higher than those experienced by humans, leaving us to wonder if the adverse effect in animals was somehow caused by the sheer magnitude of the dose. The third problem lies with estimating exposure. Most estimates of chronic risk simplistically assume that exposure occurs uniformly across a lifetime, instead of intermittently with wide variation in magnitude over time. Thus high, short-term exposure may pose higher risks than lower, long-term exposure, even though total exposure under both scenarios may be the same. Fourth, as I have demonstrated, animal tests rarely are designed to detect the special susceptibility of the young. Finally, the effects of complex mixtures of pesticides are rarely tested, although humans are normally exposed to mixtures of pesticides among a much larger universe of environmental contaminants. Regardless of governmental policies, however, it seems that given the sheer number of pesticides that are permitted to contaminate the food and water supplies, we will never fully understand the toxicological implications of our exposure to complex mixtures, any one of which may produce several types of adverse effects.

The dominant issue addressed in this chapter has been how human responses to pesticides may vary according to a person's age. Next, I turn to explore the human diet, a dominant source of human exposure to environmental contaminants, including pesticides. A gradual awareness of how variation in the diet may lead to variation in pesticide exposure evolved from 1985 to 1995, and these findings provide a basis for predicting where toxic responses or accumulation of risk in humans is likely to occur.

CHAPTER 9

The Diet of a Child

Food and water deliver a complex mixture of natural and synthetic chemicals to the human body at every meal. In 1984, an NAS committee estimated that the government then permitted 8,627 different food additives and an additional 1,800 pesticides and inert ingredients to remain as residues within the nation's food supply.[1] The synthetic additives include artificial colors, flavors, and fragrances; antibiotics, hormones, and other drug residues that may persist in meat, poultry, and dairy products; and substances that migrate to food from packaging such as heavy metals from cans or vinyl from plastic wraps. Thus the problem of managing pesticide residues lies within a much larger food safety arena, one managed predominantly by FDA, with EPA assuming regulatory responsibility to control pesticide residues and a diverse mixture of drinking-water contaminants. Moreover, on the very day it was formed, EPA was drawn into very public debates over the risks posed by chlorinated pesticides such as DDT—a controversy that demonstrated how careful pesticide management was a necessary condition for food safety, sustainable agriculture, water quality, and wildlife protection.

To cope with the more rigorous environmental health standards required by the 1972 amendments to federal pesticide law, EPA focused on regulating 325 pesticides used on food crops, under the assumption that Americans were exposed to pesticides predominantly through the diet. The agency also avoided questions regarding the toxicity of nearly 1,600 inert ingredients that constitute the largest proportion of pesticide products by weight.[2] Because residues can appear in more than five thousand food products marketed in the United States, EPA further restricted its attention to approximately seven hundred more basic forms of food, such as apples, wheat flour, and tomatoes.[3] Despite these simplifications to its regulatory strategy, since 1980 the agency has still managed nearly 8,500 separate "tolerances," which limit, but still permit, pesticide contamination of most foods. For example, in 1995, nearly 75 different pesticides were permitted by federal regulation to remain as residues on apples, and one hundred pesticides were allowed to remain in milk.[4] Thus pesticide residues in food touch nearly every human on earth every day, often as a complex mixture of contaminants.

Between 1970 and 1985, EPA assumed it could estimate pesticide exposure if it knew the U.S. average diet. To make this estimate it used only crude percentages calculated from national dietary surveys administered every ten years by USDA.[5] Slowly, however, EPA began to look more carefully at the dietary patterns of populations that differed by age, region of the country, ethnic background, season, and gender.[6] Although the agency had improved its analyses by using mainframe computers during the mid-1980s, it was exceptionally slow to adopt advanced microcomputer technologies, particularly the statistical and graphical software that can rapidly simulate food intake, exposures, and risks, providing dramatic improvements in analytical power at relatively low cost.[7]

In 1985, Congress asked NAS to convene a group of experts to try to make some sense out of the federal laws that had governed the growth of the federal pesticide tolerance system.[8] Congressional sponsors of the request were especially interested in the link between pesticides in the diet and cancer risks, as well as in the way that the Delaney amendment to FFDCA had been interpreted by EPA and FDA to apply to pesticide residues. The NAS committee focused on risks permitted by tolerances set pursuant to a risk-benefit balancing standard, because these made up nearly 96 percent of all tolerances. They quickly found that to understand pesticide risks they needed to know what foods people eat, what pesticide residues appear on these foods, and what toxic effects these residues are likely to cause. These are not simple questions.

As my students and I began working with the Delaney Committee in 1985, we sifted electronically through many national surveys of American food consumption searching for clues about likely patterns of exposure to pesticides believed to pose cancer risks.[9] By 1986, we began to understand that the most predictable variation in risk appeared age-related: children consumed very different foods than adults—or than some fictitious "average person" that EPA had fabricated by averaging food intake across all people, food by food, just as USDA had done before it. The Delaney Committee chose to assume that everyone ate the average diet, for the same reason EPA had—the sheer complexity of considering variance given so many foods and pesticides. It therefore skirted the question of higher childhood intake other than to recognize it.[10]

This chapter explores the dietary patterns that lie behind the averages. The findings are reasonably clear: differences between childhood and adult patterns of food consumption lead to distinctive patterns of pesticide exposure and risk.[11] Before this understanding, concern for children's health played little part in pesticide decisionmaking, but by 1995, most deliberations over tolerances or registrations overtly considered childhood risks. This shift in thinking demonstrates the potential political power that comes from understanding distributional patterns of risk.

Poor Data Undermine Estimates of What Children Eat

Surveys of food consumption are of intense interest to the agricultural and food-processing industries, as well as those concerned with public health and nutrition. Nearly one hundred food-consumption surveys conducted between 1970 and 1990 were evaluated prior to choosing the data presented in this chapter.[12]

Surveys conducted by USDA appeared to provide the most comprehensive and projectable data for the purposes of national pesticide exposure and risk estimation. These data sets include the 1977–78 National Food Consumption Survey (NFCS),[13] the 1985–86 Continuing Survey of Food Intake by Individuals (CSFII) for Children ages 1–5, and that for women ages 19–50.[14] Although USDA conducted an additional national survey of food intake in 1987–88, the response rate was so low that the General Accounting Office, as well as a special committee convened to judge the statistical significance of the survey results, recommended against their use for national dietary estimates without extreme caution due to probable bias.[15]

The conclusion that these data may be biased has placed EPA, FDA, and USDA in a very difficult position because they all rely heavily on these surveys for regulatory and research purposes. They have been forced to use dietary data that are ten to eighteen years old instead. Changes in food-processing technologies, marketing strategies, and the public's knowledge of nutrition and health have all influenced food consumption since 1978, but dietary experts have remarkably little understanding of precisely what shifts have occurred or how they may be relevant to pesticide exposure.

FDA also conducts a number of studies related to nutrition in the United States. These include the Total Diet Study, the Health and Diet Survey, and the Vitamin/Mineral Supplement Intake Survey.[16] The Total Diet Study is undertaken every year by FDA to assess levels of various nutritional elements and contaminants in the U.S. food supply; representative diets of specific age-sex groups are reviewed. Results from all FDA surveys are commonly used to support claims that pesticide residue levels in food constitute insignificant risks.[17] Yet, like many private-sector surveys designed to propel marketing strategies, the surveys are not suitable for estimating national dietary patterns.[18] FDA tests only a few foods, and their method of blending samples in "slurries" could easily reduce the concentration of any pesticide present to a concentration beneath a limit of detection.

There are three additional significant limitations to the data presented below. The first is simply its age. The most recent data set, the CSFII (1985–86), is already nearly a decade old, and the NFCS (1977–78) data is almost twenty years out of date. Second, EPA has normally estimated exposure and risk by aggregating food consumption data within large age classes. But only three childhood age groupings were delineated for chronic exposure and risk analyses: ages birth to one, two to six, and seven to twelve. The research described in the previous chapter demonstrates that specific organ systems evolve through periods of vulnerability that often last less than a year. Further, considerable variation in food intake and exposure within these age classes may also occur, and these may have a significant influence on the accumulation of risk.

Third, the surveys are all of short duration, which makes it difficult to estimate the "usual" intake necessary to forecast chronic exposures and toxic effects.[19] Longitudinal studies that track changes in the eating behavior of individuals across long periods of time are extremely costly and difficult to conduct; they are therefore rare.[20] None exist at a scale necessary to estimate usual intake and therefore usual exposure throughout childhood.

The Recipe Problem

The explosion in the diversity of processed foods available in the marketplace makes it very difficult to estimate intake of the basic food types regulated by EPA. When a survey participant reports eating a food, that food must be translated into amounts of basic food components. For example, pizza is broken down into wheat flour, water, yeast, tomato paste, tomato sauce, and so forth by using average recipe information provided by USDA's Human Nutrition and Information Service. Breaking "foods as eaten" into component parts is necessary because pesticide residues are regulated by tolerances—or limits—in the components, not in the processed food.[21]

Moreover, as mentioned in chapter 1, EPA's scheme for codifying tolerances for pesticide residues on foods is unique. It does not match the food-coding system used by USDA to collect dietary data, and it cannot be easily correlated with the lexicon of food codes FDA uses to monitor compliance with tolerances. These differences require computer programming gymnastics to estimate pesticide exposure—and they have significantly hampered the strategic management of pesticide risks in the United States. Joining EPA's food-coding system to the real-world USDA food-as-eaten coding system requires considerable judgment regarding the "recipes" of processed food. The objective in defining the recipes is to estimate the proportion of the processed food composed of raw foods, because residues are regulated almost exclusively in these more basic food forms. Pizza, for example, will be reported as eaten by the slice, but the current regulatory system demands an estimate of the average weight of a "slice" and the proportion composed of wheat, tomatoes, milk, and so forth. Further, although recipes for processed foods change as technology advances or tastes change, EPA's food-coding system is locked in place by the incremental nature of pesticide tolerance-setting and the cumbersome administrative procedures necessary to change federal regulations. With 325 food-use chemicals on the market, only a handful are reviewed each year, and often each review results in only modest changes in tolerance levels. The snail's pace of chemical-by-chemical reviews guarantees that the inventory of food codes that underlies EPA's 9,300 tolerances will remain inflexible. Stability in the tolerance system also stabilizes allowable patterns of exposure to pesticides. This built-in resistance to change exists despite rapidly changing pesticide toxicity data, which often demonstrates that pesticides formerly thought harmless may pose significant risks.

Children Consume More of Fewer Foods Than Adults

A careful analysis of the dietary data demonstrates that young children eat more of fewer foods than adults.[22] Of the twenty-nine foods that constituted greater than 1 percent of the U.S. average diet, children less than age seven consumed several times the U.S. average levels of some foods when intake is adjusted by body weight. Table 9–1 details the results of this inquiry. Twenty-two foods constituted more than 1 percent of the average non-nursing infant's diet. Six of these were fruits, which together accounted for 29 percent of the diet; nine were fruits or vegetables that made up 37 percent of the diet. Soybean and coconut oil, most likely from infant formula, together accounted for 5.2 percent of the total. Non-nursing infant consumption of soybean oil was 4.6 times higher than the U.S. average level, whereas coconut oil consumption was fifty times this average. Milk products constitute 21.5 percent of the average non-nursing infant's diet, 7.3 times the U.S. average for consumption of nonfat milk solids and 3.6 times the U.S. average for fat milk solids. Of the 376 foods surveyed, only 148 were positively reported as consumed by non-nursing infants, whereas 375 were reported for the entire 30,770 person sample. This latter finding is one indication of infants' relatively low level of dietary diversity. Moreover, these data immediately suggest the importance of monitoring the percentage of the total diet represented by individual foods for certain age groups. If tolerances have been set to protect against dangerous exposures based upon U.S. average food intake, then children could be at higher risk. The difference between childhood and adult exposures is dictated by how much more of a given food they consume.

For once it may be appropriate to compare apples and oranges—together these fruits constituted 16 percent of a child's diet. As I looked more care-

Table 9–1. Most Consumed Foods in the Average U.S. Diet, 1977–1978 (N=30,770)

Food	Percentage of U.S. average diet	Age class consuming highest multiple of U.S. average	Multiple of U.S. average intake
Wheat flour	7.4	Children 1-6	2.25
Beef, lean	6.9	Children 1-6	1.82
Orange juice	6.7	Children 1-6	3.08
Milk, nonfat	4.7	Infants	7.33
Potatoes	5.4	Children 1-6	2.14
Cane sugar	4.4	Children 1-6	2.49
Eggs	3.3	Children 1-6	2.30
Tomatoes	2.9	Children 1-6	1.68
Apples	2.8	Infants	6.91
Milk-fat solids	2.5	Infants	3.62
Pork	2.3	Children 1-6	1.84
Chicken with skin	2.3	Children 1-6	2.15
Beef, fat	2.2	Children 1-6	1.81
Potatoes	2.1	Children 1-6	1.79
Beet sugar	1.9	Children 1-6	2.49
Soybean oil	1.9	Infants	4.55
Apple juice	1.4	Infants	16.65
Corn	1.4	Children 1-6	2.39
Bananas	1.4	Infants	4.96
Peaches	1.3	Infants	10.60
Lettuce	1.3	Nursing mothers	1.57
Pork fat	1.2	Children 1-6	2.10
Green beans	1.2	Infants	4.65
Fish, fin	1.1	Nursing mothers	2.41
Peas	1.0	Infants	3.72
Carrots	1.0	Infants	9.05
Tomato Puree	1.0	Children 1-6	2.15
Corn Grain	1.0	Hispanics	3.37
Rice, milled	1.0	Infants	8.69

fully at childhood diets, I found that children between ages one and six consumed twenty-six different foods, each of which accounted for more than 1 percent of their diet. Seven of these were fruits or fruit juices that totaled 21 percent of their diet. Dietary diversity increased in this age class; wheat, beef, sugar, eggs, and chicken began to dominate the diet and a greater variety of vegetables were eaten. The number of foods consumed in amounts greater than the U.S. average declined, and the disparity in the amounts of these foods that were eaten narrowed. This trend is likely a response to two variables: first, the rapid increase in dietary diversity after age one, and second, the diminishing effect of the body-weight conversion factor as children's average weights increased.

Children between ages seven and twelve had slightly more diversity in their diets than the one- to six-year-old group. Wheat flour, beef, and potatoes constituted a higher percentage of the diet, whereas orange juice and apple products fell. The fifteen most consumed foods for children ages one to six and those ages seven to twelve were almost identical, although the relative rankings differed slightly. Children ages seven to twelve also approached the U.S. average for total consumption.

For teenagers (thirteen to nineteen years old), wheat flour, beef, potatoes, and eggs continued their ascendance in relative importance over fruits and vegetables with the exceptions of orange juice (which is still among the three most consumed foods) and tomatoes. Again, the dietary diversity of this group expanded with age and total food intake continued to approach the U.S. average.

Adult patterns of food consumption followed these trends. Beef was the most consumed food, whereas intake of wheat flour, orange juice, potatoes, and milk declined slightly from teenage levels. The list of the fifteen most consumed foods is almost identical for both teenagers and adults, and the variety of foods eaten increased again with age (although this tendency may be the result of the large adult sample size). Similarly, adults ate a total amount extremely close to the U.S. average.

These results are not particularly surprising to anyone who has spent time around children. In particular, the proportion of diet represented by liquids, especially fruit juices, is far higher for infants than adults—even, in some cases, when the difference in body weight is not factored in. Milk and milk products constituted an additional 10 percent of nonwater intake by infants. Nearly 45 percent of nursing infants' food was composed of fruit products, and six of the ten most consumed foods were fruits, which constituted 43 percent of their average diet. In general, nursing infants consumed over five times the U.S. average intake of twenty-two different commodities.

A Comparison of Findings Across Surveys

In 1985–86, USDA surveyed approximately 1,500 women and five hundred children, including children between ages one and five. Individuals were sampled on six occasions—at two-month intervals across for a year. If participants responded reliably, this design is more likely to have captured variance in consumption patterns than the three-day diary employed in 1977–78.[23]

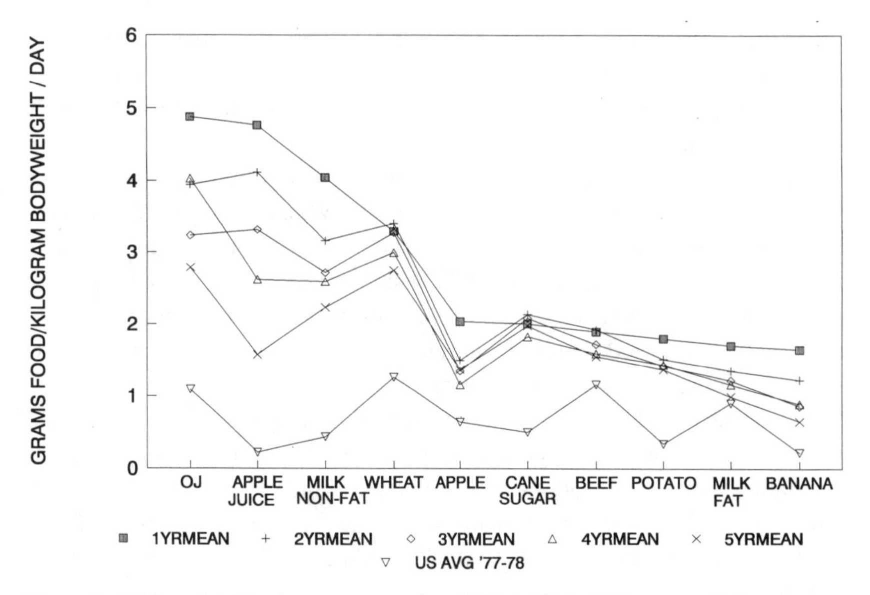

Figure 9–1. Mean food intake ages one to five (USDA 1985–1986) versus U.S. average intake (USDA 1977–1978). Mean food intake for adults is represented by the bottom line. For the foods most consumed by children, their mean intake levels are higher than those of adults. Differences are greatest for youngest children.

Figure 9–1 uses data from this study to compare young children's average intake of the most consumed foods with the U.S. average consumption of these foods in 1977–78.[24] The analysis reinforces the conclusion reached earlier that children eat more of some foods than is predicted by U.S. average dietary estimates. It also demonstrates that foods that dominate the early childhood diet (with the exception of infancy) have changed little between 1977–78 and 1985–86. Foods most consumed still include fluids, especially water, orange juice, apple juice, and milk.

The Danger of Summary Statistics

Although the majority of EPA's estimates of chronic health risks have assumed a U.S. average intake of food and water, figure 9–2 demonstrates the importance of the size of the age groupings in reaching conclusions about probable levels of exposure and risk. As the age groups become narrower and younger, the mean consumption of selected foods increases. Children ages one to two, for example, eat on average nearly 2.5 times as many apples as children ages one to six. The size of the age class may therefore be manipulated to make exposures and risks increase or decrease.

Further, summary statistics may be inadequate to represent exposure and

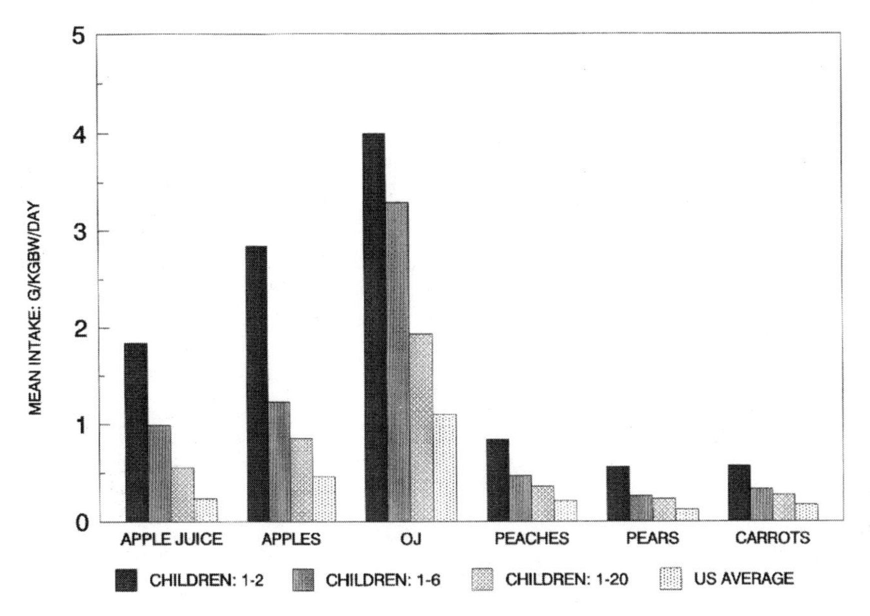

Figure 9–2. Mean intake of selected foods: comparison of age groupings. Averaging intake across short periods for very young populations produces far higher estimates for some juices and fruits. As older individuals are added to expand the age group, the mean intake level for these foods declines (USDA NFCS, 1977–1988).

risk for these age groups, because of these dietary patterns. Means or ninetieth percentile levels may differ substantially depending upon the span in age classes chosen for analysis. A comparison among intake patterns of one-, two-, and three-year-olds and women ages nineteen to fifty is presented in figure 9–3. Each of the four inset figures presents the most consumed foods for the age group and is arranged by highest to lowest mean intake, left to right. Foods most consumed by women are clearly different than those most eaten by children, and the magnitude of intake is far higher per unit of body weight for children than adults. These figures also demonstrate that median intake is often less than the mean, an indication that a considerable proportion of the sample reported eating none of a particular food. Conversely, the ninetieth-percentile intake level is roughly 1.5 to 3 times higher than the mean, whereas the maximum reported by the group is often five to twenty-five times higher than the mean for these most consumed foods. Again, it seems obvious from these data that knowledge of variation in patterns of consumption within age classes is critical when choosing among summary statistical indicators of consumption to estimate exposure and associated risks.

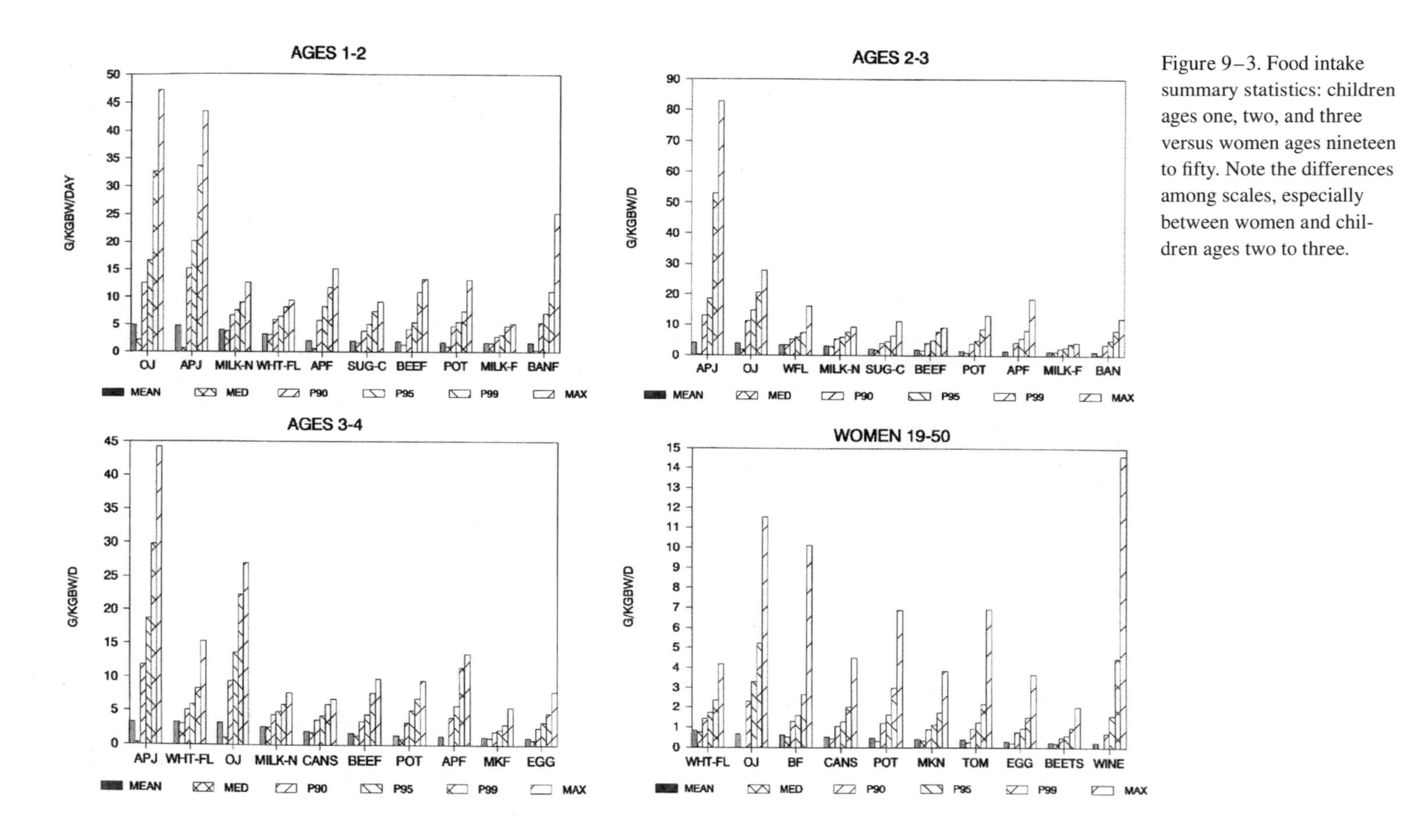

Figure 9–3. Food intake summary statistics: children ages one, two, and three versus women ages nineteen to fifty. Note the differences among scales, especially between women and children ages two to three.

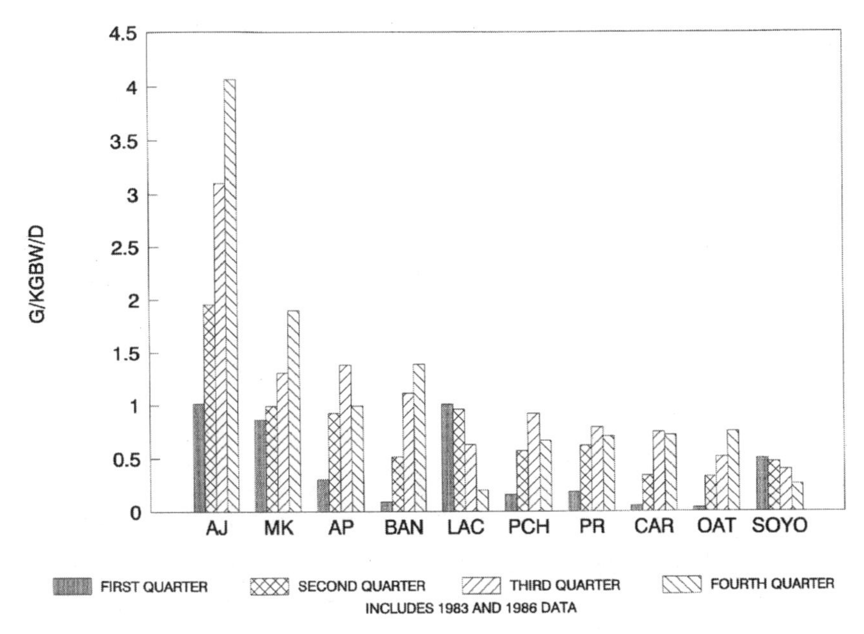

Figure 9–4. Infant mean food intake by quarter of first year. Each bar within a cluster represents a different quarter of the first year of life. The substitution of milk and juice for infant formula is demonstrated by the increasing intake for apple juice and the decreasing intake of lactose and soy oil, key components of infant formula.

The Unique Vulnerability of Infants

As explained in this and earlier chapters, from conception to age one children experience crucial periods of organ development, the most rapid rate of cell reproduction of their lives, and their lowest level of dietary diversity. During this period, the proportion of their diet composed of liquids is also higher than at any other age. All of these findings demand that special attention be given to the types of toxic residues that may exist on infants' food.

Two sources of data provide insight into infant patterns of food consumption. The first is the 1977–78 USDA survey of infants. These children have been classified by EPA into two categories: nursing and non-nursing infants. This distinction is not particularly clear, however, because most nursing infants consume foods other than breast milk. The second data set was provided by a private baby-food manufacturer and was derived from its 1983 and 1986 national surveys of infant food consumption.[25]

These studies confirm, first, that infants drink much more than they eat. Water intake is generally one order of magnitude greater than all other foods (other than breast milk during the first six months of life). Figure 9–4 ranks additional foods from most to least consumed. After water and breast milk, the next most consumed foods include the two liquids apple juice and milk.

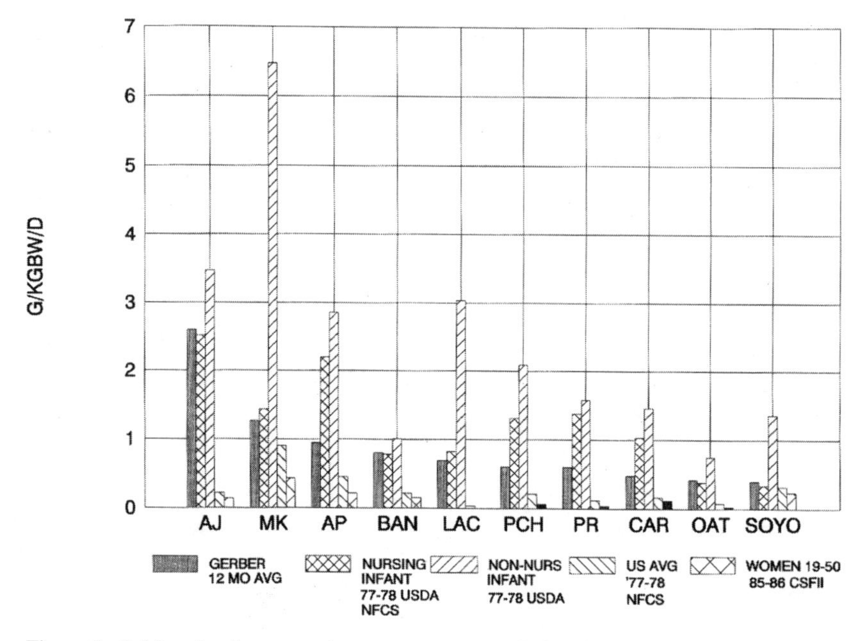

Figure 9–5. Mean intake comparison among surveys. Infant intake levels are comparable across the first three bars of each cluster. When nursing and non-nursing infant data are combined, differences with the Gerber survey are not greater than a factor of 2–3 for these most consumed foods, a difference that could easily be an artifact of variance in survey designs. Women's and U.S. average intakes are provided for comparison.

Another important finding is that the average infant diet is secondarily dominated by fruits and vegetables. As the infant matures through the first year, there is a gradual substitution of apple juice and cow's milk for breast milk. Similarly, consumption of soy oil diminishes, most likely due to weaning from infant formula. Intake of fruits and vegetables increases steadily.

The ability of any single survey to accurately mirror national consumption patterns over time is limited. Thus in figure 9–5 the Gerber data set is compared with the results of the 1977–78 USDA survey for three groups: nursing infants, non-nursing infants, and the entire 1977–78 sample (or U.S. average)—as well as with reports by women ages nineteen to fifty from the 1985–86 USDA CSFII survey.

There is some consistency between mean intake levels reported in the Gerber data and the 1977–78 nursing infant data set. The greatest similarity exists for apple juice, milk, bananas, oats, and soybean oil. Apple, peach, pear, and carrot consumption levels are roughly twice as high in the USDA data than within the Gerber data set. This difference could be partially explained by the fact that the Gerber data include nursing infants, and there is some evidence that breast milk consumption has increased since 1977–78. This theory is strengthened by the fact that the disparity in consumption is particularly evident for milk, lactose, and soybean oil—the first is a common substitute for human milk, and the latter two are components of infant formula. Infant intake is then compared with U.S. average intake within the same chart. It is clear that according to the Gerber study, infants eat much more than is predicted by the U.S. average estimates. Finally, women's mean intake of foods from the 1985–86 survey is generally within a factor of two of U.S. average intake for 1977–78.

To this point, I have presented data in the form of summary statistics—predominantly average intake values. Although these summaries may be efficient when trying to understand the meaning of complex data, they may obscure important information, particularly about what is occurring at the outer bounds of skewed distributions. For these reasons, it is important to examine intake data as full distributions. These data may then be combined with pesticide residue distributions to produce complete exposure distributions.

Intake distributions for four foods commonly consumed by young children are presented in figure 9–6. The distributions demonstrate intake for the entire U.S. sample, for one-year-olds, and for those twenty years and older—and their shape is far from normal. In fact the only food that is normally distributed is water. Although there may appear to be little distinction in the distributions among ages, if the scale is adjusted to display only high intake levels, differences among the groups emerge. The cause of the differences among mean intake levels for the various age classes, described by summary statistics above, is now apparent. An example of this effect is presented in figure 9–7 for apple juice consumption. In this figure, the top graph includes the entire distribution, and the bottom chart examines intake only at or above 4 grams/kgbw/day. The implications of skewed distributions such as these for pesticide exposure and risk assessment will be further examined in the next chapters.

All of the figures presented in this chapter show how different age groups eat different amounts of individual foods, but they give little insight into the mixtures of foods normally eaten. Understanding human exposure to complex mixtures of pesticides requires knowledge of common dietary mixtures of food. Figure 9–8 presents a food intake landscape for 170 one-year-olds, and it includes fifty of the foods they most consumed in 1985–86. If we could magnify this figure to gain a better view of each toddler's diet, we would find that most of these children consumed relatively high amounts of few foods. There are mountains of fruit, juice, milk, and wheat intake, and valleys of vegetables, meat, and fish intake.

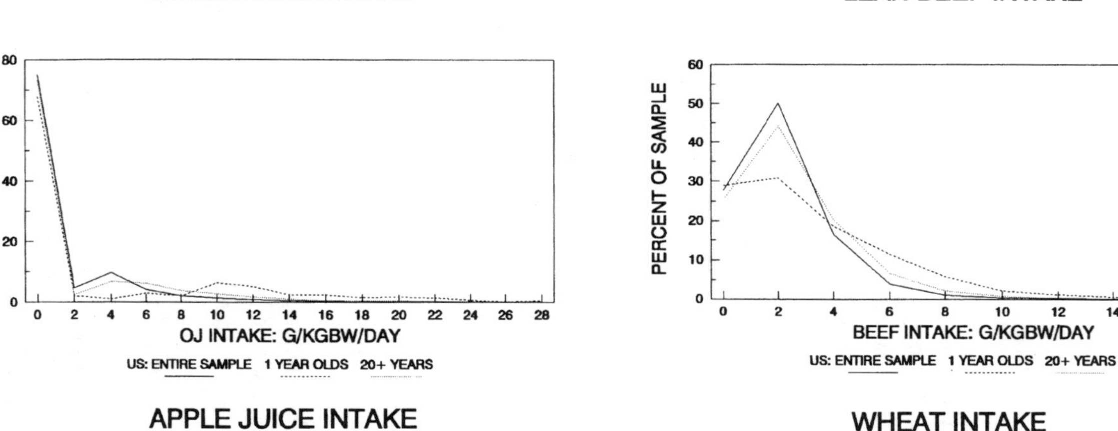

APPLE JUICE INTAKE

PERCENT OF SAMPLE

100 80 60 40 20 0

0 2 4 6 8 10 12 14 16 18 20 22 24 26 28 30 32 34

APPLE JUICE INTAKE: G/KGBW/DAY

US: ENTIRE SAMPLE 1 YEAR OLDS 20+ YEARS

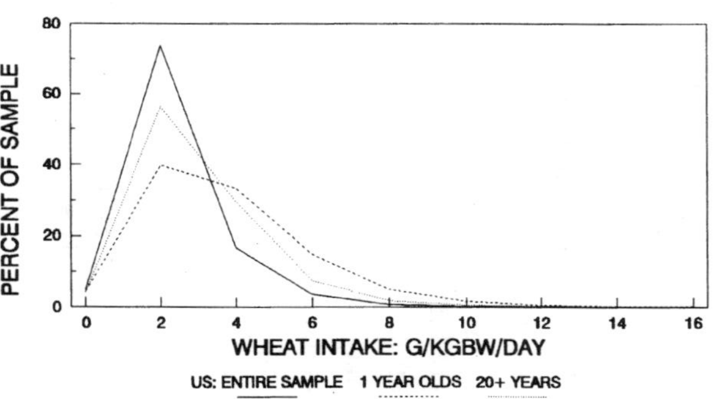

Figure 9–6. Food intake distributions: age-related differences. These distributions demonstrate that large proportions of the children surveyed consumed very low or no amounts of orange and apple juice. Those who did consume these foods tended to drink large amounts relative to adult intake. A higher proportion of the sample reported some intake of beef and wheat.

1977-78 USDA NFCS INTAKE DATA
N: 1 YR: 1473; 20+ YRS: 47,999; ENTIRE US: 87,668.

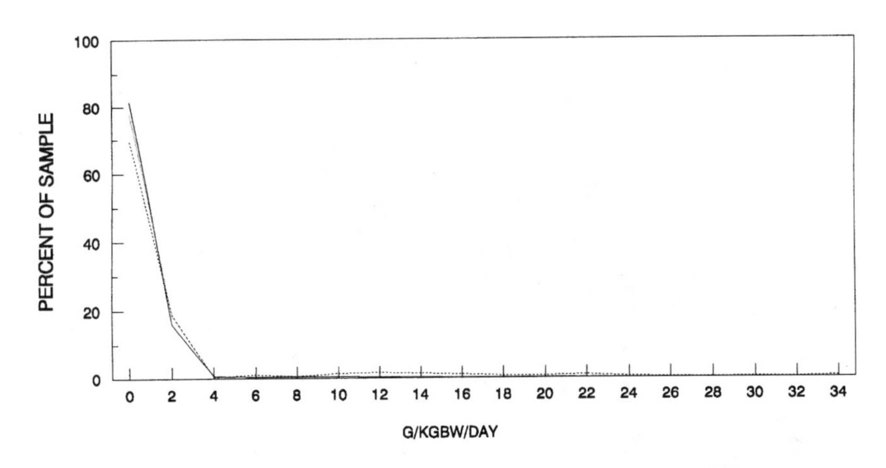

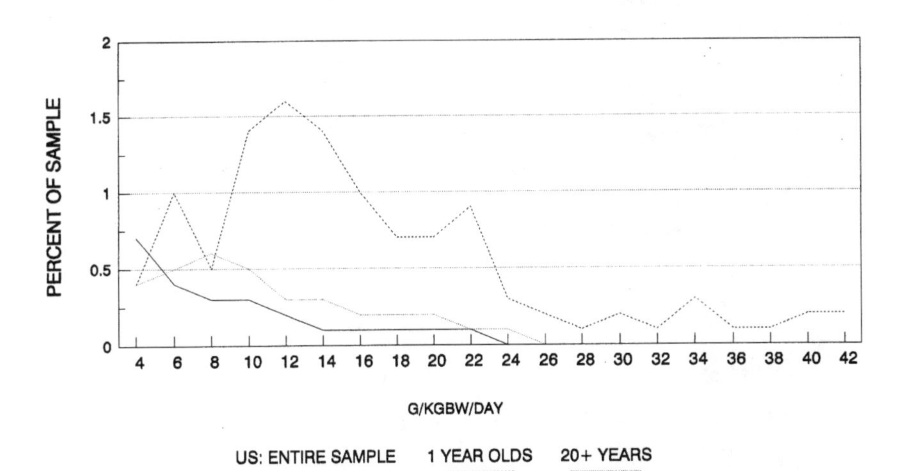

Figure 9–7. Apple intake distributions: age differences. The top figure would lead one to conclude that there are no significant differences in apple intake. Expanding the scale for those consuming more than 4 grams/kg bodyweight/day demonstrates statistically significant differences in intake when one-year-olds are compared with either U.S. average estimates or individuals older than twenty.

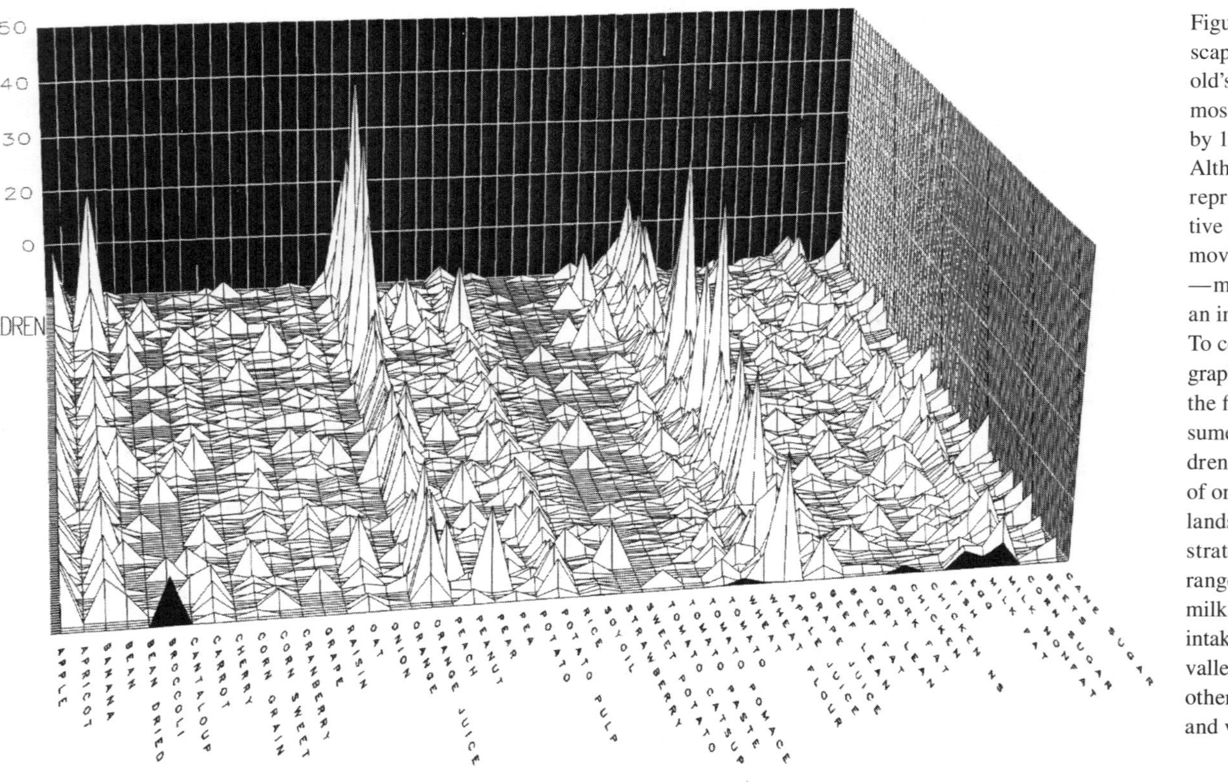

Figure 9–8. The landscape of the one-year-old's diet: the fifty most consumed foods by 170 one-year-olds. Although each column represents a distinctive food, each row—moving horizontally—maps the intake of an individual toddler. To construct the graph, I included only the fifty most consumed foods by children between the ages of one and two. The landscape demonstrates mountain ranges of fruit juice, milk, wheat, and sugar intake, along with valleys of broccoli and other specialty fruits and vegetables.

What we really want is a similar chart for these same one-year-olds that depicts levels of pesticide exposure. If pesticide exposures were known for each food, an estimate of each individual's total exposure could be derived, and from that number, a "total exposure" distribution for the entire group of children. If this method were implemented for age groups that span only one year from birth through age five and three years from ages six through eighteen, it would be easy to estimate the proportion of any age class experiencing "unacceptable" exposures to these toxins. Tolerances could then be adjusted, and safety factors included, to assure that the safety threshold is not breached. This approach is developed in chapter 10 to represent childhood exposure to benomyl on a complex of foods. It also provides the methodological foundation for understanding childhood exposure to a mixture of organophosphate insecticides, which is explored in chapter 11.

Interpretations

The primary conclusion of this research is that young children often eat more of fewer foods than adults, when intake is averaged and adjusted by body weight. Higher intake of some types of food, such as fruits, fruit juices, and some vegetables could in turn lead to a greater consumption of pesticide residues on those foods. Further, EPA's practice of grouping broad age groups of children together in their analyses may result in far lower food consumption and pesticide exposure estimates than if these age groups were more narrowly defined. Full exposure distributions for annual age groups between birth and age five, and in three-year intervals to age eighteen, should be used when considering pesticide tolerances in food. In addition, much more attention should be paid to the dietary patterns of pregnant women and infants. Within all of these childhood age groupings, ethnic, regional, and seasonal differences should be examined as well to better understand their influence on patterns of childhood pesticide exposure in the diet.

Second, outdated food-intake surveys provide a very uncertain basis for estimating pesticide exposure, especially if the estimates are used to support tolerances that are likely to remain in effect for years, or even decades. Given the long delays between national surveys of food consumption and the historic tendency for sample sizes to diminish as budgets are cut, data used by EPA to estimate childhood patterns of food intake are likely to be a poor reflection of current dietary patterns. Because the accuracy of pesticide exposure estimates is directly related to the accuracy of food consumption data, this uncertainty should be managed by using an extra margin of safety when setting tolerances. Childhood food and water intake should be nationally sampled more frequently than it has been, at least at five-year intervals, and sample sizes should support efforts to estimate pesticide exposure. This frequency would probably capture significant changes in diets due to variables such as marketing, health advisories, and advances in food-processing technologies.

Third, EPA should pay special attention to the contamination of liquids. Children, especially infants, consume much more liquid per unit of body weight than do adults. This tendency means that children could be exposed in significant amounts to pesticides in drinking water—especially because tap water is normally added to processed liquids such as concentrated fruit juices and infant formula. Families living in rural areas where pesticides and fertilizers have been heavily applied, and where water is derived from shallow wells, should therefore have their water tested for the most commonly used compounds. In addition, institutions that manage community water supplies should be certain that residue tests are employed that can detect compounds historically used within watersheds surrounding the supply: many of the pesticides used in agriculture and allowed to remain as residues on food are not monitored in community drinking water supplies. Sampling designs should also be carefully structured to detect "pulses" of contamination that may follow periods of heavy pesticide application such as preemergent herbicide use during the spring, or when heavy rains follow droughts.[26]

Fourth, the accuracy, efficiency, and effectiveness of pesticide exposure estimates—and of the regulations that are based on them—have been compromised by the fact that authority to manage pesticides has been fractured among EPA, USDA, and FDA. Each agency has used a different food coding system designed for different purposes. USDA has collected data for nutrient analyses; FDA collected residue data to find violations of allowable residue levels; and EPA has used its own coding scheme to estimate food intake and associated pesticide exposures. Promises of improved interagency coordination and improved environmental monitoring following the release of the NAS study *Pesticides in the Diets of Infants and Children* have been largely unkept because budget shortfalls increasingly constrain research efforts. EPA bears responsibility because it has never admitted the scale of monitoring required to know and manage the mixture of pesticides it has allowed to be dispersed into the food and water supplies we all encounter daily.[27]

CHAPTER 10

Averaging Games

Simplification of Exposure and Risk

Since its founding in 1970, EPA has judged the safety of individual pesticides independent from one another. Moreover, EPA has lacked the legal authority to deny registrations or tolerances based on judgments that available substitutes pose lesser risks—a constraint rationalized by an uncertain understanding of the relative toxicity among possible substitutes.[1] The complexity of estimating exposure for even single pesticides further narrowed the agency's attention and resources toward single chemicals. The result has been a highly incremental, segmented form of regulation, and one that has often focused on single crop uses of single pesticides.[2]

The averaging of risk has played an exceptionally important role in the history of EPA's pesticide regulatory behavior. As EPA granted or adjusted pesticide tolerances during its first several decades, it assumed an average level of food intake across the entire population and an average level of food contamination by pesticide residues. The agency also projected that exposure occurred evenly across a seventy-year life span, which led to its average lifetime cancer risk estimates for the entire U.S. population. The analyses became more refined during the mid-1980s as the agency developed the capacity to examine age-related food intake and exposure patterns. But, as explained in chapter 9, even by 1995 the age groupings EPA examined were broadly defined—children ages one to six and ages seven to twelve—and these groupings were not related to shorter and earlier periods of physiological vulnerability such as those described in chapter 8.

By 1992, EPA acknowledged that childhood exposure and risk estimates had never caused tolerances to be adjusted or registrations to be revoked.[3] By neglecting to consider variations in diet and pesticide residues across age groups, EPA had little understanding of whether the thousands of tolerances it oversees afforded protection for children. In fact, the legal standard for tolerance-setting—which permitted them to balance risks against benefits—did not require that a limit guarantee protection.[4]

This chapter explores the effects that averaging methods may have on the ways that risks are characterized. As this discussion reveals, many forms of averaging have been used by the agency, each of which has effectively lowered estimates of risk. Averaging together food consumption and detected residue statistics trivializes most risks. Further, these averages cast public attention away from minorities bearing significant risks.

An alternative method of exposure and risk estimation that does account for variance in the distribution of risk is presented here, using the widely used fungicide benomyl as an example. This approach considers the distribution of exposure and risk among children in yearly age classes and provides a necessary introduction to the problem of managing pesticide mixtures, which follows in chapter 11.

Benomyl: A Case Study

Benomyl, or Benlate, is a systemic fungicide manufactured by E. I. du Pont Corporation. Since its introduction in 1972, it has become one of the most widely used fungicides within a chemical family known as benzimidazoles. Benomyl is effective in preventing more than 190 different fungal diseases. It acts as a protective surface barrier and penetrates plant tissues to arrest infections. Benomyl is applied as a seed treatment, transplant dip, and foliar spray, and it is registered for use on more than seventy crops in fifty countries, including imported foods such as bananas and pineapples. In the United States, benomyl is registered for use on a wide variety of crops, and nearly seventy-five food and feed tolerances for it exist.

Benomyl captured considerable public attention during the early 1990s because it was one of several compounds that EPA proposed to remove from the market due to its suspected cancer-causing effects. Because the compound concentrates in some foods as they are processed, it is also prohibited from use on these crops by the Delaney clause of FFDCA.[5] Also during this period du Pont was the target of hundreds of lawsuits by farmers and other plant growers who claimed that a reformulation of the fungicide was responsible for widespread plant death and damage. The company has paid more than $500 million in compensation based upon these claims.[6]

EPA based its judgment on benomyl on toxicological evidence that suggests that benomyl is both mutagenic and carcinogenic in laboratory animals. EPA concluded that hepatocellular carcinomas or combined hepatocellular neoplasms occurred in male and female mice at all doses.[7] Tests including a metabolite of benomyl, carbendazim,[8] caused hepatocellular tumors in male mice and hepatocellular adenomas (benign tumors), carcinomas, and combined hepatocellular neoplasms in female mice.[9] The manufacturer contended, however, that increased incidences of liver cancer in mice are confused by a high rate of spontaneous tumor development in both the control group and those exposed. It also points to the absence of a carcinogenic response in two strains of rat and in one strain of mouse, as well as the absence of any indication of a tumor-inducing effect in humans.

Several studies have explored the effects of benomyl on the fetal development of laboratory animals. When fetal rats were treated with benomyl in one study, many craniocerebral anomalies were reported, including hydrocephalus and the growth of cell masses that overgrew and sometimes obliterated subcortical structures of the brain.[10] High doses (125 mg/kg of maternal body weight) produced late fetal death. Other observed systemic malformations included cleft palate and misshapen tails. Finally, benomyl has been linked to an inhibition of the formation of microtubules in the brain that are important to normal early brain development in rats.[11]

There is also evidence that benomyl damages male rats' reproductive organs. Rats exposed over seventy days in another study were claimed to experience a dose-dependent decrease in testicular weight, depressed sperm counts, and lower fertility.[12] By 1984 there was some evidence that these effects may be age-related, because animals treated prior to puberty showed no reproductive organ effects, whereas those treated during puberty experienced at least one of the following: decreased testicular or epididymal weight, decreased epididymal sperm counts, or testicular lesions.[13]

In 1989 carbendazim was found in still another study to have hormonal effects. As exposure to MBC increased, more follicle stimulating hormone (FSH) and pituitary luteinizing hormone (LH) were produced by the body. And following subchronic exposure to benomyl, gonads changed by the contaminant appear to upset hormonal balances by acting on the central nervous system.[14] Reduced fertility may have been caused by testicular swelling; occluded, or blocked, ducts; duct or tubular atrophy; or tissue reduction. These conditions generally appeared to be dose-dependent.[15] In another study, benomyl appeared to cause changes in the chromosomes of somatic cells, and when tested in mice the compound induced chromosome changes in oocytes—immature female reproductive cells.[16]

Estimating Childhood Exposure

Methods of estimating exposure and associated risks have long been confused by the absence of current, high-quality data and by the dispute over appropriate ways of combining food, residue, and toxicity data. Given these limitations, EPA has relied upon computer models that project the magnitude of exposure, its distribution in the human population, and its toxicological significance. The choice of data sets for food intake, residue, chemical use, and toxicity can thus have dramatic effects on exposure and risk estimates. Similarly, the selection of a method for combining and presenting these data may have broad implications for exposure and risk forecasts.

Benomyl provides a good case with which to explore these problems, because dietary exposure may come from as many as seventy-five different food sources, each of which is permitted by tolerance to contain benomyl residues in the marketplace.[17] The method presented below considers the exposure of children ages one through five to benomyl on numerous foods that they reported eating. Intake values for each child are combined with average residue values for each food to create an exposure estimate for each child. Rather than assigning some average level of exposure to a broadly defined age group, it identifies the probable exposure of each child.

As demonstrated in the previous chapter, children consume more of fewer foods than adults, especially during the first few years of life, and particularly fruits, juices, and some vegetables.[18] For any given estimate of pesticide residue levels, this means that differences in exposure among individuals will be driven exclusively by differences in what is eaten. Yet the problem is not this simple; the risk assessor must choose among the different residue data sets that commonly exist for each pesticide-food combination. In many cases, no residue data exist for important foods such as juices or milk, and the analyst is forced to make many assumptions.

This exposure analysis includes far fewer foods than the total number permitted by law to contain benomyl residues. The reason for this simplified approach is that residue data available from FDA's residue sampling program are often of insufficient quality to estimate exposure.[19] FDA does not test all foods that have tolerances, and when it does test, the number of samples is often far fewer than necessary to estimate national exposure. Benomyl posed an additional problem in that it required its own special residue detection test; it was not picked up by common multi-residue scans.[20] Compounds that require extraordinarily expensive individualized tests are far less likely to be tested by FDA than compounds detectable using less costly multi-residue scans. In defense of FDA, they perceive their mission to be the policing of EPA tolerances in the marketplace, not natural exposure analysis. For this reason, along with budgetary constraints, FDA has not restructured its sampling designs to support the types of exposure analyses suggested in this book.

Those foods chosen for the exposure analyses had FDA residue data of sufficient quality existed and are normally eaten by children. These foods include apples, apple juice, bananas, cherries, grapes, oranges, orange juice, peaches, pears, pineapples, plums, strawberries, and tomatoes.[21]

The Danger of Averages

Exposure to a pesticide residue is normally computed by multiplying the amount of food intake by the expected concentration of the contaminant. For example, to estimate the national average exposure to benomyl on apples, one would simply estimate the average intake of apples (0.45 grams/kilogram of body weight/day) and then calculate the average residue level on those apples (0.16 ppm). These two figures can then be multiplied to produce an average exposure estimate from a single chemical and a single food. This simplified approach is the one used by EPA for the past two decades to estimate average dietary exposure to carcinogens. The product, however, captures none of the variance in either food intake or residue concentration. As demonstrated in chapter 9, for example, one-year-olds may consume on average as much as fifteen to twenty times more apple juice than do adults.

The problem is far more complex than that just described, because we are really interested in estimating exposure to benomyl residues from all potential food sources. In this case, mean intake of each food may be multiplied by mean detected residue levels for each food; the sum of the resulting exposures for each food will give an average exposure to benomyl for all foods. This method, however, still does not capture any of the differences in either the food-intake or residue data.

To account for variance in dietary patterns and levels of contamination, distributions can be summarized by simply ranking all values between lowest and highest. Given any distribution, the ninetieth-percentile value is exceeded by only 10 percent of the sample; a 99 percent percentile value is exceeded by only 1 percent of the sample. Food intake and residue data have been combined in different ways to produce exposure estimates by EPA, depending on the type of toxic effect which is being explored. To estimate exposure to a carcinogen, for example, EPA commonly chose to average food intake values and multiply them by average residue values for each food. The resulting exposure estimates are then summed across foods as just described.

An even more conservative approach would be to combine ninetieth-percentile food-intake levels with ninetieth-percentile residue levels. One would initially suspect this method to produce even higher exposure estimates than those produced by multiplying both mean food intake and mean food residue values, and that the product would significantly exceed normal exposure. This approach might be appropriate if the quality of residue or food intake data has been diminished by sampling problems. For example, confidence in the FDA residue data is diminished both by the small sample sizes of residue tests and by the purpose for the data's collection. If FDA is legally responsible only for determining if a tolerance has been exceeded, it has no incentive to report levels beneath the tolerance. The FDA data sets contain many "nondetections" that therefore demand careful interpretation.[22]

Summary statistics of food intake and residue levels may be combined in different ways to produce exposure estimates. Mean food intake for children between ages one and six are combined with mean and ninetieth-percentile residue values to produce the exposure values presented in figure 10–1. A

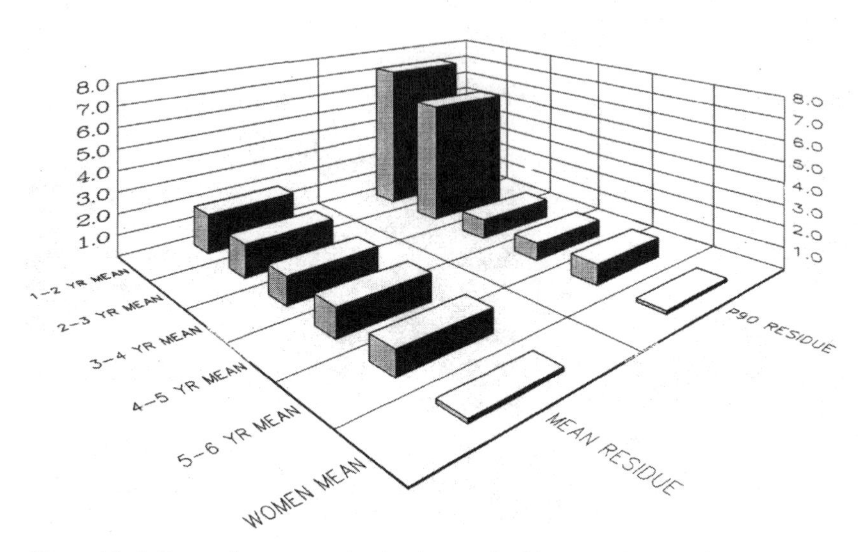

Figure 10–1. Benomyl exposure: the simple case. In this case only mean food intake is combined with mean and ninetieth percentile residue values to produce exposure estimates. Residue values are derived from FDA residue data sets from 1988 and 1989. Food intake data are derived from the USDA CSFII, 1985–86.

modest decline in exposure is seen as children reach age six, whereas women's exposure is significantly less, almost by a factor of ten.

The effect of using ninetieth-percentile food-intake levels to estimate exposure is demonstrated in figure 10–2, which also contains the data presented in figure 10–1. If we assume that the probabilities of eating more than the ninetieth-percentile level for each food considered is a physical impossibility—simply because no one can eat that much food—and that the probability of benomyl contaminating *each* of these foods at the ninetieth-percentile level is very close to zero, these estimates might be thought of as extremely conservative.

Surprisingly, the combination of ninetieth-percentile residue levels with ninetieth-percentile food intake values may produce results, not anticipated by the earlier figures. For some single foods consumed by three-, four-, and five-year-olds, for example, ninetieth-percentile exposure levels are *zero,* lower than the combination of mean residue with mean food intake. This phenomenon occurs in cases where a high proportion of residue testing results are reported as zeros. If more than 90 percent but less than 100 per-

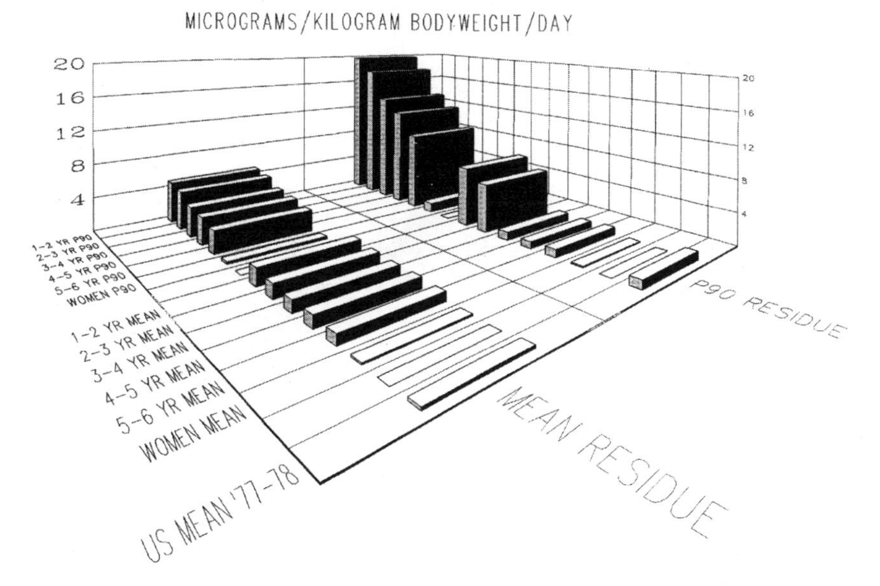

Figure 10–2. Benomyl exposure scenarios, including ninetieth percentile residue levels. These exposure estimates include those contained in figure 10–1. In addition, ninetieth percentile food intake values were used to estimate exposure.

cent of the tests are reported as zeros, then mean residue levels will be higher than ninetieth-percentile residue levels. The same effect may also be common for foods that are rarely consumed. The figures presented above do not capture this possibility, because children consume moderate or large quantities of most of the foods analyzed. Thus the use of summary exposure statistics such as ninetieth-percentile values to represent the upper limit of exposure levels may offer a false sense of security.

Tolerance levels for residues are then used to calculate the exposure allowed by regulation, presented in figure 10–3. Whereas tolerance-level contamination is rarely detected in any residue monitoring surveys, this figure demonstrates that exposures at the legal limits are ten to one hundred times higher than what is predicted using detected FDA residue data. It is important to emphasize at this point that all of the exposure analyses presented so far provide no information concerning the magnitude of risk posed, because they remain divorced from any toxicological data.

A Richer Perspective: Exposure Distributions

An alternative to combining summary residue and food intake statistics is to use detailed data about individuals' diets. This method permits exposure to be estimated for each individual and the expression of age-class exposures

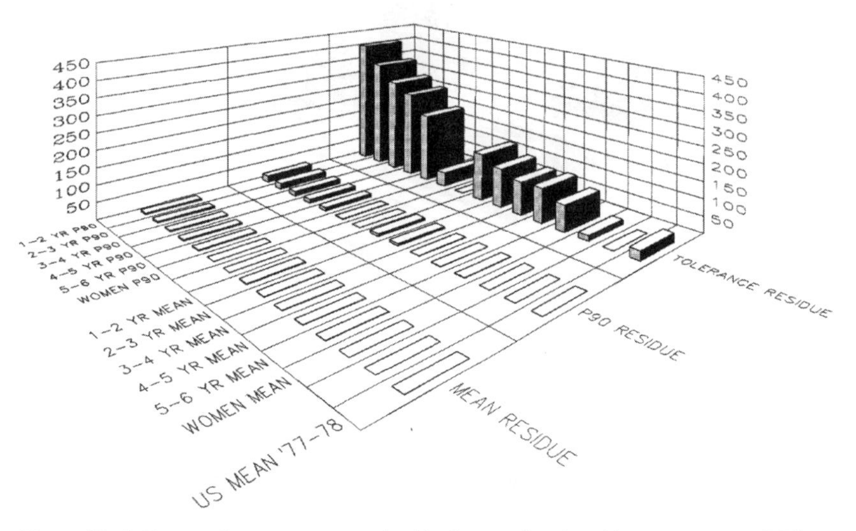

Figure 10–3. Benomyl exposure scenarios if tolerance-level residues are assumed. The exposure estimates of figures 10–1 and 10–2 are dwarfed by the estimates produced if legally allowable, or tolerance-level, residues are assumed. Although the Delaney Paradox study was severely criticized for estimating risk by assuming that residues remained on foods at the tolerance level, it seems reasonable to ask whether tolerances protect children and others from exposures that pose significant health risks.

to be presented as a distribution.[23] I prepared this analysis for each of five age groups (each of which spanned one year). The results were then plotted as the distribution presented in figure 10–4.[24]

This method fully captures the differences in eating habits among people, although it neglects the variance within individuals that occurs.[25] In addition, because this distribution was prepared for a compound that EPA believed poses some cancer risk, residue data were summarized as mean values.[26] Representing exposures in this manner facilitates rapid identification of the sample proportion estimated to fall above any known toxicologically significant dose. The exposure scale may be adjusted by a potency factor and may be time-weighted to estimate the cancer risk accumulated during the year of exposure.[27]

Interpretations So Far

It is clear that summary statistics can provide an incomplete and potentially deceptive representation of pesticide exposure patterns. However, each alternative presented above is valuable for different reasons. The combination of

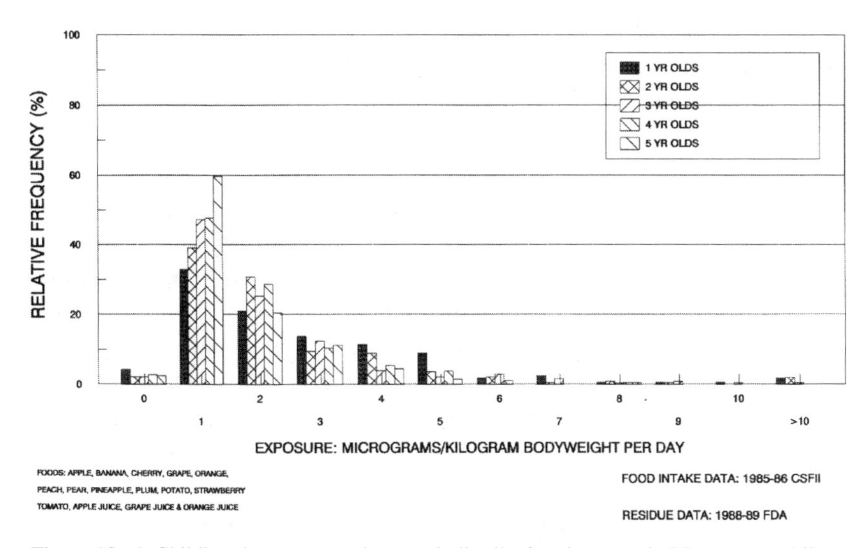

Figure 10–4. Childhood exposure to benomyl: distributional approach. Many more children were estimated to be exposed to benomyl than not. Less than 5 percent of children ages one to five received no benomyl exposure, according to this very conservative estimate.

summary statistics provides a fairly efficient method for detecting rough magnitudes of exposure and risk. One can clearly determine if residues at the tolerance level pose health risks, and depending upon the quality of data, combining food-intake data and pesticide residue information in summary statistics provides a general outline of likely exposure distributions. Yet these statistics can be deceptive; both the food intake and residue distributions tend to be highly skewed, creating a condition in which the ninetieth-percentile is often either less than the mean value or zero.

The population distribution method, by contrast, provides a more complete picture of possible exposures, particularly when the groups studied are delineated by year of age. One can quickly see if thresholds of acceptable exposure have been breached, and exposures can be translated into annualized risk estimates or proportions of the acceptable daily intake. But one should not conclude that the exposure estimates presented in this case are accurate. Use of other residue data sets, including field trial and market-basket data, can yield significantly different results. Also, considerable uncertainty surrounds the estimates of residues in processed foods such as juices, which are consumed in great amounts by small children. The sensitivity of exposure and risk estimates to varying assumptions regarding the magnitude of residues is tested in the following section, where I again rely on the distributional method just described.

Residue "Monkey Business"

Given the problems associated with available pesticide residue data described here and in chapter 4, I have chosen to compare six different scenarios of childhood exposure. In all six I use the same food intake data, but then I explore the effects of incorporating six different residue data sets. These residue data include (1) du Pont field trial results, (2) the results from a du Pont Corporation market-basket survey, (3) FDA compliance and surveillance data for the years 1988 and 1989, (4) National Food Processors' Association (NFPA) data provided by corporate members, (5) data collected in raw food testing by a private residue-testing laboratory, and (6) exposures legally permitted by tolerances in place as of 1993. The sampling designs vary significantly among these types of residue data, as do the sample sizes and analytical methods. Sample sizes and the number of positive detections are compared in table 10–1.

The primary purpose of the comparisons is to analyze the differences in exposure estimates that result from the use of alternative residue data sets. The analyses that follow test the hypothesis that exposure estimates—along with associated risk estimates—are sensitive to the source of residue data. Each of the residue data sets is described briefly below.

Food Industry Residue Levels

The National Food Processors' Association routinely collects data from its member organizations. These data include the sample, chemical, food, detected residue if any, and the limit of detection. The suitability of this information for national exposure and risk assessment, however, is questionable for several important reasons. First, the data were not collected according to any uniform sampling plan; instead, sampling strategies were defined differently by member corporations. Moreover, because the randomness of the survey methods is uncertain, the data's representativeness of residues likely to be found in the nation's food supply remains ambiguous. The only cases of positive benomyl detection used in the analysis were values detected by NFPA in apples and apple juice. It is important to recognize that since this study was completed, NFPA has standardized its testing protocols, providing for more reliable data for the purpose of risk assessment.

Market-Basket Residue Levels

Du Pont conducted a market-basket survey to examine residue levels in the marketplace. The sampling design of a market-basket survey is particularly important because the results can be dramatically affected by regional pat-

Table 10–1. Residue Scenarios: Number of Samples (S) and Number of Positive Detections (D)

	FDA		DP FT		DP MB		NFPA		Private lab	
	S	D	S	D	S	D	S	D	S	D
Apple	134	35	138	122	26	5	67	25	127	65
Apple juice	—	—	—	—	—	—	29	16	—	—
Apricot	—	—	—	—	—	—	6	0	19	5
Banana	72	8	—	—	—	—	4	0	—	—
Bean	5	0	35	29	30	3	19	0	38	10
Blueberry	—	—	—	—	—	—	—	—	14	3
Carrot	—	—	—	—	—	—	—	—	12	1
Celery	—	—	—	—	—	—	—	—	24	4
Cherry	21	5	—	—	—	—	7	0	4	1
Cucumber	—	—	—	—	—	—	6	0	4	1
Grape	27	12	71	65	11	4	—	—	11	5
Nectarine	14	8	18	5	—	—	—	—	39	14
Orange	6	0	18	13	12	12	1	0	6	1
Orange juice	1	0	—	—	—	—	2	0	—	—
Peach	26	13	82	72	15	1	15	0	81	44
Pear	23	1	15	14	24	6	—	—	—	—
Pineapple	25	18	—	—	—	—	1	0	7	4
Plum	21	10	—	—	—	—	—	—	28	18
Raisin	—	—	—	—	—	—	13	0	—	—
Raspberry	14	0	—	—	—	—	2	0	17	6
Rice	—	—	—	—	—	—	6	0	—	—
Squash	4	0	—	—	—	—	4	1	—	—
Strawberry	30	2	—	—	—	—	6	0	16	11
Tomato	20	0	35	23	25	1	—	—	—	—
Watermelon	5	0	—	—	—	—	3	0	—	—
Wheat	—	—	—	—	—	—	12	0	—	—

Key: FDA = Food and Drug Administration; DP FT = Du Pont Field Trial; DP MB = Du Pont Market Basket; NFPA = National Food Processors Association

terns of chemical use and food distribution. A properly designed market-basket survey is very valuable because it obviates the need to make complex assumptions regarding the effects on residues of food processing, as well as regional variations in pesticide use and food distribution.

Among seven foods analyzed, a total of 143 samples were tested; of these, thirty-two (22 percent) had residue levels above the limit of quantitation. Juices were not sampled. This low percentage is similar to that found by the NFPA, but approximately 50 percent lower than the percentage positive found by the more focused sampling design of the private testing laboratory.

FDA Surveillance Residue Levels

Only FDA surveillance sampling results were used in this analysis, and these data were used only if sample sizes exceeded twenty for any individual food. These small sample sizes make any attempt to estimate exposure highly suspect; this problem also often confronts EPA when it attempts to judge pesticide risks. Whereas benomyl is registered for use on more than seventy foods, FDA residue data—pooled for 1988 and 1989—had sample sizes that exceeded twenty for only ten foods listed in table 10–1. Among the twenty-six foods, 448 samples were tested, and 112 of these (25 percent) had residues reported above the detection limit.

These data demonstrate that FDA focused its scarce monitoring resources on fresh rather than processed foods. For example, apple juice was not sampled, and orange juice was sampled only once. This observation is particularly important because these two juices are among the foods most consumed by young children. These findings also raise an important question about the reasonableness of testing only a few samples of single foods—for example, five of beans, six of oranges, and four of squash. Although there may be a rationale for focusing monitoring resources on suspected violators, the usefulness of FDA's surveillance data for developing national estimates of pesticide exposure from food is limited at best.

The disadvantages of these data for estimating exposure to pesticides include (1) extremely limited sample sizes for most foods for which pesticide tolerances have been established, (2) the absence of regional or seasonal stratification in sampling design, (3) the fact that residue concentrations are diminished by FDA's procedure of blending slices from different pieces of food, (4) the realization that detection limits in some cases may have been set only to detect tolerance violations, and (5) the fact that sampling of processed foods is extremely limited compared with raw agricultural commodities—a situation that forces complex assumptions regarding the effect of processing on residues for each food. The primary advantage of FDA data is that it is the only sampling scheme that is applied uniformly across the nation's food supply.

Residue Levels Provided by a Private Laboratory

A private testing laboratory certifies for some grocery manufacturers—and, in turn, grocers and consumers—that marketplace produce is free from detectable levels of certain pesticides. The lab tested 447 samples for benomyl residues (almost identical to FDA's 448 samples for 1988 and 1989). Among these, 193 (43 percent) had residues above the detection limit. By contrast, the NFPA tested 203 samples and found only 42 cases (20.6 percent) where residues exceeded the limit of quantitation. The distinction in findings may have several explanations. First, the private lab focuses its sampling on produce and sources of produce where a compound of concern has been detected. Benomyl's status as a B2 (probable human) carcinogen has caused them to test extensively for the compound. Second, the private lab tested a different but overlapping set of foods than did NFPA. Residues were detected most frequently on apples, peaches, and plums. The highest percentage of residue detections above the limit of quantitation in NFPA sampling was found on apples (37 percent) and apple juice (55 percent). The private lab did not test any apple juice, which raises questions regarding the appropriateness of using any single residue data set for estimating national levels of pesticide exposure.

Residue Levels Detected in Field Trials

Du Pont Corporation submitted a substantial body of residue data in 1989 to EPA to support continued registration of benomyl. These data are extremely important because the application rate and detected residue level for each sample of raw agricultural commodity were carefully recorded and sample sizes for single raw commodities were often large enough to permit statistical analyses. Unfortunately, residue tests were much less frequently conducted for processed foods made from benomyl-treated raw foods; this absence of information has forced complex assumptions to be made regarding residue fate. Another limitation of field trial data is that detected levels are likely to be far higher than those found in market-basket surveys due to uneven use of the compound around the country. Du Pont tested a total of 412 samples. Of these, 343 (83 percent) contained residues above the detection limit. This discovery suggests that if treated crops are tested, residues will likely be detectable. Moreover, du Pont's analyses were conducted under the assumption that only ten foods are consumed.[28]

Residues at Tolerance Levels

Although benomyl is allowed by law to be used on more than seventy foods, exposure analyses were conducted assuming tolerance level residues on only fifteen foods in order to permit reasonable comparison with the residue data sets described above. Tolerance-level residues permit exposures roughly one to two orders of magnitude above those residue levels detected using these other residue data sources.[29]

Interpretations

Exposure estimates for children based upon the six different residue scenarios are presented in figure 10–5.[30] Variation in the shape of the distributions demonstrates the importance of using judgment when selecting residue data to estimate exposure. Several conclusions may be drawn from the analyses. First, the level of childhood exposure varies widely depending upon the residue data set chosen for analysis. In this case, field trial data yields the highest exposure estimates, followed by data from private testing labs, FDA data, market-basket data, and finally, NFPA data. The comparison across exposure estimates is made difficult by the fact that there is some variation in the foods that were included in the scenarios (as determined by the availability of adequate sample sizes). As mentioned earlier, this problem routinely confronts EPA as well.

Second, the interpretation of "nondetections" is most important when they comprise the majority of any residue data set, as they do in the case of the FDA and NFPA data. The influence of nondetection interpretations also depends on the level of consumption of the food in question, as well as on the potency of the chemical of concern.[31] Third, assumptions regarding food-processing effects have a dramatic influence on exposure estimates for children—especially when one considers that children consume large amounts of liquids such as apple juice, orange juice, grape juice, and milk. EPA's requests for data on residues in processed foods should thus target foods consumed most by children, including processed juices, which are increasingly supplied by corporations in tropical nations.

Fourth, EPA's use of "percent acreage treated data" to adjust exposure and risk estimates often reduces estimates by 80–90 percent. This approach is another form of exposure and risk averaging. It is extremely important that these data be explicitly excluded from the exposure estimates and examined openly and critically before their use. If employed, one must acknowledge that exposure and risk are averaged across the country, even though regional patterns of pesticide use and food distribution could cause local populations to be regularly exposed well above average levels. Fifth, any conclusion regarding which foods are "riskiest" may be falsely governed by the pool of foods used to conduct the risk assessment. For example, if only five foods from a market-basket survey are deemed to have sample sizes large enough for exposure and risk assessment purposes, whereas fifteen foods from a field-residue survey are considered sufficient, the relative ranking of riskiest foods will differ significantly between the two data sets.

Finally, and perhaps most importantly, childhood exposure to pesticides should be represented as distributions, calculated from individual food intake

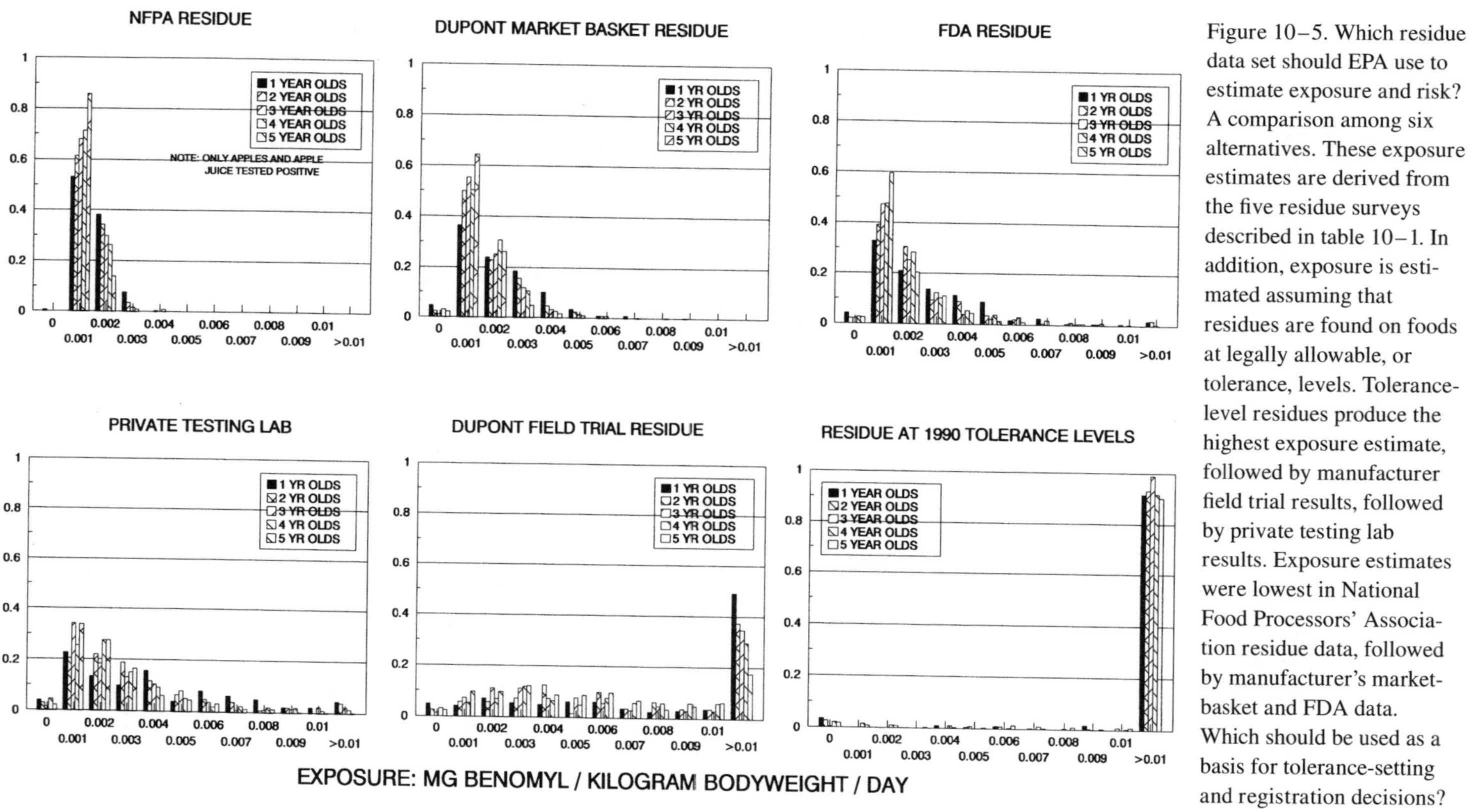

Figure 10–5. Which residue data set should EPA use to estimate exposure and risk? A comparison among six alternatives. These exposure estimates are derived from the five residue surveys described in table 10–1. In addition, exposure is estimated assuming that residues are found on foods at legally allowable, or tolerance, levels. Tolerance-level residues produce the highest exposure estimate, followed by manufacturer field trial results, followed by private testing lab results. Exposure estimates were lowest in National Food Processors' Association residue data, followed by manufacturer's market-basket and FDA data. Which should be used as a basis for tolerance-setting and registration decisions?

data: summary statistics can obscure important high levels of exposure. Although in the case presented above I employed individual food intake data, I combined it with average residue data to estimate exposure. In the following chapter, a method to estimate exposure to complex mixtures of pesticides in foods is suggested—one that relies on full distributions of both dietary and residue data and thereby provides the richest possible image of exposure and risk distribution.

CHAPTER 11

The Complex Mixture Problem

Since its inception, EPA has been overwhelmed by questions concerning the toxicity, exposure, and risks posed by single pesticides. This situation has prevented the agency from examining the distribution and effects of pesticide mixtures in the diet and other environments. By constraining questions to consideration of single pesticides, and by averaging their risks as demonstrated in the previous chapter, pesticide dangers often appeared trivial—and this conclusion has rationalized EPA's choices to grant tens of thousands of tolerances and registrations to manufacturers. If EPA instead asked how a single pesticide contributes to risks posed by other pesticides—or, better still, other types of toxins—we would build a system of knowledge that would permit the most significant threats to environmental health to be identified and managed.

This licensing behavior in the absence of understanding the collective risks posed by mixtures has created a risk assessment problem of extraordinary proportions. To effectively protect public health, the agency must know the likelihood that combinations of pesticides might appear as residues in the marketplace. Because a single pesticide may contribute to several types of toxic effects, forecasting the risks posed by complex mixtures of toxins in the diet is one of the most perplexing analytical problems facing the agency.

In this chapter, I summarize research on the health effects of a class of insecticides that have often damaged the human nervous system. I also describe a method to estimate childhood exposure to mixtures of organophosphate insecticides, which was reported in the 1993 NAS publication *Pesticides in the Diets of Infants and Children.* The method was developed by Richard Jackson of the U.S. Centers for Disease Control, Daniel Krewski of Health and Welfare Canada, and me.[1]

During the last several decades of the twentieth century, EPA permitted roughly fifty organophosphate insecticides to remain as residues in the food supply. Many of these chemicals have the common neurological effect of inhibiting enzymes known as cholinesterases (ChE's), which are necessary for the normal transfer of signals among nerve cells. If EPA regulates these insecticides individually, what is the collective risk they pose? The purpose of this chapter is to develop an approach to answering this question, rather than to provide a definitive estimate of the hazard posed by the specific mixture considered. By trying to answer the question, we learned a great deal about how impoverished EPA's understanding of risk has been as well as the magnitude of the research effort necessary to provide a reasonable knowledge base for effective risk management.

The Effects on Human Health

Organophosphate insecticides increasingly replaced organochlorine insecticides such as DDT and dieldrin during the 1960s and 1970s. These substitutions resulted in the production and distribution of not only new compounds, but also a very different mixture of health and ecological risks. The organochlorine insecticides were persistent, mobile, bioaccumulative, and induced widespread reproductive failure in wildlife, especially birds. Organophosphate insecticides degrade at varying rates; they are most persistent where the climate is dry and temperatures cool.[2] Moreover, although many organophosphates are fat soluble, they tend not to bioaccumulate. They are, however, acutely toxic to the central and peripheral nervous systems of humans, and they can cause irreversible damage and death.[3]

The potential for organophosphates to inhibit cholinesterase was first discovered by Willy Lange and Gerhard Shraden in 1932. Bladan was the first organophosphorus pesticide to be sold, in 1944.[4] The neurological effects of these compounds were intentional; many of the organophosphate insecticides are close relatives of nerve gases developed during World War II. It is important to note, too, that although the potency of the nerve gases was reduced one thousand to ten thousand times for agricultural applications, the organophosphate insecticides are still responsible for the majority of poisoning cases reported to the Centers for Disease Control and the World Health Organization.[5] In 1989, organophosphate and carbamate insecticides, either alone or in combination with other pesticides, caused nearly twenty thousand pesticide poisonings (as reported to the American Association of Poison Control Centers), with nearly 2.5 million farmworkers at highest risk.[6] The primary route of exposure in these cases was through the skin, but these insecticides may also be absorbed quickly through the lungs, eyes, or stomach.[7]

The quality of evidence linking pesticide exposure to acutely toxic effects such as severe cholinesterase depression is quite strong when compared to evidence linking exposure to delayed, chronic effects such as cancer. In the case of organic phosphorous compounds, the type, intensity, and duration of human illness are all well predicted by levels of acetylcholine. In general, accumulation of acetylcholine in the central nervous system due to ChE inhibition can also lead to anxiety, restlessness, insomnia, headache, emotional instability, nightmares, apathy, and confusion.[8] The victims' complaints are often non-specific, however, and may easily be mistaken for influenza, alcohol toxicity, or simply fatigue. Childhood exposure to anti-cholinesterase insecticides may lead to acute pancreatitis.[9] In the worst case, death can result from depression of central respiratory control reflexes or from paralysis of the respiratory muscles resulting in asphyxiation. Statistics on organophosphate and carbamate poisoning are therefore believed to significantly underrepresent actual occurrence due to diagnostic errors by both victims and clinicians.

Although bioaccumulation of these insecticides does not appear to occur either in food chains or in the human body, accumulation of the cholinesterase-inhibiting effect can occur, and it may lead to signs and symptoms of poisoning following small, repeated doses.[10] This type of accumulation is of special interest to me and others concerned with food contamination because it could result from repeated dietary exposure to complex mixtures of these insecticides.[11]

Some organophosphorus compounds may induce other toxic effects such as damage to DNA. The insecticide Imidan, for example, was found to cause single-strand breaks in human DNA and to be mildly mutagenic.[12] There is also some suspicion of the potential for organophosphates to induce non-Hodgkins lymphoma.[13] One well-established chronic effect from exposure is organophosphate-induced delayed peripheral neuropathy (OPIDN).[14] Neuropathy is a general term for functional disturbances caused by lesions in the central nervous system. More than forty thousand cases of OPIDN in humans have been documented. Improvement or recovery follows most mild cases, but long-lasting neurological dysfunction from spinal cord damage results from severe exposures.[15]

A prolonged type of organophosphate toxicity is known as intermediate syndrome. This condition appears to be caused by prolonged exposure, and it results in persistent cholinesterase inhibition. Clinical symptoms include incomplete respiratory paralysis, weakness in several territories of cranial motor nerves, and depressed tendon reflexes. The duration of the syndrome is variable; it lasts at least days or weeks and may persist for many months.[16]

Long-term neurobehavioral changes have also been confirmed by recent animal and human studies.[17] Polyneuropathy—disease involving several nerves—and abnormal EEG records were found in 1990 in 22.9 percent of workers employed at an organophosphate production facility.[18] In a study of 187 patients diagnosed with organophosphate insecticide poisoning between 1981 and 1986, 10 percent developed delayed neurologic effects;[19] some neurobehavioral and psychiatric effects were also found in adult humans.[20] In another study, aggressive and violent behavior was believed to be caused by organophosphate exposure.[21] In general, these neurobehavioral effects appear to be pesticide-specific, rather than characteristic of all organophosphorus pesticides.

A Closer Look at Cholinesterase Inhibition

Compared with many types of delayed, toxic effects such as cancer, toxicologists have a reasonably good understanding of the biochemical mechanisms that produce the acute effects described above. In healthy humans, normal neurological function is maintained by the body's careful regulation of acetylcholine. Acetylcholine is a principal neurotransmitter found in neurons—or nerve cells—located at neuromuscular junctions, at synaptic junctions of the autonomic nervous system, and throughout the central nervous system. Importantly, high concentrations of acetylcholine are normally found in areas of the brain that are responsible for cognitive functions such as learning and memory. Excess acetylcholine is controlled by a family of enzymes known as cholinesterases, which break the neurotransmitter into choline and acetic acid. This process prevents nerves from constantly firing.

Organophosphorus and carbamate insecticides upset the regulatory process just described. They bind with and inactivate cholinesterases, an act that in turn allows the neurotransmitter acetylcholine to remain or increase in concentration. As mentioned, excess acetylcholine hyperstimulates neuronal receptors in both the central and peripheral nervous system. Organophosphate insecticides irreversibly inactivate the enzyme. Carbamate insecticides similarly depress cholinesterase, but this effect is generally reversible at low doses.[22]

Hyperstimulation of neural receptors following exposure to carbamates is generally of shorter duration than that experienced after contact with organophosphates. The process of recovering from carbamates occurs within hours of exposure, whereas the recovery mechanism from organophosphate-induced inhibition is complex and may last for weeks—or until new enzyme is synthesized and normal activity is thereby restored.[23] This may take some time because cholinesterase is replaced in the blood only through its synthesis in bone marrow. This rate of recovery is potentially important to estimate risk, because daily exposure (even exposure several times a day at different meals) to low doses of cholinesterase-inhibiting pesticides is probably common.

The Toxicity of Mixtures

Organophosphates and carbamates may also undergo conversion in the environment or in the human body to form metabolites of potentially greater toxicity than the parent compounds. Malathion and parathion, for example, degrade into maloxon and paroxon when exposed to heat and air, and both degradation products are much more toxic than the parent compounds. A case in point is the extensive maloxon poisoning occurred among malaria workers in Pakistan following their exposure to malathion that had been improperly stored.[24] One study has concluded that the combined effect of parathion and methyl parathion exposure is more likely to induce intermediate syndrome than parathion poisoning alone, despite the fact that malathion is a far weaker anti-cholinesterase agent than parathion. The authors added, however, that as with most mixtures the mechanisms of interaction between the two compounds remain uncertain.

Synergism among organophosphate compounds, such as that demonstrated among the insecticides malathion and EPN, may be an important toxicological variable when considering exposure to numerous compounds.[25] In a study of dose additivity, synergism and antagonism in various combinations of thirteen organophosphorus insecticides were tested:[26] Toxicity was dose-additive in twenty-one pairs of compounds and less than additive in eighteen pairs; and a synergistic effect occurred in four cases.[27] Synergism appears to depend upon many factors. One of these is the rate of metabolic detoxification, which may be regulated by exposure to other compounds. The NAS Panel on Complex Mixtures offered this general conclusion: "One might predict that synergism will occur only when the dosage exceeds the threshold where metabolism becomes a rate-limiting factor in toxicity. Of course, that dosage becomes smaller as critical pathways of detoxification are inhibited by other compounds."[28]

It is possible as well that both potent and weak inhibitors compete for the same active catalytic sites on acetylcholinesterase molecules. If the weak inhibitor occupies these sites at the expense of the strong inhibitor, exposure to both compounds will produce an antagonistic effect rather than an additive one.[29]

Defining Adverse Effects

There is little controversy regarding the accuracy of acetylcholinesterase as a biological marker of exposure to anti-cholinesterase compounds such as organophosphate insecticides.[30] This relation is well established primarily through laboratory analyses of blood samples from farmworkers,[31] pest-control applicators,[32] greenhouse workers,[33] pesticide manufacturers,[34] household users who have misapplied compounds,[35] and other individuals accidentally or purposely poisoned.[36]

"Neurotoxicity" was defined by EPA in 1988 as an adverse change in the structure or function of the nervous system following exposure to a chemical agent. Thus, considerable regulatory pressure exists to define that level of ChE depression which is adverse. The level of exposure that produces an adverse effect will often influence how EPA sets a regulatory ceiling for allowable exposure—and thereby allowable crop uses for the insecticides. Thus, the debate over what is an adverse effect and how to measure it has been energetic, especially during the early 1990s. Within EPA, scientists argue over the validity of measuring cholinesterase inhibition in peripheral tissues—that is, in plasma and red blood cells, as a surrogate for measuring inhibition in the central nervous system.[37] Despite these conflicts, there is already a substantial body of epidemiological and clinical evidence that damage to the central nervous system, particularly delayed neuropathy, is associated with moderate to high level of exposure to organophosphates.[38]

Variance in Susceptibility

Susceptibility to neurotoxicants appears to be related to the stage of brain development. Because in humans brain growth occurs most rapidly until nearly age six, early childhood exposure to neurotoxicants deserves careful consideration. Studies on lead, irradiation, fetal alcohol syndrome, and oxygen deprivation generally indicate a higher vulnerability of the developing brain. In 1993, the NAS Committee on Pesticides in the Diets of Infants and Children concluded: "The data strongly suggest that exposure to neurotoxic compounds at levels believed to be safe for adults could result in permanent loss of brain function if it occurred during the prenatal and early childhood period of brain development. This information is of particular relevance to dietary exposure to pesticides, since policies that established safe levels of exposures to neurotoxic pesticides for adults could not be assumed to adequately protect a child less than 4 years of age."[39]

Normal levels of cholinesterases seem to vary considerably both among and within individuals, and these differences affect susceptibility to the neurotoxic effects of organophosphate pesticides.[40] Nearly 3 percent of the population is genetically deficient in producing cholinesterase, which could make them more susceptible to organophosphate poisoning.[41] Also, pregnant women and women taking oral contraceptives had ChE levels approximately 12 percent lower than others tested.[42] ChE levels among pesticide sprayers in Egypt were significantly lower among the population that smoked than among non-smokers.[43] Thus a variety of genetic and environmental factors could influence background levels of cholinesterase and therefore susceptibility to anti-cholinesterase insecticides.

Animal tests also provide evidence suggesting age-related susceptibility.

For example, acephate, leptophos, and methidathion were more toxic to weanling rats than to adults.[44] In 1965, F. C. Lu and colleagues found significant differences in the toxic effects on newborn and adult experimental animals of a variety of pesticides. In 1963, J. Brodeur and K. Dubois determined that eleven of sixteen anti-cholinesterase insecticides were more toxic to the weanlings than the adult rats.[45] In a more recent study, C. Pope and T. Chakraborti administered one of three commonly used organophosphate pesticides—methyl parathion, parathion, and chlorpyrifos—to both neonatal and adult rats. They concluded that developing mammals are "markedly more sensitive" to acute toxicity from exposures to these insecticides.[46]

In an important 1990 study, M. Stanton and L. Spear evaluated data on adverse effects detected across species. They compared the responses to seven toxicants of rodents, non-human primates, and humans across several categories of neurobehavioral function. Although this type of comparison is extremely complex, they concluded that laboratory animals generally experience the same cognitive, motor, and sensory deficits as humans when they are exposed to the same compounds at high doses. Effects of low doses were more difficult to analyze, but similar effects were found across species when similar measures of outcome were employed.[47]

How Risky Are Complex Mixtures?

Perhaps the greatest challenge to experts in environmental health is to evaluate risks associated with complex mixtures of toxic substances. Human exposure to toxins normally occurs in mixtures. In any single day, we may encounter diesel exhaust, gasoline vapors, cigarette smoke, pesticides in the diet, and drinking-water contaminants. Toxicity data, when available, normally exist only for single compounds rather than for mixtures of environmental contaminants. Variance in the quality of data available among compounds further divorces risk estimates from reality.

This problem was first publicly addressed by EPA in 1986 when the agency attempted to estimate risks from the diverse forms of dioxins.[48] Because each form appeared to have a different level of cancer-causing potency, there was an obvious need to develop an indicator of relative potency, or toxicity equivalence. According to this method, the toxicity of any single compound is compared to the toxicity of a reference compound, normally that compound with the broadest consensus regarding level of toxicity. This ratio is known as a "toxicity equivalence factor" (TEF), and it has been used repeatedly by EPA to study mixtures containing dioxin and dibenzofuran congeners.[49] TEFs were also used to estimate carcinogenic risks from a mixture of radionuclides and to estimate risk equivalence among radionuclides and other hazardous substances.[50] Other methods used to judge the risks of complex mixtures have also been reported in the literature.[51]

For the analysis that follows, only those organophosphate insecticides that induce the same type of cholinesterase-inhibiting effect were chosen. It is important to estimate the relative toxicological potency of each compound as accurately as possible. To accomplish this, the "lowest observed adverse effect level" (LOAEL) for numerous ChE-inhibiting insecticides was researched. The chemicals finally chosen for analysis were acephate, chlorpyrifos, dimethoate, disulfoton, and ethion.[52] Chlorpyrifos was chosen to be the reference compound for estimating toxicity equivalence, because its effects on cholinesterase inhibition have been most thoroughly examined. Thus all other compounds were compared to it to estimate their relative potency.[53]

Several important assumptions lie at the heart of this analysis. First, it was assumed that exposures to numerous pesticides could be added across sources to estimate a cumulative dose. Thus neither synergism nor antagonism were assumed. Second, it was assumed that the adverse effect will persist only for a twenty-four-hour period, even though longer durations of action are possible. Thus it was assumed that ChE rebounds to a normal level prior to the next day's exposure. This simplification did not account for the possibility that exposure during day one may come during dinner, whereas exposure during day two may come at breakfast. In fact, there is little evidence to justify the assumption that ChE rebounds to normal levels at the end of each day.

Estimating Childhood Exposure from the Model

Information about food intake by two-year-old children was combined with residue data to estimate exposure. The choice of specific foods for analysis was made based on the availability of residue data and the amount of each food consumed by two-year-olds, and an attempt was made to include foods most highly consumed by young children.

Food consumption data were derived from the 1977–78 USDA National Food Consumption Survey.[54] These data were chosen due to the limitations of the 1987–88 National Food Consumption Survey[55] and the smaller sample sizes within one-year age groups of the 1985–86 Continuing Survey of Food Intakes of Individuals.[56] The data were provided by EPA and were still used by the agency in 1995 to estimate exposures for registration reviews. Residue data were obtained from FDA for the years 1988 and 1989.[57]

One of the key limiting factors for understanding the magnitude of public exposure to complex mixtures has been the absence of data demonstrating multiple residues on single food samples. FDA, for example, has not kept track of cases where this effect occurred. The existence of multiple residues might be predicted if one knew how often and where farmers used combinations of cholinesterase-inhibiting compounds. Their failure in this regard demonstrates how single-chemical recordkeeping can come to support an inadequate policy of single-chemical regulation.[58]

The residue data employed in this study were collected by FDA from around the United States—and predominantly from samples obtained through its warehouse surveillance program. Because these data are generated using a random sampling design (in contrast to the compliance program's sampling design, which strategically targets probable violators), each data set is presumed to accurately reflect the distribution of residues in the nation's food supply. The high proportion of zero residue values in each residue data set may reflect the percentage of the crop produced without the use of the compounds.[59]

Exposure was estimated using a computer simulation model that combined food intake and residue data for every child. An exposure estimate was calculated for each food consumed by each individual. Exposures for individual foods were then summed to estimate every child's total exposure to the mixture of organophosphate residues, all in chlorpyrifos equivalents. The model was designed as a Monte Carlo simulation—it pulled random residue values from the residue distributions available for each food. Thus five thousand separate exposure scenarios were developed for each of the 1,831 child-days, which produced a total of roughly 9 million exposure estimates. The figures that follow simply summarize these exposure estimates by demonstrating the percentage of the group falling at different levels of exposure.[60]

The sensitivity of these exposure estimates to assumptions regarding the transfer of residues from raw to processed foods was then tested. First, processed juices were excluded from our analysis because of the uncertainty of processing effects on residues. This exclusion could lead to an underestimate of exposure if residues did in fact transfer to juices, because apple juice and orange juice are so highly consumed by young children. The effect on exposure of assuming that residue levels in juices were the same as those found in raw fruits was then tested. Two other assumptions were also tested regarding the meaning of zeros reported in the FDA data set. First, it was assumed that all nondetections represented pesticide-free food. For the second analysis, it was assumed that the nondetections meant that residues were at the analytical detection limit.[61] Because FDA normally reports more than 90 percent of its samples as "non-detections," this interpretation could have a significant effect on exposure and risk estimates.

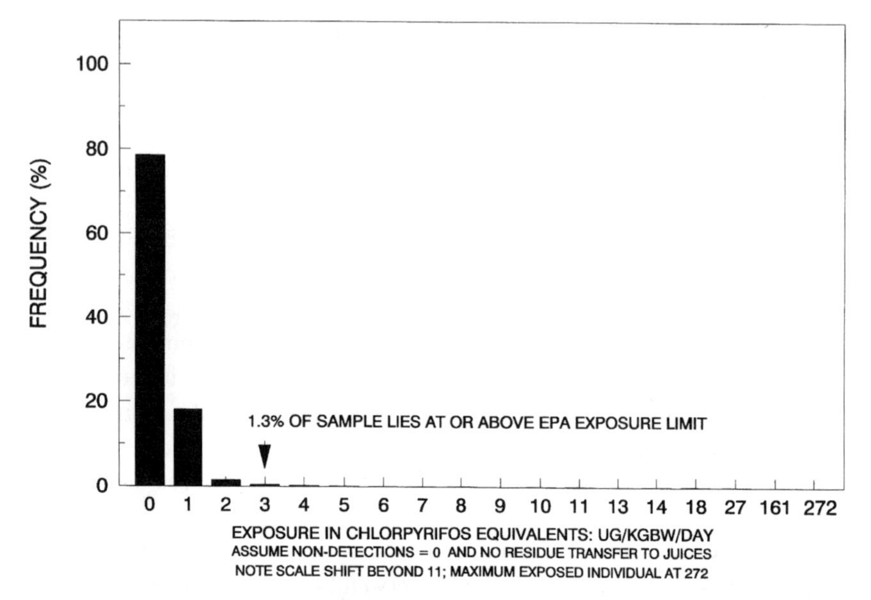

Figure 11–1. Two-year-olds' exposure to organophosphate pesticides: frequency distribution of simulated exposures. Using childhood food intake survey results and FDA detected residue data, only 1.3 percent of the simulated exposures for two-year-olds fell above EPA's acceptable daily intake level for chlorpyrifos. Nearly 80 percent of the sample fell at or below the limit of detection, assuming that all of the FDA reported non-detections were literally zero residue values and that no residues transfer from raw foods to processed juices.

Findings and Interpretations

The results appear in figure 11–1. Approximately 1.3 percent of the population of children fell at or above EPA's definition of the threshold of safe exposure. In this case the threshold lies at 3 micrograms of residue per kilogram of body weight per day. Here it was assumed that "nondetections" in the residue data really meant no residue remained on the food tested, and that no residues transferred from the raw fruits to the fruit juices.

The effect of assuming that residues transfer from raw fruits to juices is demonstrated in figure 11–2. The first case assumes that no pesticide residues transfer from raw fruits to juices; results are represented by dark bars. The second case assumes that 100 percent of residues transfer to juices, and these findings are represented by light bars. If it is assumed that residues transfer to juices, the sample percentage at or above EPA's threshold of safety increases from 1.3 percent to 4.1 percent. The inset to figure 11–2 presents the same data; the X-axis scale, however, begins at 3 μg/kgbw/day to highlight the distribution at higher exposure levels. In this simulation, the "maximum exposed individual" experienced 272

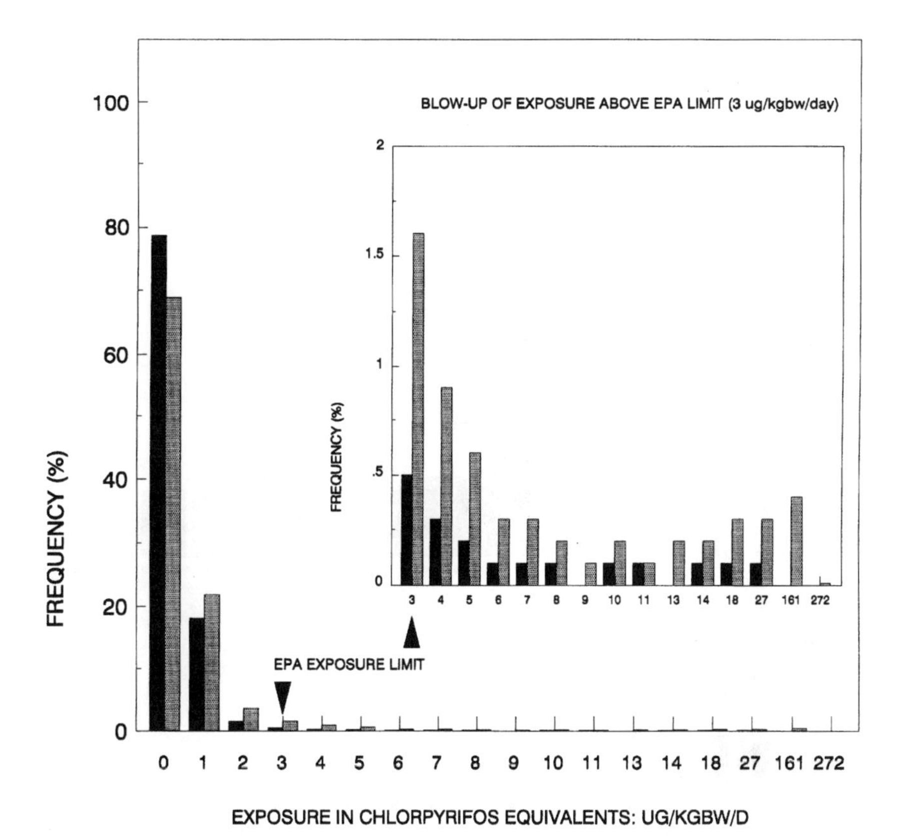

Figure 11–2. Sensitivity of exposure estimate to assumptions regarding residue transfer to juices. If it is assumed that 100 percent of the residues detected in raw foods transfer to juices, then 4.1 percent of the simulated exposures fell above EPA's acceptable daily intake for chlorpyrifos. The inset figure presents the same data, providing a clearer image of higher exposure levels by beginning the scale at 3 micrograms/kg bodyweight/day.

μg/kgbw/day of chlorpyrifos equivalents, nearly one hundred times above EPA's safe level of exposure.

Figure 11–3 compares the assumption that the nondetections are equal to zero with the premise that they are actually at the smallest detectable level (estimated by FDA to have been—on average—0.01 ppm). The only significant difference between the two distributions is the percentage of the sample falling either at zero exposure or within the second bin (defined as "greater than zero but less than or equal to 1 μg/kgbw/day"). The modest effect occurred under both scenarios depicting the transfer of residues from raw foods to juice. This type of analysis may be especially important if the threshold of safe exposure is quite low—that is, if the compound is highly

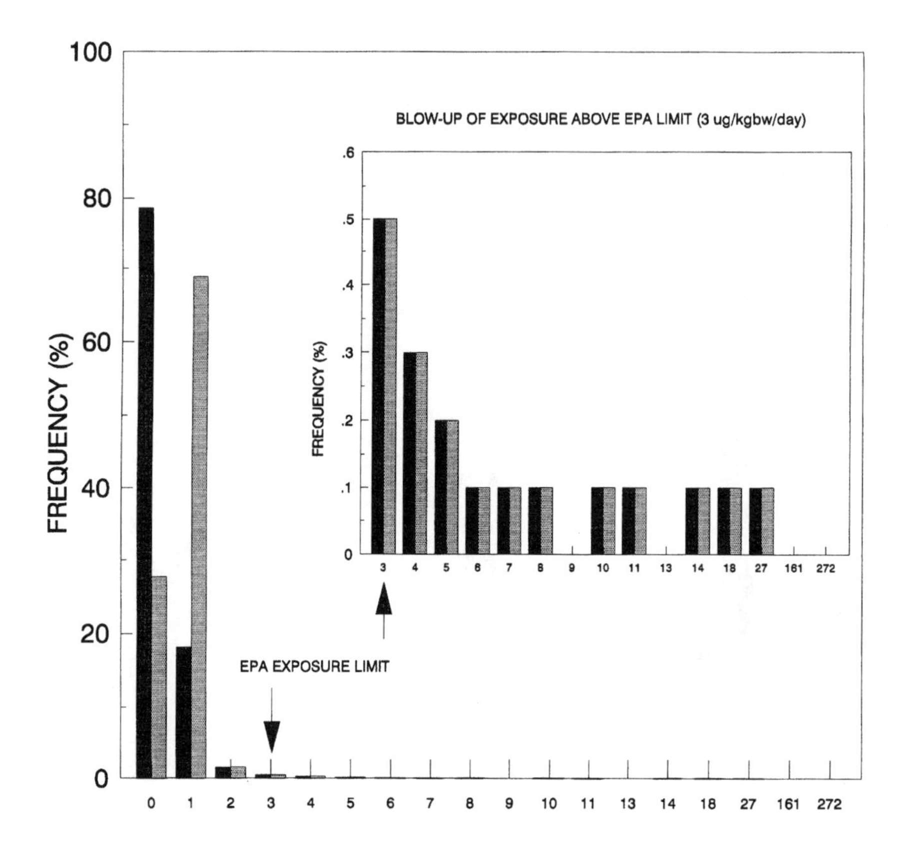

Figure 11–3. Interpreting the meaning of "non-detection." This final scenario tests the effects of the assumption that zero values in the data set could be at or just beneath the analytical limit of detection. Although a shift from the first to the second cluster resulted, the proportion of the sample exposed above the acceptable daily intake level did not change. Again, the inset figure demonstrates that the highest exposure levels fell at nearly one hundred times the EPA acceptable limit.

toxic—so that slight increases in exposure may place some individuals at risk.

Interpreting these findings demands caution. Should we extrapolate from these percentages to conclude that roughly between 1 and 5 percent of the nation's 4 million two-year-olds are at risk? This makes sense only if we are confident that the food intake data accurately reflect all two-year-olds' eating patterns; that the residue data represents the degree of actual contamination of the nation's food supply; and that the toxicity indicators are reasonable estimates of chemical potencies. The first assumption is suspect due to the age of the food intake data, but there is some evidence, presented in chapter 9, that other than an increase in fruit juice consumption children ate essentially the same foods in 1977 as did children of the same age in 1985. The second assumption—that FDA residue data is sufficient—is particularly suspect given the analyses presented in the previous chapter, which show that the sampling design chosen to collect residue data has a significant effect on the percentage of food tested found to contain residues, and the amount of contamination on each food. Finally, the belief that toxic potencies are well understood appears to be reasonably defensible—at least for these chemicals and for this type of neurological effect.

The threshold of safety, or "reference dose," was chosen by selecting the most sensitive and credible "no observed adverse effect levels" (NOAEL's) among animal toxicity studies. This NOAEL is then adjusted, or diminished by a safety factor (of normally one hundred), to arrive at the safety threshold. It is generally believed that this safety factor was designed to account for variance in susceptibility among species and among humans, but not to consider the special physiological susceptibility of children to neurological damage.[62]

At this point it is important to recall that these analyses were not conducted for all pesticide compounds known to be blood plasma cholinesterase inhibitors. Nor were all foods that these compounds are allowed to contaminate considered. Also, as mentioned before, neither previous- nor subsequent-day exposures were assumed; cholinesterase levels were presumed to have rebounded to "normal" levels at the beginning of each day. Variation in background cholinesterase levels within the population due both to the natural and environmental conditions described earlier were not included. And finally, nonfood exposures to cholinesterase-inhibiting compounds were not considered in this study, even though they would contribute to dietary exposure.[63] Given these assumptions, there may well be a significant percentage of our children who are regularly at considerable risk from pesticides that inhibit cholinesterase.

Carcinogens pose a similar type of problem in that cancer risk comes not only from numerous pesticides allowed to remain as food residues, but also from hazardous air pollutants, drinking water contaminants, and intentional food additives. If a pesticide is a carcinogen, the relevant question to ask is not only what cancer risk a single compound poses, but also how this individual chemical risk contributes to our total risk of developing cancer.

Pesticides are managed in the United States predominantly by controlling their concentration within various environmental media such as the air, water, food, or soil. Within-medium control has evolved incrementally through chemical-by-chemical reviews. This discussion revealed, however, that single-chemical and single-medium analyses are necessary building blocks for understanding exposures and risks from mixtures of pesticides. This case of cholinesterase inhibition suggests that government should demand the scientific evidence necessary to understand our exposure to these mixtures of toxins, especially those that appear to contribute to the same type of adverse health effect. It also suggests that a chemical-at-a-time and environmental medium-at-time risk assessment and regulatory process will continue to fail to protect the public health.

PART 4

Managing the Coevolution of Knowledge and Law

CHAPTER 12

Fractured Law, Fractured Science

Restating the Pesticide Problem

I began this book by suggesting that pesticides create a management dilemma: If they pose risks while providing benefits, how should these effects be managed? Although benefits have normally been portrayed as relatively certain and immediate, risks have been represented as uncertain, distant, and somehow manageable—primarily by foreseeing and preventing dangerous exposure.[1] The effect has been the registration of tens of thousands of pesticide products and the construction of an elaborate but ineffective tolerance system to manage human exposure in foods.

The early benefits of pesticides were enormous and indisputable. Following World War II, the gradual eradication of such insect-borne diseases as malaria and yellow fever in many temperate and subtropical parts of the world saved probably tens of millions of lives. Moreover, because young children are most susceptible to malaria, many of those saved were under age five. By the end of the war, pesticides were widely respected, if not revered, for their role in containing epidemics.

But these benefits are no longer provided by pesticides in most of the temperate world, where many insect-borne infectious diseases have been eradicated. In the United States, for example, pesticides are now used predominantly to ensure plentiful crops and beautiful lawns. Judging the net benefit of pesticide use for these purposes is difficult, because the costs of contaminating water, food, soil, and fields, as well as workplaces and other indoor environments, need to be accounted for. Pesticide use has thus often left behind a cascade of environmental monitoring, contamination, and health problems. Thousands of farmworkers are poisoned annually in the United States and hundreds of thousands are poisoned abroad.[2] Millions of residents of the midwestern United States now drink water contaminated with pesticide residues.[3] Our food supply is allowed by law to contain residues of nearly 325 different pesticides and roughly 1,500 "inert" ingredients at levels that were *not* set to protect public health, because government has always had the authority to weigh health risks against benefits when setting tolerances.[4]

Compounding this problem is the inadequacy of our systems for monitoring pesticides in the environment. In the United States alone, we spend billions of dollars annually to monitor residue levels in food and drinking water, with little assurance that the sampling designs are sufficient to control hazardous contamination levels.[5] Pesticide concentrations in indoor air remain completely unregulated, even though most homes and offices contain detectable levels of pesticides from paints, rugs, curtains, clothing, and the intentional spraying of floors or houseplants. Pesticide residues in schools, cafeterias, offices, and even swimming pools remain largely unmonitored and uncontrolled. Unfortunately, it seems all too clear that during the last half of the twentieth century the institutional structure for controlling the distribution of pesticides has remained insensitive to new information about the risks and benefits of these compounds.

USDA's decisions to license pesticides between 1945 and 1970 created complex and enormous management problems that EPA has never been able to fully comprehend, let alone control. EPA's dominant strategy has been to try to manage exposure. Interestingly, pesticide management poses some problems similar to those encountered in the pharmaceutical industry. The toxicity of both drugs and pesticides is controlled by regulating dosage. But pesticide exposure is far more difficult to manage than drug dosage. We can see drugs—count pills or measure tablespoons of elixirs—whereas pesticides are invisible. Moreover, the invisibility problem is compounded by the great diversity of ways we may be exposed to pesticides. Accurate prediction of exposure requires understanding how pesticides are used, how they move through the environment, and the ways that people encounter residues through air, water, food, or the skin. Given the thousands of pesticides licensed for use in the world, their invisibility, and the diversity of sources of exposure, we are unlikely to ever fully understand the intensity or implications of our encounters with them.[6]

In this chapter, I make three basic claims. First, I suggest that U.S. law governing pesticides is highly fractured by different statutes that regulate contamination of various environmental media such as food, water, and air. Legal attention is further narrowed within each medium to control individual pesticides and their specific uses. Second, I argue that the science demanded by these laws has been uncoordinated and has produced a narrow and incomplete understanding of how pesticides contaminate the environment and affect human health. Third, I contend that the entanglement of broken law and specialized but uncertain science has served to promote and sustain pesticide registration and use.

Environmental Law Is Fractured

Pesticides are regulated by a diverse set of statutes. That which has had more influence than any other in governing the dispersal of pesticides in the United

States is the Federal Insecticide, Fungicide, and Rodenticide Act (FIFRA).[7] FIFRA requires that pesticides be registered for specific uses and requires EPA to apply a risk-benefit balancing standard to judge pesticide suitability for registration. FIFRA can be thought of as a gatekeeping statute, in that the decision to register a compound creates an endless set of questions concerning residue fate and environmental health effects. Once released into the environment, control is attempted by setting standards for the maximum allowable contamination in different environmental media. The decision to register each compound has set off a broadscale scientific detective game that is structured by federal environmental law. If residues are found in food, water, soil, or the air, experts converge to judge the significance of possible effects.

After FIFRA, the most significant statute regulating pesticide use is the Federal Food, Drug, and Cosmetic Act (FFDCA), which requires that tolerances be set to govern maximum allowable pesticide residues in food. A risk-benefit balancing standard is applied to set tolerance levels in raw agricultural commodities, whereas a separate zero-risk standard is applied to carcinogenic pesticides if they concentrate during food processing. If pesticides are carcinogenic and they do not concentrate, maximum residue levels are set under the risk-benefit balancing standard. Approximately 97 percent of all food tolerances have been set for raw commodities pursuant to the risk-benefit balancing standard.

Congress passed the Safe Drinking Water Act (SDWA) in 1974, in response to a concern that drinking water had been contaminated by the underground injection and storage of toxic wastes and the widespread application of agricultural chemicals. This statute requires EPA to set maximum allowable contamination levels for a number of toxic substances. By 1994, pesticides and heavy metals constituted well over half of EPA's priority list of drinking-water contaminants.[8] The SDWA requires that EPA establish *maximum contaminant level goals,* which are set to ensure that no adverse effect on health results, and which provide for an "adequate margin of safety."[9] EPA must then set *maximum contaminant levels* as close to the applicable goal as is "feasible with the use of the best technology, treatment techniques and other means." Choice of treatment technology must consider costs and technology effectiveness under field conditions.[10] The statute was amended in 1986 over frustration at EPA's delay in setting standards under §1412. The law went so far as to prescribe the contaminant levels for eighty-three separate compounds, most of which were pesticides, organic solvents, or metals. EPA has set the maximum contaminant level for Class A and B carcinogens at zero, whereas Class C carcinogens have been allowed to persist at levels judged to present insignificant risks.[11]

Pesticides can also be regulated under the Clean Air Act as "hazardous air pollutants." The airborne transport of pesticides came as an unwelcome surprise to scientists, and exposure to these compounds through the air may pose health risks if the duration or intensity is prolonged. Exposure may come from several sources. Drifting sprays from phenoxy herbicides applied in the 1970s to national forests in the northwestern United States infuriated owners of private, adjacent lands, who called attention to the reproductive and possible cancer-inducing effects of 2,4,5-T and its dioxin byproducts. In addition, many pesticides are volatile, and as they evaporate into the atmosphere they may attach to water droplets or dust particles. DDT and its metabolites have bound to organic matter in the forest floor of New Hampshire mountains, deep within a national forest never sprayed.[12] DDT has also been found in Antarctic ice, penguins, and most species of whales. In addition, pesticides may attach to the water droplets in fog, and fog can often hover for weeks at a time during winter in the Central Valley of California. Toxaphene, chlordane, and heptachlor are examples of pesticides currently listed as "hazardous air pollutants" in §112 of the Clean Air Act, and all are banned for agricultural purposes.[13] It is still possible, however, that they may be released to the atmosphere as byproducts of manufacturing. Indoor air may contain pesticide residues from use of pesticide sprays, treated pets, paints, or fabrics. These indoor air-residue levels remain unregulated under the Clean Air Act.

The control over pesticides in water is also influenced by the Federal Water Pollution Control Act, which requires that any manufacturer or formulator of pesticides receive a permit prior to discharging effluent into any body of water.[14] Pesticides can be regulated as "toxic substances," which would require a special discharge license; however, they are generally classified as "non-point source" pollutants, which makes them fall under the jurisdiction of state and local governments. States, for example, are charged with protecting water from agricultural uses of pesticides and fertilizers.[15] But the more recent discovery of widespread contamination of groundwater aquifers and surface water from agricultural runoff has caused intense scrutiny by environmentalists and health advocates of how this responsibility should be delegated. The efforts of state and local governments have been further hampered by insufficient federal funding for forecasting how pesticides move from soil into water resources.

The Toxic Substance Control Act (TSCA) does not include any chemical substance "when manufactured, processed, or distributed in commerce for use as a pesticide."[16] TSCA, however, does apply to "inert ingredients" or other "intermediate substances" used to manufacture a pesticide.[17] Chemicals that are intended for use as pesticides fall under TSCA rather than

FIFRA until the producer proposes to use the compound for pest control.[18] Further, as many as 1,800 "inert ingredients" may be added to "active ingredients," and most of the inert compounds have not yet been evaluated by EPA for their health effects. Some chemicals that the agency permits as "inert" include compounds whose former classification as active ingredients has been canceled by the agency.[19] Other compounds permitted to be used as inert ingredients, and therefore likely to exist as residues on food, include substances listed as "hazardous" by the Resource Conservation and Recovery Act. These include potent toxins such as 1,1,1-tricloroethane, 1-butanol, 1,2-dichloropropane, and cyclohexane.[20]

The Resource Conservation and Recovery Act (RCRA) regulates the generation, treatment, storage, and disposal of hazardous wastes, including most pesticides.[21] Pesticides make up many of the regulated chemicals.[22] If any waste listed as hazardous is mixed with nonhazardous waste, then the entire mixture is classified as hazardous, and the mixture must be restored to its precontaminated condition. RCRA was amended by the Hazardous and Solid Waste Act, which requires EPA to identify facilities and sites where hazardous wastes have been improperly stored or released into the environment. The universe of contaminated sites requiring clean-up, or "corrective action," has continually expanded and by 1995 included 4,300 sites. As of 1992, however, intensive clean-up efforts had been initiated at only forty-three. And the number of facilities requiring cleanup seems likely to expand as detection technology improves.[23]

Pesticides are regulated as "extremely hazardous substances" by the Comprehensive Environmental Response, Compensation, and Recovery Act, known more popularly as "Superfund." The statute was originally intended to provide funds for rapid cleanup of abandoned, contaminated sites, efforts that otherwise would be paralyzed by litigation initiated to establish fault. The law established "retroactive, strict, joint, and several liability" without fault, on owners, transporters, and generators of hazardous substances for "response costs."[24] The theory behind this strategy is that liability will deter behaviors such as property transfer that pass cleanup costs on to others. Superfund has resulted in enormous unanticipated technical and legal costs associated with case-by-case site assessments and litigation, however—costs that have consumed a significant proportion of the trust fund established, and continually replenished, by Congress.[25]

In 1987 pesticides were found to constitute approximately 15 percent of the hazardous substances contaminating sites on the Superfund national priorities list.[26] This list had expanded by an average of 112 sites per year through the 1980s. By 1990, 1,200 sites were on the national priority list, and EPA then estimated total cleanup costs to be $26.5 billion. (More realistic cost estimates range between $26 billion and $125 billion and higher.[27]) Between 1981 and 1991, EPA restored only sixty-three sites to their precontaminated condition.[28]

The history of federal efforts to protect farmworkers from pesticides is one of the more shameful U.S. environmental health stories, and it clearly deserves its own book. The Bureau of Labor Statistics estimates that between eighty thousand and three hundred thousand U.S. farmworkers are injured annually by occupational exposure to pesticides, and of these between eight hundred and one thousand die.[29] Effective, health-protective policy has been impeded by the overlapping jurisdictions of EPA and the Occupational, Safety, and Health Administration (OSHA).[30] OSHA attempted to adopt farmworker protection standards in 1973 when it issued emergency standards for reentry of fields after application. But these guidelines were quickly overturned by the Court of Appeals, which found that no "emergency" justified the action.[31] OSHA failed to issue new farmworker protection standards until 1987, when it was forced to respond to a court order and adopt a "hazardous communication standard" for the nonmanufacturing sector. In the interim, OSHA simply deferred to EPA's authority, which had been granted by FIFRA.[32] A memorandum of understanding designed to coordinate the regulation of pesticides was finally signed by EPA and OSHA. The responsibilities remain confused, however, because OSHA has recently adopted standards to control airborne pesticides.[33] OSHA has set permissible occupational exposure limits for seventy-five chemicals classified as potential human carcinogens, far fewer than are regulated by EPA. Lois Gold and her colleagues found in 1994 that the permissible exposure levels were within a factor of ten of the carcinogenic dose in rodents for nine of these chemicals, and between a factor of ten and one hundred for seventeen others. They concluded that the cancer risks allowed by these exposure limits is far higher than is allowed by other federal agencies that regulate carcinogens.[34]

EPA's regulations are relatively short and simple, and many contend that they offer little protection to farmworkers.[35] They require, for example, that farmworkers not be exposed to pesticides without their knowledge; that unprotected workers leave the field prior to spraying; that workers stay out of the fields until sprays have dried or dust has settled (or longer for twelve highly toxic pesticides); that workers be notified before entering fields treated or to be treated with pesticides; that warnings be posted, and, if there is reason to believe that workers cannot read, given orally; and that warnings be given in languages understood by workers, "when required."[36]

Some pesticides, but not all, are also regulated by the Hazardous Materials Transportation Act (HMTA).[37] The primary purpose of this law is to prevent the accidental discharge of hazardous materials into the environment, and in the event of a discharge, to ensure that everything possible is done to protect health and the environment. In July 1991, a train derailment in Dunsmuir, California, caused nearly twenty-six tons of the herbicide metam sodium to spill into the Sacramento River and to eventually drift into Lake Shasta. Metam sodium sterilized nearly forty miles of the river, killing at least a hundred thousand fish and numerous songbirds, along with many of the trees and brush on the riverbank. Although EPA had previously received a study demonstrating that metam sodium caused birth defects in 1987, it had not reviewed the study by the time of the spill, and the chemical was not listed as a hazardous material.[38] This resulted in hundreds of people being exposed to toxic fumes that formed as the herbicide reacted with air.[39]

The incident caused a federal interagency committee to recommend the "cross-listing" of hazardous substances so that toxic substances identified under one law would automatically become listed as potentially dangerous under another. It was later discovered that many pesticides were not on the Department of Transportation hazardous materials list, and that 226 hazardous air pollutants listed under 112 of the Clean Air Act were also unlisted. Spills of any unlisted compounds would therefore not trigger an emergency response under HMTA.[40]

The segmented structure of law just described is mirrored by a fractured administrative structure within EPA, which has separate subdivisions responsible for regulating air; water; solid waste and emergency response; and prevention, pesticides, and toxic substances. As mentioned earlier, pesticides are managed according to whether they contaminate air, water, or land. Different regulatory missions, agendas, and schedules among these divisions have led to generally uncoordinated data production. In turn, an incremental, division-specific pattern of decision-making and standard-setting has evolved that makes assessing comparative and cumulative risks across compounds and environmental media nearly impossible. Thus we are left with little understanding of whether government is concentrating on the most significant risks to public health and environmental quality.[41]

Why is environmental law related to pesticides organized by the medium of exposure: food, water, air, or the workplace? The answer may lie in an understanding of the ways that scientific complexity is reduced and problems are narrowly defined, both of which are necessary conditions for the passage of legislation. It is normal for individuals and groups to break complex problems into components that seem simpler. Managing air, water, and food quality each independently is easier than managing ecological systems that include all of these components.

This medium-based approach is also promoted by highly specialized academic disciplines, professions, and institutions. Specialized knowledge appears to have resulted in the specialized definition of problems. For example, although carcinogens appear in air, water, and food, our law does not regulate human accumulation of carcinogenic risks across these spheres.

Institutions and their subdivisions, which themselves are structured to manage problems of specific types, are often key actors in defining future problems so that the process of narrow problem definition becomes self-perpetuating. Interest groups—industrial and environmental—also support this effect because they are better able to influence regulation if the purpose of law is narrowly defined. Focusing law on problems that cut across a broad array of interests limits the potential of any one group to control the outcome. Also, environmentalists have commonly sought to disengage their problems from other social issues in order to avoid competition for scarce management resources. Similarly, the more recent conservative emphasis on cost-benefit analysis to justify regulation is an attempt to reconnect environmental issues to broader problems of economic health and social welfare.

Politicians also tend to engage in problems that they may gain credit for "solving" through the passage of law. It is easier to seem responsible for a narrowly defined law than for one that requires consensus across a broad coalition of interests (and therefore many sponsors). Thus air-quality control is a more likely target of law than regulation of neurotoxic chemicals that contaminate all environmental media.

Scientific Inquiry Is Fractured

The questions that scientists and regulators have posed to understand the environmental health effects of pesticides have been narrowly defined for political and scientific reasons. This phenomenon partially explains the series of surprises narrated in the early chapters describing evolving pest resistance, transport of residues worldwide, childhood susceptibility to toxic effects, and childhood patterns of exposure. Government attention has focused within individual environmental media (such as food or water), on specific compounds (for example, DDT and Alar), and on specific types of toxic effects of each compound (such as cancer or neurological risks). But understanding pesticide use, environmental transport and fate, ecological effects, and human health effects for even just a single compound is an enormous scientific and analytical task. If EPA is responsible for the registration and control of tens of thousands of pesticide products, what knowledge is necessary to identify significant health risks? What are the right questions to ask? Several categories of knowledge that seem most relevant are explored below.

Understanding Pesticide Use

Within the past half-century, nearly 250 billion pounds of pesticides have been released into the global environment, primarily to protect cultivated plants and control insect-borne infectious diseases.[42] I argued earlier that pesticides are perhaps the most difficult of toxic technologies to control, because their effectiveness normally depends upon their dispersal into the environment. Knowing where and when pesticides are released to the environment is thus critical. But keeping track of all uses around the world is virtually impossible given the diversity of compounds, their many purposes, and the scarcity of resources to manage this information. Because it is impossible to check for all pesticides in all foods and all sources of drinking water, knowledge of where compounds are released to the environment is necessary to direct scarce environmental monitoring resources. Thus those compounds that appear to pose the greatest risk, registered or unregistered, domestic or foreign, should be tracked most closely.

Further, it is important to acquire and maintain this information for foreign uses of pesticides not registered in this country, because the United States imports tremendous amounts of food from abroad. Unfortunately however, sales data may be of little use to predict the flow of residues in the international marketplace; compounds sold this year may not be used until next year, or they may be mixed and applied with other compounds purchased previously. Worse, foreign use may include experimental pesticides unregistered anywhere in the world.

Understanding Residue Pathways and Fate

Since the 1940s, our understanding of how pesticide residues flow through the environment has been facilitated by advances in analytical chemistry. Before these advances, insensitive detection tests had led to the broadly held opinion that pesticide residues normally dissipated and that there was no need to test for toxicity or ecological effects. These innovations permitted the detection of pesticide residues at extremely low levels—often parts per billion, and occasionally parts per trillion—where previously none were thought to exist.

In the history of pesticide use, scientists have been dismayed by discoveries about residue pathways and residue fate on at least three occasions. The first unfortunate surprise concerned the persistence and widespread distribution of lead in the environment (lead arsenate was heavily used as a pesticide during the early part of the century). The accumulation of lead in agricultural soils, the persistence of lead on fruits and vegetables, and the buildup of lead in human tissues, especially bone, was unexpected. As we now know, lead has profound neurological effects; even low levels of exposure can diminish cognitive function.[43] A second unanticipated problem concerned the persistence and bioaccumulation of the chlorinated compounds that had been quickly substituted for lead and arsenical pesticides during the 1940s and early 1950s.[44] Even very low background levels of DDT contamination accumulated along food chains, resulting in high levels of human contamination—greater than 10 parts per million in human fat tissue during the 1960s. The third major surprise was the extreme mobility through ground and surface waters of water-soluble pesticides, especially herbicides and soil fumigants. Low levels of herbicide residues in the soil were originally attributed to dissipation, but in fact significant proportions of amounts applied volatilized and moved through the air, were washed off of fields into surface water, or leached into groundwater. In each of these three situations, an understanding of the environmental persistence, mobility, accumulation, and fate of these compounds evolved only after fragments of evidence were pieced together from inconclusive data.

Ecological Risks: Defining Relevant Endpoints

Knowledge of how pesticides move through our environment naturally led to new and difficult questions concerning their effects *en route* and where they came to rest. Understanding the pathways and fate of DDT, for example, especially its tendency to bioaccumulate, made it possible to correlate high levels of fish and wildlife contamination with their reproductive failure and to recognize the rapid evolution of resistance among insects to individual pesticides. Farmers and health officials have commonly responded to evidence of insect resistance by increasing the pesticide's concentration, applying it more frequently, or substituting a different class of pesticide.

Ecologists trying to measure the health of an ecosystem have been challenged by the enormous complexity of interactions among numerous species within highly variable environments.[45] Species diversity, the comparison of current net primary productivity (increases in biological mass) and the theoretical maximum net primary productivity, as well as the biotic regulation of biological, geological, and chemical fluctuations have all been advanced as theoretical metrics of ecosystem health.[46] Defining ecosystem health seems to depend on how we value ecological changes such as the extinction of species due to pesticide contamination. Thus health seems entangled with the ways that we value resources within the system, as we transform trees into building materials, rainfall into drinking water, plants into food, or the wilderness into recreational areas.

The concept of ecological risk therefore seems closely tied to the likelihood of losing some valued element within the system such as biological diversity, productivity, or high-quality water. Pesticides have long been implicated in the loss of fish, wildlife, and water quality. This perception that something highly valued is at risk focuses the attention of lawmakers, regulators, and scientists.

Defining "Adverse" Human Health Effects

What is an adverse health effect? This question, especially as it pertains to regulatory policies, has had a powerful influence on the direction of scientific study. Both science and politics play an important role in determining what is an "adverse" effect. For example, public fear of environmentally induced cancer has commanded extraordinary attention by both scientists and regulators since the 1950s. The Delaney clause was a legislative response to perceived cancer risk. This law alone has commanded a very large proportion of resources available for regulatory science.[47]

EPA has slowly expanded the number of adverse effects that it considers before registering a pesticide. It now deliberates over carcinogenicity, mutagenicity, neurotoxicity, and effects on the reproductive, immune, and endocrine systems. In considering the influences on human health of pesticides, EPA has historically been most concerned with cancer and with acute neurological effects, especially those from organophosphate and carbamate insecticides. Only recently has the agency asked probing questions about chronic neurological damage, immunological effects, and disruptions to the endocrine system. This means that for most of the thousands of pesticides now registered EPA has not secured relevant data to judge these types of risks.

Acute effects are relatively well known for most pesticides early in their product lives; this information is derived primarily from accidental or occupational human exposures.[48] By contrast, knowledge of delayed human health effects appears to have lagged far behind knowledge of residue pathways, ecological effects, and acute human health effects. The slow pace of understanding has resulted from many factors, especially the difficulty we have had in knowing the fate of pesticide residues within the human body and the specific biochemical or physical mechanisms involved in producing damage such as a tumor or a birth defect.

Delayed outcomes such as cancer are especially difficult to understand given the number of compounds believed to play some role in the etiology of the disease. The link between chemical exposure and a delayed effect normally remains hypothetical, given the fact that the person afflicted may have been exposed to a complex array of other compounds during the intervening period. The longer the delay, the more difficult it has been to establish causality between an exposure to an adverse effect. In addition, when disease occurs far from where pesticides have been applied, and long afterward, there is much less likelihood that earlier exposure to residues will even be suspected. This explains the relatively small number of compounds classified as "known" human carcinogens, in comparison with "suspected" human carcinogens, which have caused tumors in laboratory animals.

Understanding Thresholds for Adverse Effects

Once consensus is reached that a compound produces an adverse effect, at least two basic questions normally follow: (1) is there a threshold of exposure beneath which no adverse effect is detectable? and (2) how does the magnitude of the adverse effect vary with the level of exposure? For most types of toxic effects, toxicologists presume that exposure must exceed a threshold before the problem appears. This supposition underlies the use of the lowest "no observed adverse effect level" (NOAEL), discovered through animal studies, in setting allowable human exposures to any single compound.

Yet significant uncertainty surrounds the accuracy of animal study results for estimating human health effects. For this reason, safety factors are commonly employed to lower the level of acceptable human exposure beneath the NOAEL from animal studies. The appropriate size of the safety factor is the subject of considerable debate among interest groups and scientists. Whereas EPA has formally declared a policy of dividing the NOAEL by one hundred to determine a safe level of exposure, in 1993 a NAS panel recommended increasing this by an additional factor of ten to account for children's increased susceptibility and potential for exposure to pesticides. The stated policy of EPA also differs substantially from the reality of existing standards, which permit human exposure at levels that employ no additional safety factor from NOAEL levels found in animal studies.[49]

Moreover, the entire process of setting allowable contamination levels, or acceptable levels of exposure for individual pesticides, neglects the probability of everyday exposure to numerous *other* compounds that may contribute to the same type of adverse effect. Cancer risk may accumulate across compounds in this way, for example, as may the risk for disruption of the nervous system through cholinesterase inhibition.

Relative Potency

As described in chapter 8, Paracelsus's claim in 1567 that the dose makes the poison has become the foundation for modern toxicology.[50] Thus discoveries about the relationship between dose, or exposure, and the severity of the adverse effect have often been used to distinguish between trivial and significant chemical risks. For acute effects, the threshold between an exposure that produces an effect and one that does not is the primary interest of health scientists. The lower this threshold, the more potent the compound is considered to be. For carcinogenic effects the problem is different, because there is a presumption that even the slightest exposure increases the probability of cancer. Attention in cancer risk assessment is therefore directed to understanding how the incidence of tumors increases as dose increases. One compound is considered more potent than another if it produces more tumors for any given dose. Because many compounds produce tumors following heavy exposures, those that produce tumors at low doses quickly draw the attention of regulators. Similarly, those compounds that produce tumors following the shortest duration of exposure are of greatest regulatory concern.

Knowledge of how much of a toxin is needed to produce an adverse effect is essential for administrators trying to allocate scarce regulatory resources toward compounds likely to pose the greatest risks. But the task of these managers is further confounded by research-related problems, such as inconsistency in the quality of studies, differences in the relative certainty of their conclusions, and EPA's slowness in interpreting already submitted studies. Together, these circumstances diminish the likelihood that potency and risk comparisons among pesticides and their substitutes will occur.

Understanding Variance in Exposure and Risk Distribution

Variations in human health risk may be the result of differences in exposure to a toxin, in susceptibility among the exposed population, or both. As explained more fully in chapter 8, the developmental stage of specific target organ systems or functions may be an important factor. And chapters 9 through 11 demonstrated that exposure may vary according to where and when pesticides have been used, or differ in relation to what people eat and drink. This poor understanding of variance in risk distribution has led EPA decisionmakers to rely on statistical summaries of risk. Yet when these summaries are used to set standards for allowable exposure, highly exposed or highly susceptible groups such as children have historically been ignored.[51]

Understanding Cumulative Effects of Complex Mixtures

Each day, we are all exposed to a complex mixture of contaminants in food, water, and air. Methods of estimating our exposure to mixtures and their toxic effects, however, are often primitive and are likely the result of laws and regulations that generally support the single-media, one-pesticide-at-a-time regulation of chemicals. Real-world risk reduction will require formal scientific and legal consideration of human exposure to complex mixtures of toxic substances across the artificial environmental compartments defined by current laws.

Laws that demand analysis of single compounds within a single medium cause scarce analytical resources to be used to explore narrow questions and to identify minuscule risks. As risks approach zero, the marginal costs of risk reduction tend to increase exponentially—and to consume available resources before a search for systemic or cumulative effects can even be started. Further, the specialized knowledge produced promotes more technical regulation, which in turn directs EPA's attention to questions even more limited in scope.

The problem of estimating and managing cumulative effects from diverse compounds has many dimensions. Exposure to different compounds that each contribute to the same adverse effect may come from numerous sources, each managed independently. One example is childhood exposure to neurotoxic compounds such as heavy metals or pesticides: these chemicals are found in air, water, and food, each of which is controlled without regard to the other. Cancer risks appear to pose another type of cumulative effect problem whereby a single cell may move through several stages before a tumor begins to grow and promotion from one stage to the next may be induced by different compounds at different points in time. A final example may be our exposure to diverse chlorinated organic compounds that mimic estrogen. In these cases, each individual release of the compound into the environment seemed to pose little threat so we permitted release of minor amounts—but in an uncontrolled manner, with no one keeping track of amounts released or their distribution in the environment, accumulation in the body, or adverse health effects.

Understanding Sources of Uncertainty in Risk Estimates

Uncertainty in any risk estimate may be caused by a variety of factors including (1) poor sampling design, (2) systematic or random error associated with measurement instruments, (3) poor data management, (4) the absence of trend data when the problem is one of projecting past trends into the future; (5) analytical methods that fail to distinguish between inherent variance and statistical error, and (6) our limited ability to control risks through exposure management. The quality of a risk analysis may often be judged by how carefully and honestly these potential weaknesses are considered and expressed, as well as how they are translated into a statement of the likelihood, or probability, that specific types, magnitudes, and distributions of damage will occur.

Scientific uncertainty can easily become a black hole for agency attention and resources. Confronted with the statutory demand to estimate the effects of hundreds of active ingredients—and thousands of their combinations with inert ingredients—EPA has focused on incremental choices such as how to manage a single pesticide, or worse, how to manage a single type of effect of a single pesticide. Given this overwhelming responsibility, as well as the uncertainty and complexity of nearly every question raised, few resources are available to explore broader issues of cross-compound and cross-media risk assessment and management.

Uncertainty is normally a part of each component of a risk estimate. These component variables include (1) the magnitude of contamination released to the environment, (2) the distribution and fate of the contaminant in the environment, (3) human exposure to the contaminant (via air, water, food, and so forth), and (4) the human response to the exposure, or the dose-response relationship. Each of these parameters has inherent variability—which we may or may not well understand. Knowing the sources of uncertainty among these components of a risk estimate may be important to direct both science and regulation. For example, if scientists generally agree that a pesticide is highly toxic but the exposure is uncertain, then exposure assessment would deserve immediate scientific attention. If the level of toxicity is ambivalent, with best available estimates demonstrating only moderate or low potency, then resources would best be directed toward toxicology, rather than toward better estimates of exposure, to resolve the uncertainty.

Absence of Standards to Judge the Quality of Evidence

Conflict over appropriate regulatory policy often results from the general absence of clear standards to judge the relative quality of competing interpretations of uncertain evidence. Instead, standards emerge within highly technical fields such as reproductive toxicology or cancer biochemistry, and as a result they are very difficult for laypersons to interpret.[52] EPA has gradually articulated guidelines for the quality of toxicological and ecological effect studies submitted to justify pesticide product registration. Yet different institutions, such as EPA, EPA's Scientific Advisory Board, NAS committees, academic departments, peer-reviewed journals, industry, environmental interest groups, and the popular press all have different standards by which to judge the quality and meaning of evidence. Each relies on different bodies of evidence, different analytical methods, and most fundamentally, different values and ideologies when crafting interpretations of uncertain research results. Patterns of interpretation among the groups are quite discernible, often predictable, and they define pesticide politics.

EPA learned how difficult it is to manage public perception of risk when it lost control over the regulation of Alar, a growth regulator formerly produced by Uniroyal Corporation. The Natural Resources Defense Council (NRDC) developed their own interpretation of the toxicity evidence and concluded that children were at significant risk from Alar and many other neurologically active pesticides. By reporting the NRDC claims on its *60 Minutes* program, CBS became an interpreter of science and the regulatory process, and EPA appeared confused and unreasonable.[53] Meanwhile, the NAS panel on pesticides in the diets of infants and children, which had been convened prior to the controversy, chose to avoid interpreting the Alar evidence.[54] As a result, the institutions interpreting the risks associated with Alar included the producer, Uniroyal, EPA, NRDC, and following the *60 Minutes* broadcast, virtually every major television network and daily newspaper in the nation. All struggled to understand the magnitude and distribution of risks in the absence of any consistent set of standards regarding quality of evidence or methods for its interpretation. The dominant lesson from the Alar controversy is that the government failed to develop consensus over the estimates of risk among manufacturers, government, interest groups, the media, and the public. It was the public that eventually regulated Alar by refusing to purchase Alar-treated apples and apple juice due to fear of cancer. Although the blame for this effect is most often directed toward NRDC, it should be targeted at EPA, which has the responsibility to produce credible, timely interpretations of evidence.

Understanding Relative Risk

If our knowledge of the relative risks posed by different pesticides is limited, our understanding of pesticide risks relative to other types of environmental health hazards is primitive. Where does the greatest environmentally induced cancer risk come from: the diet, drinking water, or air pollution? Answering this question would seem necessary to justify the allocation of scarce scientific and regulatory resources. But obtaining a clear response is inhibited by the all-too familiar problems of poor quality data; inconsistency among analytical methods; uncertainty in understanding patterns of exposure to complex mixtures; uncertainty in understanding variance in the distribution of risk; ambiguous standards of acceptable risk; and incremental, chemical-at-a-time regulation in each regulatory arena. Rational comparison of risks across these arenas is therefore nearly impossible.

The Value Dimension: Estimating Benefits and Costs

Surprisingly, estimating benefits is limited by many of the same factors that make assessing risk so troublesome. Still, some benefits appear to be beyond disagreement. Large areas of the world, especially temperate and subtropical regions, are now free from deadly insect-borne diseases such as malaria, yellow fever, and typhus primarily due to the systematic and repeated use of pesticides. The control of these diseases has unquestionably saved millions of human lives and has dramatically improved the quality of life for hundreds of millions of others who have avoided illness. Also without question is the important role that pesticides have played in reducing pest damage to food and fiber crops. Even with their use, today nearly one-third of cultivated plants are lost to insects, diseases, or competition from weeds.

Yet a full accounting of the value of pesticides demands careful consideration of their environmental and health costs. For example, the costs of controlling insect damage appear to be increasing, and the rapid evolution of insect resistance to pesticides demands the application of ever more potent compounds. Increasing the intensity of application, however, does not preclude insects from evolving resistance to the new treatment, and this strategy may carry additional risks of environmental contamination and human exposure.

In reality, balancing of risks and benefits is normally reduced to crude comparisons—for example, between hypothetical cancer risks and increased crop yields. These diverse effects are expressed in units that are very difficult to compare quantitatively. Worse, little if any attention is devoted to explaining the enormous uncertainty that exists in all estimates of net welfare; it is seldom clear how projected benefits and costs are distributed in society. Benefit forecasts used to justify pesticide registrations normally fail to account for the societal costs of managing the risky technology or its unintended consequences.

When in Doubt, Delay

In summary, our knowledge base is constantly changing, and it tends to be fractured medium-by-medium, chemical-by-chemical, and by type of adverse effect. It expands systematically, in a pattern driven by EPA's data-call-in and chemical-by-chemical review process. This process inhibits the joint consideration of comparative risk among chemical substitutes. The effect has been one of decade long delays in reregistering individual compounds, and a failure to strategically substitute low-risk for high-risk pesticides. Worse, the one-at-a-time review process could cause higher risk compounds to be substituted for those posing lower risks. Because full review of any single compound may not occur again for another decade or two, these types of mistakes may pose long-term risks to environmental health. During this lengthy review period, scientific evidence of transport, fate, human exposure, and health effects may evolve dramatically and make former judgments of risk obsolete. In most instances, additional study does not appear to result in consensus on the magnitude or distribution of risks. Instead, more questions are raised.

When risks are ambiguous, waiting for better evidence is a strategy used with equal effectiveness by both environmentalists and manufacturers. Environmentalists commonly demand more proof of safety prior to the federal licensing of *new* products, whereas manufacturers use uncertainty as an argument to delay the removal from the marketplace of *old* products newly found to be risky. Delay is suggested until the evidence of damage becomes clarified. These strategies have frustrated efforts to strategically reduce risks by substituting new, low-risk compounds for older, high-risk ones.

The Feedback Effect: Fractured Law, Fractured Science

Fractured environmental law has defined the structure of EPA, the organization of its expertise, and the boundaries surrounding the questions agency scientists have normally asked. Law has shaped regulatory science by demanding answers to technical questions that are chemical-specific, medium-specific, and effect-specific—all of which are necessary for setting standards. How much benomyl, for example, will likely remain in heat-processed apple juice, and what if any cancer risk might the residues pose? Answers to these types of questions normally cause regulations to be further elaborated or qualified. Since 1970, attention has imploded on narrow topics, a trend that has caused a proliferation of regulations. It is an unfortunate irony that the comprehensive demand of current law—that each chemical be studied with equal intensity—has necessitated an incremental regulatory process that moves at a glacial pace. Quite simply, law has not directed attention or resources to examine the bigger questions, such as how are we accumulating risks across chemicals and environmental media—risks we face in daily life? The potential for contamination to flow across environmental compartments, and for people to amass exposures and risks as they move from one environment to the next, have been largely ignored.

Current law seems to reflect a highly simplified conception of nature, one that neglects important ecological processes. Although breaking nature into compartments is a necessary basis for pursuing improvements in environmental quality and risk reduction, the boundaries among the compartments are necessarily artificial. Many lessons from ecology—some of which are detailed in earlier chapters—have long warned us to consider the natural processes that connect the units. More worrisome, this entanglement between law and science gives little hope that we are even aware of the magnitude and distribution of significant threats to environmental health.

CHAPTER 13

Toward Reform

In a world of constantly changing knowledge of relative risk—and one of highly stable law, agency structure, and expertise—what principles should reorient scientific and legal attention to ensure that government recognizes and controls the most significant hazards?[1] Reform is necessary in three areas: (1) how we produce knowledge of risks, benefits, and costs associated with pesticide use; (2) how we distribute this knowledge; and (3) how we choose to manage the allocation of risks. Collectively, the guidelines that follow provide a new architecture—admittedly a normative one—for pesticide risk assessment and management.[2]

Producing Knowledge of Risk

Quality of Evidence

Articulate clear standards of quality for evidence used to estimate risks.

Types of evidence relevant to this process include data regarding the production, distribution and use of pesticides; the residue fate of complex mixtures in diverse environmental media; human exposure to pesticides within single and diverse media; human health effects from exposure to single compounds and complex mixtures; and ecological effects.

Management efforts have been severely hampered by data of poor quality, although there is little consensus about how good the data need to be. Quality is determined by numerous factors, which may vary by type of data. The most important limitations of current data include their age, the non-representativeness of samples, insensitivity of measurement methods, the absence of information on trends, and heavy reliance on animal tests rather than epidemiological studies to estimate human health effects. If experts hold conflicting views or are confused in their definitions of quality evidence, how can the public be expected to accurately discriminate among these explanations? Many environmental health debates involve the interpretation of complex evidence. Groups that commonly attempt to interpret government regulatory decisions for the public—such as the visual and print media as well as special interest groups—apply widely varying standards for interpreting uncertain evidence. The disparity between expert and public opinions regarding the relative dangers that diverse contaminants pose to health and the environment may be partially explained by the absence of data quality standards.[3]

Variance in Risk Distribution

Clearly express variance in the magnitude and distribution of exposure and risks. Demonstrate the effect of using alternative assumptions on these estimates.

Risk has traditionally been expressed through the use of summary statistics such as average or ninetieth-percentile values. This practice, however, obfuscates how risks are truly distributed in society. Averaging distracts attention from highly exposed or susceptible individuals and causes nontrivial risks to be imposed on unknowing individuals. Understanding variability in the magnitude and distribution of risk is essential to the control of health and may be one of the most neglected subjects in environmental science and law.[4]

Variability in risk can result from many factors, including differences in patterns of pesticide use around the world; variance in residue levels within different environmental media; the wide range of human exposures to these residue levels; and variance in the toxic potency of a pesticide, which in turn may depend upon differences in human susceptibility to the effect. Averaging ignores variance and has justified EPA's licensing of risky technologies such as pesticides that may pose low risks on the majority, but higher risks on minority populations. Simple assumptions that risks are distributed evenly among individuals and across time and space are not likely to promote health-protective policies.

Expression of Uncertainty

Clearly articulate standards for the expression of certainty for both adverse and beneficial outcomes.

Because risk estimates are normally produced by complex computer models that link numerous variables—such as contamination levels, food, water or air intake, and chemical toxicity—it is essential to understand the uncertainty contributed by each variable. Uncertainty derives from ignorance, which may have many sources. Sampling designs, for example, often fail to accurately measure the effect or phenomenon of interest. For decades, no one sampled herbicide runoff from fields into surrounding ground and surface waters when heavy spring rains fell soon after application. Researchers therefore missed the pulse of contamination that is normally transported to surrounding waters. The absence of evidence was interpreted simply as dissipation, and no problem was recognized.

A similar sampling-design defect surrounds estimates of food and water intake. For example, if you asked me to estimate your average exposure to pesticide residues from contaminated food for any given year, the answer would depend on the sample interval and the time of year. If I asked you what you eat on one day in July as a basis for calculating your average daily intake of pesticide residues for the entire year, the estimate would be very different than if I sampled your diet over twelve months at two-month intervals. The latter approach has a better chance of capturing variance in your food intake over time and is more likely to produce a more accurate estimate of what you eat on the days that were not surveyed. Thus the sampling design used to produce data for each component of any risk equation must be carefully structured to capture the natural variance of the component. Designs that poorly capture variance contribute uncertainty to risk estimates.

Uncertainty surrounds estimates of the toxic potency of pesticides as well, and it is derived from our need to rely on animal bioassays to predict delayed human health effects. As discussed in chapter 5, human variance in susceptibility to toxins appears to be age related, with windows of vulnerability opening and closing for key organ systems and functions at different stages of development. Chapter 10 demonstrated that reliance upon summary statistical estimates of exposure and risk can easily mask significant variance among populations such as very young children.[5]

Exposure

> Greater understanding of the legal, organizational, educational, and cultural conditions surrounding pesticide use is necessary for effective management of pesticide exposure to ensure that risks do not exceed trivial levels.

The history of pesticide management has really been one of attempts to control exposure. Whereas great effort is normally expended to understand the toxic potency of pesticides, relatively little effort is spent analyzing patterns of exposure and our institutional ability to control it. By contrast, other highly toxic substances are managed with great care. Radioactive material, for example, must be monitored from point of production to disposal, and exposure is tightly controlled through highly skilled technicians, protective equipment, and routine detection to ensure against contamination. Similarly we control access to prescription drugs, many of which are specialized biocides, by administering them through two layers of experts, physicians and pharmacists, to manage the risks that they carry.

It is striking, then, to realize that pesticides are commonly dispersed into the environment by untrained workers who have little technical education and inadequate protective clothing, and who are rarely monitored for contamination. Imagine the migrant farmworker in the Central Valley of California during summers when field temperatures often exceed one hundred degrees Fahrenheit. The worker commonly moves from farm to farm, and crop to crop, facing very different types of pest problems; therefore he or she is exposed to a constantly changing mixture of pesticides. The complexity of managing exposure under these conditions is staggering.

If controlling risk depends on how we manage post-release exposure, how should we judge the potential effectiveness of both public and private management systems to control this exposure? EPA, for example, has pursued a variety of regulatory tactics, including limiting contamination levels in food, water, air, and soil; labeling pesticides with cautionary instructions; requiring protective clothing; instructing applicators; and establishing field reentry intervals following spraying. But the potential for these restrictions to prevent dangerous exposure rests on at least four pillars. The first is accurate knowledge of the toxicity of pesticides, especially how toxicity is related to dose, another word for exposure. The second is knowledge of how people are exposed through contaminated food, water, air, or soil. The third pillar is an understanding of how exposure and susceptibility vary among the population. The final pillar is knowledge of probable contamination levels, which underlie exposure estimates. Failure to fully understand any of these issues may lead to a dangerous underestimation of risks, and these failures have historically been common.

Exposure management has been a potent underlying force shaping the structure of environmental law as it gradually evolved to control different contaminants in various media. Although Congress crafted the patchwork quilt of environmental laws, legislators never seemed to fully understand the enormous burden this approach places on scientists—private, public, and nonprofit—and regulators. Given the limited resources devoted to health and environmental science, comprehensive analysis of each chemical and all possible mixtures has been infeasible due to cost and time. The Achilles' heel of exposure control as our dominant tactic to govern risk is the complexity and cost of knowing the movement and effects of even a single pesticide once it has been released into the environment.[6]

Cumulative Effects

> Estimate the potential for accumulation of exposures across compounds and environmental media which could produce additive or synergistic effects.

Toxic effects may result from the accumulation of exposures to diverse compounds, each of which is regulated under different laws, statutory standards, and administrative units within EPA. As we have seen, this fractionalization appears to have exacerbated the problems inherent in estimating risks from compounds that are carcinogenic and neurotoxic. Moreover, this regulatory focus on single compounds rather than complex mixtures reveals how narrow science can drive narrow regulation.

The problem of multiple-media, multiple-compound exposure seems best modeled probabilistically. In other words, we need to ask: What is the probability of being exposed to compounds from different media that together may produce an adverse health effect? The quality of the estimates produced are of considerable concern, because accounting for additional sources of exposure will likely compound error, variance, and uncertainty. The case presented on organophosphate insecticides in chapter 11 demonstrates one approach to the complex-mixture problem within a single medium—the diet—an approach that could be extended to other media. The analysis used to judge the potential for significant cholinesterase inhibition was severely constrained by the absence of high-quality residue data.

Within the past five years, Congress has demanded that industry pay the costs of registering pesticides. Manufacturers have long funded toxicity and ecological effect studies, which EPA regularly reviews as a basis for registration and tolerance-setting decisions. Government should require manufacturers to also fund analyses of complex mixtures. If performed by independent testing laboratories, these studies could provide a higher quality database to judge the potential for risks to accumulate. Currently, no individual manufacturer has any incentive to fund complex mixture studies, which would normally involve products produced by other firms.

Strategic Attention and Expert Judgment

> Direct scientific attention strategically toward understanding the effects of single pesticides and mixtures that appear to be most potent, and to cases where public exposure is believed to be highest. Any hope of meaningful understanding of the relative risks posed by environmental contaminants will require new forms of mixture analyses, and greater reliance upon the opinions and judgments of experts in their interpretations of uncertainty.

Given the uncertainty of our knowledge of toxicity and exposure, the allocation of scarce scientific resources should be guided by expert judgment. In general, acute effects are better understood than delayed effects, largely due to knowledge gained through the occupational exposures of manufacturers, formulators, and farmworkers. Thus delayed effects such as cancer, neurological damage, and reproductive failure deserve special attention.

EPA has considerable experience in the use of expert judgment to explore uncertain effects and to establish regulatory priorities. The agency's advisory committees and boards have often been used to interpret conflicting evidence important to regulatory decisions. But expertise should also be directed toward strategic planning. The complexity of pesticide risk assessment demands a standing committee of government, academic, consumer, and industry experts to plan for the allocation of scarce scientific resources.[7] Others have called for a separate federal agency to judge the relative risks of diverse hazards and to estimate the costs and benefits of their management.[8] This type of comprehensive and comparative analysis is crucial given the incremental nature of regulation and science.

Although EPA would certainly respond that it is already conducting this type of expert review, it has severely underestimated the scale of the effort necessary and ignored the limitations of its current decision-making process. If EPA expanded the role of its scientific advisory board and gave its members more authority to define key research questions, a credible scientific foundation might finally emerge that would direct public attention toward the most significant environmental health risks we face. Funding for this effort could come from pesticide manufacturers—through registration fees or a progressive tax based on a product's use, volume, and toxicity.

In addition, government needs to establish protocols for research that ensure high-quality data, schedules for the production of these data, agreements regarding the use of appropriate analytical methods, and comparisons of the relative risks posed by substitutable compounds. No single quantitative indicator will evolve to permit comparison across all pesticides, but defensible rankings should be established for separate toxic effects—carcinogenicity, neurotoxicity, and estrogenicity—using the judgments and inferences of those most knowledgeable about each. The rankings would change as new, higher quality evidence emerged, but they would still provide a logic for the regulatory agenda.

Burden of Proof

> A continuing burden of proof of safety should lie with the producers of risky technologies such as pesticides.

As scientific protocols for testing become more sophisticated, producers should be required to regularly update their tests of environmental fate and effects so that EPA may continue, revise, or terminate product registrations. Although FIFRA requires registrations to be renewed every five years to allow for advances in our knowledge about these compounds, EPA has normally revisited individual chemicals within only a ten-year period. The reasons for this noncompliance include EPA's delay in requesting additional studies and industry's sluggish production of analyses. Further, when studies are finally submitted to EPA, the agency often requests additional information. The general delay is also understandable given the agency's limited resources, the enormous volume of evidence they must request and review, and the diversion of their attention toward highly public debates about the safety of individual compounds such as DDT, dieldrin, EDB, and Alar. Industry normally favors delay in the review of products already registered, unless the costs of additional studies are expected to exceed anticipated profits. Manufacturers voluntarily canceled registrations for thousands of products during the early 1990s when they were faced with the costs of providing updated environmental health data.

It is not uncommon for a pesticide to be reviewed for fifteen years while waiting for "high quality" data to be submitted regarding its likelihood of causing different types of damage. A chronic-effect animal bioassay may take five years to complete, and during that time scientific protocols for conducting the study may change and render the conclusions suspect. Moreover, because each pesticide has a unique data-production schedule, comparison of relative risks among possible substitutes has rarely been possible. Reforms should require that classes of substitutable compounds be reviewed simultaneously, with data requests synchronized among pesticides and companies. Manufacturers are clearly more comfortable with the current chemical-at-a-time review, because this method is unlikely to discover a cost-effective, less risky substitute for their products.

The universe of new compounds introduced in international commerce is continually expanding; nearly seventy thousand compounds are currently listed as hazardous in a registry kept by the National Toxicology Program. The subset classified as harmful steadily expands as we test their carcinogenic, reproductive, neurotoxic, and other effects and as advances in analytical chemistry permit us to trace the transport and fate of compounds formerly believed to have disappeared. It seems fair to conclude that the pace of our understanding of compound fate and toxicity has been far slower than the speed at which new compounds are introduced.

Accounting for Costs and Benefits

The same standards of quality suggested above for risk estimates should be required for cost and benefit estimates.

The standards contained within FIFRA and FFDCA have required that the costs of regulation be considered during the registration and tolerance-setting processes, which demand risk-benefit balancing. But technical problems inherent in choosing effects to value, methods of valuation, treatment of probabilities, and uncertainty together create a black hole for scarce analytical resources that is at least as deep as that engulfing risk assessment.

What costs and benefits should be considered, how should they be measured, and how should we predict their future value? Accounting for benefits is simpler than accounting for costs, because manufacturers are well informed about the effectiveness of their products in improving crop yields, preserving wood from rot, and controlling disease-bearing or nuisance organisms. Estimates of jobs likely to be lost from canceled product registrations, crop losses from using less effective substitutes, or effects on foreign trade of chemicals and foods are far less certain.

Current methods of cost accounting are tied directly to estimates of risk. Uncertainty in risk estimates is therefore transferred to some types of cost estimates. Uncounted costs of pesticide use have included expenses from monitoring the environment for residues; costs of health care for pesticide-related illnesses; outlays for federal, state, and local regulation; costs of cleaning up contaminated soil, ground, and surface waters; costs of managing genetic resistance (often involving more intensive application or demanding new pesticides); and the costs of losing nontargeted animals, plants, birds, insects, and fish.

Many of these costs are incurred simply by attempting to manage pesticides; it is expensive to run a regulatory program or to monitor residues in drinking water—a cost borne by most municipalities in the United States. Some costs are easier to identify, such as the medical expenses incurred by acutely poisoned farmworkers. But the uncertain role of these compounds in causing delayed effects such as cancer or neuropathy has left overall cost estimates ambiguous and suspect. Finally, understanding the distribution of the costs of pesticide use and management is as relevant to decision-making as is understanding the distribution of risks. If the risk-benefit decision standard for pesticide management is retained, a more careful and accurate accounting of costs and benefits—along with an understanding of their distribution—will be necessary for responsible public choices.

The Distribution of Knowledge

Property Rights to Knowledge of Risk

> Knowledge of pesticide risks should be considered international common property.

Claims of proprietary knowledge of pesticide toxicity have further narrowed what is known about pesticide risks. The U.S. government has deemed that data about the health effects of these compounds should be protected as a private property right. Manufacturers have thus insisted that this information remain confidential to prevent manufacturers of generic brands of pesticides from reusing this research to justify registration of their generic versions. The government has also played a significant role in restricting public understanding of the risks from pesticide use by limiting access to the data sets and computer programs they use to estimate exposure and risks. Data on chemical use, residues in the environment, ecological effects, and health effects—along with computer programs necessary to perform analyses and interpretations—should all be freely available.

One of the benefits of requiring full disclosure of chemical use, residue, exposure, and toxicity data is that bias would be easier to identify.[9] Bias may creep into risk estimates in numerous ways—in the definition of the problem to be addressed, in the choice of evidence, in the selection of methods for analysis, in the interpretation of uncertainty, and in the choice of conclusions to be conveyed to the public. Real improvement in the quality of risk estimates could come from a requirement that data, methods of analysis, computer programs, and assumptions be formally exposed for public review.[10] Given current low-cost computer technology, it would be easy to make these materials publicly available electronically, a process already begun by EPA and FDA but threatened by budget cuts.

Communicating Risk

> Purchasers should be forewarned of products containing pesticide residues—including food and drinking water.

Most humans are exposed to pesticides daily without their knowledge or consent. One principle for the management of invisible hazards, such as radioactive material or pesticides, is simply public warning, or risk communication. Although the concept is simple, its implementation in the case of pesticides is difficult. To ensure that the consumer is given an accurate image of the compounds used on a product, and the risks they carry, would require keeping records of the use of pesticides on crops throughout the world and tracking these foods through global crop and processed food markets. Labeling could be based upon simplified hazard classifications. This field-to-grocery store tracking system is conceptually similar to the cradle-to-grave tracking of hazardous waste required by the manifest system of the Resources Conservation and Recovery Act. Opposition to this idea may be vocal, however: pesticide manufacturers and food distributors have long argued that the risks are not significant enough to warrant such an information-intensive solution.

In this era of soaring costs of risk assessment and management, support for market solutions to environmental problems has become fashionable. But truly free markets are hardly likely to diminish risk, largely because exchange is facilitated by advertising. The purpose of most advertising is to exaggerate a product's benefit and appeal; warnings of risk are rarely included without government requirement. One way to diminish this market defect would be to improve and control the information given to purchasers about pesticide risks. There is ample precedent for this strategy in the sale of drugs, which must be accompanied by warnings of possible side effects. Public opinion surveys demonstrate considerable anxiety over pesticide residues in food and drinking water and at least the stated willingness to pay more for pesticide-free food. If consumers were given more complete information about the risks of pesticides, their buying decisions would likely encourage the safer use of these compounds or the substitution of other insect-control strategies.

Another important policy yet to be employed in pesticide management is hazard-based taxation. Taxes could be levied based upon the relative risks posed among pesticides, which would internalize the costs of damages believed associated with pesticides. Prices, influenced by hazard-based taxation, would communicate risk to the consumer.[11] Yet another option for risk communication is the opposite of labeling by hazard classification. Products *not* containing pesticide residues could be labeled, avoiding the entire tracking and relative hazard evaluation problem. Labeling of products as "organic" could accomplish this, although the word "organic," like "natural" has been misused in the past. California has made the greatest progress in controlling the quality of organic food labeling; it has required certification that pesticides have not been used for at least three years in the fields where organic produce is grown and that no post-harvest use of pesticides has occurred. Similarly, a designation such as "no residues detected" has been applied in California to products that have been screened using sensitive multiresidue-detection methods, even if they had previously been treated with pesticides in the field. This too has become an effective marketing strategy among grocery store chains.

Whatever form this communication takes, it will be important that the government explain its risk evaluation in relation to other hazards that the public is more likely to understand. Further, the expression of risk should include a full discussion of how some people may be affected differently than others, especially if there are some individuals or groups who are likely to be highly exposed to the hazard or some who may be particularly susceptible to an adverse health effect. One analogy is the warnings to the ill and asthmatic to stay indoors during periods of severe air pollution. This type of warning is particularly useful because the warning includes a safer option: stay indoors and turn on your air conditioners. In the case of pesticides, alternatives like this would be difficult to recommend because the government does not even know which fields are treated with which pesticides, let alone their probable level of contamination.

Education

> Government should ensure that risks which flow from their regulatory decisions are understood, not just communicated.

If government licenses the use of toxins that may persist as contaminants in food, water, air, clothing, and so forth, it has an obligation beyond simply warning the public of possible risks—it should do so in a way that has some relationship to everyday choices. Publication of quantitative estimates of exposure and risk for single compounds in the *Federal Register* hardly meets this obligation. This understanding is also necessary for effective public participation in decisions about pesticide management.

Government, as the primary financial supporter of public education, similarly should require that school curricula include enough of the natural, social, and health sciences so that students will develop an understanding of environmental health risks. This knowledge would empower citizens to participate in public policy debates and to become more informed, discriminating, and demanding players in markets. Without this knowledge, it is very difficult to interpret the legitimacy of competing claims of benefits and risks made by manufacturers and environmental groups.

A government with scarce communication resources often assumes that the public is not competent to understand risk estimates. By not taking the time to explain its analyses and justify its policies, the job of government becomes simpler, because the concentration of knowledge among technical experts constrains participation. Fewer people are competent enough to contest government and industry claims of safety and to challenge the tens of thousands of former government licensing decisions. In this way, concentration of knowledge reinforces the status quo, diminishes the public's ability to influence regulation, and undermines the regulator's accountability to the public.

International Disclosure of Use

> International agreement to maintain and share records of pesticide use, and to label products containing pesticides, is a necessary condition for the effective domestic management of pesticide risks.

International trade in pesticides results in the international exchange of risks. Because regulation relies upon knowledge of where risks are coming from, full disclosure of information on chemical use, in combination with tracking products containing pesticide residues, should first be attempted. Without an effective information management system, regulatory efforts have little chance of being health protective.

International trade in pesticides, crops, and processed foods make disclosure and education especially difficult. An active ingredient produced in the United States may be shipped to Switzerland where it is combined with other ingredients and shipped to Egypt. In Egypt, it might be applied as an insecticide to cotton. Cotton seeds may then be harvested and sold to a commodity broker in Israel, who then sells them to a manufacturer of cottonseed oil in Italy. The Italian firm may then sell the oil, perhaps mixed with other oil from seeds grown in Guatemala with the help of another pesticide, to an American food processing company. This firm may use the oil to produce many different products, from cookies to pretzels to cereals, and then ship the processed foods around the world.

The global nature of these markets demands that other nations agree to disclose which pesticides are likely to appear in specific products. This knowledge is fundamental to the principle of informed consent. If society is serious about warning the consumer, the complexity of international pesticide and food markets could easily create a nightmare of recordkeeping. But the problem could be simplified by creating a few categories of hazard and insisting that all nations agree to be most vigilant about pesticides classified as most hazardous. The United Nations already keeps a list of pesticides that have been banned or severely restricted around the world. Products containing these compounds, for example, should be clearly labeled. EPA has already identified a set of the most hazardous pesticides that they have not banned but have targeted for special, priority review due to their suspected hazard. Other compounds have been designated as "restricted use pesticides," and these require applicators to receive special training and be certified prior to sale. Perhaps of greatest concern are those compounds that are used on food crops but are not registered in any country; in these cases, importing nations do not even know what to look for. Hazard categorization and product labeling could cause consumer demand to shift toward the less risky products and thereby send a clear message to growers and food processors.

The Management of Risk

The Delaney Clause

> Replace the Delaney clause with a negligible risk standard applied uniformly across all pesticides and health effects.

The Delaney clause of FFDCA seems unreasonable for several reasons. First, it applies only to a small proportion of pesticides—those that are both carcinogenic and concentrate during food processing—and neglects all other health effects. Second, scientific evidence supports the claim that some compounds pose cancer risks that approach zero, but are not definitively zero. Forcing these compounds off the market may cause more risky substitutes to be used. Third, the uncertainty in cancer risk assessment is often substantial due to its reliance on animal studies. Fourth, the costs of achieving a zero-risk pesticide environment are enormous and difficult to justify given the possibilities of reducing other types of more certain health risks at a lower cost. Fifth, the Delaney clause has caused scarce regulatory resources to be disproportionately allocated to manage cancer risks rather than other types of effects such as neurological and reproductive damage.

Risk-Benefit Balancing Standard

> Replace the risk-benefit balancing standard in FIFRA and FFDCA with a health-protective negligible risk standard.

The predominant standard for pesticide registration—risk-benefit balancing—has been implemented in a way that permits humans to experience greater-than-trivial risks without their knowledge or consent. These outcomes have been justified not only by claimed benefits, but also by employing "risk averaging," which avoids considering highly exposed or susceptible groups such as children. The act of balancing risks and benefits is often justified by analyses that conclude that average risks are trivial. Under these conditions, benefit claims are not challenged by EPA, and products are registered. Yet this policy allows the imposition of greater-than-average risks on many, because most distributions of dietary exposure and risk from single chemicals are quite skewed—often because of the heightened exposure or susceptibility of some. In these cases, the majority of a potentially exposed population experiences little or no exposure, whereas often less than 5 percent of the population has significant exposure. Thus a policy that ensures protection of 95 percent of those exposed still leaves 5 percent at more significant risk. Considering that there were approximately 4 million two-year-old children in the country in 1995, a regulation that protects 95 percent of them still leaves two hundred thousand children at risk.

What other justification is there for a public policy that allows some to experience greater-than-trivial risk without their knowledge or consent? Some types of pesticide management decisions, for example, may be characterized as risk-risk tradeoff problems. In these cases, the magnitude of the risks involved, their distribution, the certainty of outcomes, and the manageability of the risk are all relevant to decisionmaking. Intensive indoor use of pesticides in tropical countries commonly reduces the risks of disease and death from insect-borne parasites. These effects are relatively certain and provide ample justification for the use of pesticides. In this case the same individual may be exposed to either the risk of a toxic reaction to applied pesticides or the threat of insect-borne disease. Health risk to the individual is thereby lessened, providing a clear moral defense for the public choice to apply pesticides.

Averaging and summary indicators of risk also play an important role in managing pesticide substitution. Suppose that a new chemical is being considered as a substitute for an older one, and that the newer one will pose less risk to 90 percent of the population potentially exposed, whereas 10 percent will experience substantially increased risks that experts agree are greater than trivial. Also suppose that many in this group will be children. Registration of the new pesticide has often been allowed under these circumstances, and has relied on summary risk-reduction estimates. In this case, risks would be reduced for 90 percent of the population—neglecting those more highly exposed. In situations such as these, alternatives should be explored that offer additional protection for the 10 percent at greatest risk from the new compound. The pesticide, for example, might be restricted from use on foods that comprise a high proportion of the diet of children. If it is impossible to contain the higher risks beneath an acceptable limit, the substitution appears to be indefensible.

Childhood vaccinations offer another example. In this case, probable adverse reactions to the vaccine are compared with probable disease incidence if the vaccine is not administered. Although the risk of an adverse reaction may be greater than trivial, the individual bearing this risk also experiences a reduced chance of coming down with the disease.

By contrast, consider the example of broadscale aerial spraying of pesticides in Los Angeles to control medfly infestation of fruits and vegetables. In this case the population sprayed assumes the risks, whereas the benefits accrue to farmers who avoid crop damage. Although the net benefits may be extraordinary to the farmers, wholesalers, exporters, and consumers, most would agree that these benefits are irrelevant to the spraying decision unless the risks imposed are deemed trivial by experts. Thus it is difficult to defend a government decision to improve the net welfare by imposing biocidal risks on one population when benefits accrue to another.

The case becomes even thornier when risks to a minority are greater than trivial—but seemingly controllable. Our management of farmworker exposure to pesticides falls within this category. It is theoretically possible to manage farmworker risks so that they do not exceed a trivial level, but the extraordinary vigilance, training, and care necessary to do so make health protection costly, uncomfortable, and highly unlikely. Statistics on farmworker hospitalizations and their incidence of chronic disease confirm that farmworkers and other applicators often unknowingly assume the highest degree of risk from pesticides. Because the individual farmworker experiences no offsetting risk reduction—as occurred in the case of insect-borne disease control and vaccination—reliance on an exposure-protection strategy seems defensible only if it has a high probability of success.[12]

A final example demonstrates the importance of political boundaries to the argument. When we ban compounds from use within the United States due to excessive risk, we still allow their export. This freedom permits domestic corporations to accrue the benefits of pesticide sales; it also creates jobs and a favorable balance of trade for the United States. But risks are imposed on other individuals, often those living in nations that do not have the resources to effectively manage technological risks. In addition, as with farmworkers those most heavily exposed rarely have full knowledge of the nature of the risks they face. The 1984 pesticide plant tragedy in Bhopal, mentioned earlier, provides an excellent example.

Negligible Risk Standard

> Risk standards should be set to ensure health protection for the entire population of individuals likely to be exposed to pesticides.

Maximum allowable contamination levels of any medium (air, water, food, or soil) should be set to ensure protection of the entire population of potentially exposed persons against significant risks, unless those exposed would otherwise face higher relative risks. Thus both the Delaney clause and the current risk-benefit balancing standards contained within FIFRA and

FFDCA should be replaced with a negligible-risk standard, because neither requires protection of public health. Acceptable risk ceilings should be clearly articulated and should prevent the accumulation of risk above thresholds of significance for all probable adverse health effects. As the benomyl case demonstrates, if any replacement for the Delaney clause is to be effective, both the quality of data and the analytic methods used to estimate exposures to single compounds or complex mixtures need to be controlled. For example, should risk be estimated for each year of childhood, or should childhood exposures be averaged in five-year intervals? The latter method would certainly reduce estimates of how much carcinogenic risk children face from their diet. Past failure to control data quality, analytical methods, and the expression of uncertainty has led to seemingly endless debates over magnitudes and distributions of risk. The regulatory system, long stagnant from its nonstrategic approach to risk assessment, may soon be paralyzed.

Precaution in Standard-Setting

> Set contamination limits that are precautionary in the face of uncertain evidence of risk.

As earlier chapters have shown, our history of pesticide management has not been cautious. The current system has a number of defects that lead to this conclusion: (1) tens of thousands of pesticide use registrations have been issued by the federal government, (2) most of the toxicity data used to support these licenses is now considered insufficient for a complete health-hazard assessment, (3) residue monitoring of food, water, and air is inadequate to accurately predict human exposure, (4) most of the standards for allowable exposure have been set by weighing economic benefits against risks, and are therefore not necessarily health protective, (5) risk-averaging methods employed by EPA have neglected the heightened exposure and susceptibility of groups such as children, (6) once tolerances have been established, they normally remain in force for at least a decade, despite evolving knowledge of chemical toxicity and exposure that may change estimates of risk, and (7) maximum contaminant levels for single pesticides have been set without considering additive or synergistic effects with other chemicals also permitted as residues in food or other environmental media. For all of these reasons, risks permitted by the current complex of registrations and tolerances are highly uncertain. Given this background, it seems wise to be precautionary when setting standards for allowable pesticide exposure in all environmental media. This should be accomplished by using ample margins of safety to protect against all suspected adverse health effects.[13]

Cross-Media Law

> The potential accumulation of risks across toxic compounds and across environmental media should be controlled for each type of adverse health effect.

This proposal requires comparative evaluation of risks across media, across types of pollutants, and among pollutants such as pesticides and demands a significant additional commitment to environmental science. EPA should establish cross-media and complex-mixture risk ceilings for each type of possible toxic effect, such as cancer.

Risk management by toxic endpoint should cut across laws and EPA programs and be intelligible to the public. Many presume that EPA keeps track of total risk from all possible sources of contamination. It does not, nor does any other federal agency. Pesticides provide an excellent example. EPA has had an unwritten rule that allows a one in 1 million cancer risk for any single pesticide across all crop uses. There are 325 food-use pesticides licensed by EPA, and nearly one-third of these are now suspected of being human carcinogens. Thus the one in 1 million ceiling has become—on average—an acceptable risk of nearly one in ten thousand. This allowance does not consider cancer risks from exposure to pesticides from sources other than foods (such as indoor pest control, lawn care, contaminated drinking water, or pet flea collars). Nor does it consider cancer risks from compounds other than pesticides, such as hazardous air and water pollutants. And because these are average estimates, some are most likely accumulating far higher risks.

A comprehensive risk accounting system would at first be of little use for decisionmaking, due to the current range in the quality of knowledge about the fate and effects of different types of toxins. Gradually, however, such a system would restructure and broaden the questions posed to scientists and would expand the context within which individual chemical management decisions would be made. Congress and EPA would slowly improve their understanding of the relative significance of the environmental health risks we face. Legislators and the public would then better understand how to achieve the highest level of risk reduction at the least cost.[14]

Avoiding High-Risk Technologies

> Prohibit the licensing of technologies that pose extreme risks, unless compelling evidence of risk throughout the product life cycle is presented.

Some pesticides are so extremely toxic that there is little confidence that their risks—throughout their product life cycle—can be effectively managed by traditional methods such as packaging, labeling, tolerance-setting, applicator training, protective clothing, and disposal requirements. Or perhaps the compounds can be managed effectively only in nations with vigilant regulatory systems. To control significant pesticide risks may require the careful training of pesticide applicators, a relatively sophisticated legal and administrative structure for enforcing restrictions on a pesticide's use, and a network of agricultural extension agents who may provide expert knowledge of risks in the field. In many poorer nations, these approaches to risk management have little hope of being employed. This partly explains why nearly 95 percent of pesticide induced deaths occur in low-income nations, even though most pesticides are applied in high-income nations.

Strategic Risk Reduction Plans

> Allocate regulatory attention to reduce the greatest risks in the shortest possible time.

When pesticides pose risks that exceed trivial levels, reform should encourage the substitution of less risky alternatives. Substitution, for example, is now severely limited by the fact that alternatives are regulated independently. This situation has led to the staggered and uncoordinated production of health-effect data among possible pest-control substitutes, making their comparison impossible.

Strategic risk reduction will require concentrating regulatory attention on several key components of risk estimates: those compounds that are the most potent; the most susceptible; those most exposed; primary routes of exposure; and the most vulnerable ecological areas. Given similar estimates of the magnitude, distribution, and certainty of risks posed by competing technologies, attention should be directed toward those risks that may be reduced at the least cost. In particular, food regulations should consider the risks of substitutable pesticides for individual crops and encourage farmers to select less toxic compounds. Because knowledge of risk is continually changing, as is the available pool of pesticides for any pest problem, the system should reward the continual substitution of ever-less-harmful compounds. Current, high-quality data on the fate and effects of pesticides used on each crop should be employed to assist these decisions. In cases where current registrations permit nontrivial risks, the government should design and implement risk-reduction plans, organized by toxic endpoint (for example, cancer or cholinesterase-inhibiting effects).

Pollution Prevention

> Apply and manage pesticides so that they produce the least possible contamination of the environment.

Where similarly effective substitutes are available, pesticides that are the least persistent, least mobile, least likely to bioaccumulate, and least toxic to nontargeted species should be chosen. Special concern should be given as well to where these compounds are used; some ecological regions increase the potential for pesticide mobility. These include areas where permeable soils overlie groundwater aquifers; lands adjacent to streams or waterways; sensitive plant and wildlife habitats; and agricultural lands lying adjacent to residential or recreational facilities where spray drift could result in significant human exposures.

Pollution may also be prevented by reducing pesticide use and by employing integrated pest management and organic farming practices. Commodity price supports should be revised where they encourage high levels of pesticide application; subsidies should instead be shifted to promote "low-input" agriculture, which achieves pest control through practices such as crop rotation, intercropping, and mechanical cultivation.

Because it is possible that most of pesticides applied may never reach their target pests, intensive investigations should be conducted to determine whether pests are present prior to chemical application. Agricultural extension agents already play an important role in this work and their efforts could be further supported.

Funding Risk Assessment and Management

> Funding for scientific investigation of pesticide risks and their management should come entirely from the producers of the technology.

As regulatory responsibilities continue to shift from federal to state authorities, consistent and effective risk assessment and management will become ever more difficult. But there is widespread industry interest in uniformity of regulation, given the interstate and international character of pesticide product and food markets. It would be especially helpful if state governments collected land-use and pesticide-use data and monitored residues in food and water supplies. Funding for these tasks should come from producers, and it should be adequate for states both to accomplish any delegated tasks and to ensure that uniform, high-quality monitoring, analytical, and recordkeeping practices are maintained.

Epilogue

I began this project by wondering how society should best balance risks and benefits from pesticide use given the diversity of problems that pesticides both solve and create. I have come to believe that this is the wrong question to ask, because the discretion to balance has resulted in choices that give great weight to estimated benefits but neglect the distribution of risks among susceptible populations such as children.

Since World War II, two powerful Western philosophical traditions have given legitimacy to these choices. The first is utilitarian thought, which suggests that public decisions are best if they maximize collective satisfaction. The second is liberalism, especially the assertion that a primary purpose of government is to protect broadly defined individual liberties and rights. Each philosophy is in turn supported by faith in the human capacity to make rational choices, especially the belief that humans can understand and predict the effects of human behavior and technology.

Utilitarianism rests on the premise that science and analysis may provide reasonably accurate predictions of the future effects of collective choices—both benefits and harms—and reliable estimates of how society will value those outcomes. The utilitarian is directed toward choices which maximize aggregated social values. Liberals—those who broadly define and defend individual liberties and rights—have depended upon science and rational analysis in a very different way. They are skeptical of the predictive accuracy of analytic efforts, especially claims that specific behaviors or technologies have caused damages—claims often made by environmental health advocates in their attempts to restrict private rights. Together, utilitarian and liberal beliefs have provided a mutually reinforcing moral logic for the granting of tens of thousands of pesticide use entitlements.

The Limited Rationality of Environmental Science

The global scale and intensity of pesticide use ignite questions of such enormous diversity regarding chemical movement, fate, and biological effects that we may never fully understand the consequences of past licensing and use behaviors. Former efforts to predict the adverse effects of pesticides have produced a history of errors. Scientists misjudged the rapid evolution of insect resistance to pesticides and parasite resistance to drugs, which together fueled the market for new or more concentrated biocides. Experts also failed to forecast the persistence, mobility, and bioaccumulative potential of many chlorinated pesticides and their byproducts. These miscalculations in turn led to low estimates of human exposure to mixtures of pesticides that have contaminated diverse environmental media. Finally, government officials have long overlooked how heightened childhood exposure and susceptibility to adverse effects place children in greater danger than adults. Further, although children happen to be the primary example used in this book, government has similarly neglected variance in exposure and susceptibility that may be associated with ethnicity, other age groups such as the elderly, and medical conditions such as pregnancy or illness.

Although these misunderstandings occurred independently from one another, they had the collective effect of distorting knowledge of the magnitude and distribution of human exposure and risk. These errors resulted in part from the ways in which scientists and regulators narrowly framed research questions in space and time. Oversimplified assumptions, such as rapid pesticide degradation and uniform public exposure, further diminished the accuracy of risk forecasts. Finally, these distortions grew from the presumption that no evidence of contamination meant no contamination existed. If no residues remained after pesticides were applied, why should society bear the expense of testing for toxicological effects? Increasingly sensitive residue-detection technology, however, proved these conclusions to be premature and prompted the government to demand a wave of new toxicity tests. In many instances, when older pesticide residues were carefully tested they were found to pose risks to humans. But because the compounds were already licensed and markets were established for thousands of products, the stage was set for a regulatory nightmare even before EPA was formed.

The limited vision surrounding pesticide research is perhaps best illustrated by findings that chlorinated pesticides have continued to move from one human generation to the next in contaminated breast milk. Policy responses to these initial findings were delayed for decades while human exposure continued. Government scientists knew by 1951, for example, that chlorinated insecticides were accumulating in human fat tissues and contaminating breast milk. Yet USDA and FDA failed to respond effectively through regulation for nearly a quarter-century, exposing tens of millions of additional people in the United States to these residues. This is not a claim of damage, but it is one of contamination and risk imposition. Our attention should be similarly captured by the strength of evidence concerning the scale and breadth of pesticide contamination of foods, water, and human tissues and their adverse effects demonstrated among highly exposed people and other species. Although available evidence does not constitute proof that humans are normally damaged by low-level exposures to pesticide residues, it does suggest that the potential for human damage is significant—and uncontrolled by government management efforts that are fractured and in disarray.

Government continued to license the use of old and new pesticides through the last half of the twentieth century, according to the logic that exposure could be controlled to protect against toxic effects. All humans are regularly exposed to pesticide residues as complex mixtures in a variety of environments, especially the food supply. The sheer number and volume of pesticides used around the world, along with the expense of environmental monitoring for residues, ensure that government has very little understanding of how we are exposed. Thus controlling risks by governing exposure could not help but fail, simply due to the enormity of resources required to understand the pathways that residues follow through the environment and their fate in our water, food, and bodies. This failure has left government, especially EPA, in the insupportable position of claiming the safety of our food, water, air, and workplaces without the evidence to demonstrate these reassurances.

Under these conditions, pesticide registration was and continues to be an act of human experimentation. Following World War II, the experiment grew quickly, as markets and chemical use spun well beyond government control. Before EPA was created, not only was the experiment uncontrolled, it was conducted on unknowing and unconsenting participants. This is not a condemnation of science or scientists. Nor is it an attack on the manufacturers who tended to concentrate their attention on the immediate benefits and sales potential of their products. It is a criticism of government, which has allowed the massive experiment to evolve through tens of thousands of disconnected, incremental decisions. The international character of the experiment has grown far beyond the ability of any national government to manage it without expending enormous public resources to track down the fate and effects of all possible pesticide mixtures that humans encounter.

Science in the Liberal Tradition

Within most liberal democracies, efforts to regulate market externalities are viewed as threats to individual rights, especially property rights. If government has a fundamental purpose of protecting individual rights, how does it justify their constraint? Science plays a crucial role in providing a rationale for and against the limitation of these rights. Early perceptions that pesticides posed few health or ecological risks, or that exposure was manageable, justified a pattern of government licensing that in turn fueled the rapid expansion of domestic and global markets. As evidence mounted that pesticide residues were more difficult to control or that risks were higher than earlier presumed, manufacturers began treating licenses and market shares as rights, or, in effect, as a form of property. They demanded a high degree of certainty in damage forecasts before agreeing to remove pesticides from the market. Much of this history, told earlier, is one in which the inability to produce reasonably convincing estimates of future damage has supported government decisions to license pesticides to produce relatively certain near-term benefits.

The personality of the pesticide industry has been innovative, aggressive, and experimental, and impressive in both scientific and legal talent. This expertise has been used with great effectiveness to secure a variety of advantages from governments: to register admittedly risky products; to provide an image of risk manageability (if not safety); to open and maintain access to domestic and international markets; to establish and enforce product patent claims; and to set allowable contamination limits within a variety of environmental media. Government's primary effect in pesticide management has been one of promoting and protecting a global exchange, rather than ensuring that human exposure and risk is maintained beneath a health-protective threshold.

Manufacturers became highly expert in the use of scientific methods to continually challenge claims that pesticides posed unreasonable risks. The pursuit of certainty through science became a way of protecting the rights to use risky technologies and of securing or expanding trade. It was especially effective in postponing the prohibition of technologies that were suspected of inducing delayed effects but which left no unique signature. Causality under these conditions is extremely difficult to prove. Without proof, suggestive evidence of damage normally motivated government to demand more research rather than to adopt precautionary regulations.

The sheer enormity of the environmental health questions posed by pesticides creates a management dilemma. The evidence produced to respond to these questions has suffocated management efforts and distracted government from identifying and controlling the most significant threats to environmental health. By the mid-1990s, EPA was overwhelmed with responses to its comprehensive request during the 1980s for more current environmental fate and toxicity data. At the same time, comparative risk analysis provides the only logical approach to identify and reduce the most significant dangers.

To break the pattern will not be easy. At a minimum, scientific inquiry must become simultaneously comprehensive and strategic—in order to direct the attention of both government and the public to the most significant threats through an analysis of relative risks. And for government to protect environmental health, it must squarely face the enormity of the analytical problem that their past licensing choices have created. Given the sea of uncertainty created by former choices, new technologies should be licensed only if they will replace riskier alternatives. Government and industry research should be redirected to speed up these substitutions.

The Insufficiency of Utilitarianism

> The main idea (behind utilitarianism) is that society is rightly ordered, and therefore just, when its major institutions are arranged so as to achieve the greatest net balance of satisfaction summed over all the individuals belonging to it. —**John Rawls, A Theory of Justice**

Utilitarian logic is a form of moral reasoning often used to define just government. To a utilitarian, a collective choice is just if it yields the greatest possible satisfaction of desires, and fair decisions are those that account for all possible outcomes and maximize social value. Knowing if a decision is just depends on our ability to forecast outcomes and to understand how all potentially affected parties may desire or value those outcomes. These are hardly trivial demands on the natural and social sciences, and our use of this standard in pesticide control demonstrates considerable faith in our ability to conduct rational analyses from impoverished and loosely relevant data.

Utilitarians have historically had little concern about the *distribution* of satisfaction or dissatisfaction. Policy is justified if the gains of some are greater than the losses of others, even if the losses are incurred by a majority or by those already disadvantaged. To the utilitarian, what is rational for society—selecting policies that produce net gains—is similarly rational for the individual.

Relying on *aggregated* estimates of benefits and costs, rather than requiring an understanding of how they are distributed in society, also provides enormous analytical relief. But relaxing the need to consider variance in the distribution of negative outcomes has had important implications for the least advantaged in society. If the concern is benefit distribution, such as providing a safety net for income, health care, or education, avoiding knowledge of who is most needy averts the social costs of both finding them and improving their condition. In terms of how risk is distributed—such as in the case of childhood susceptibility to pesticides—avoiding the identification of the most exposed and susceptible makes policies that achieve an average level of protection seem reasonable.

The decision standard that USDA, FDA, and EPA have applied to pesticide registration and tolerance decisions during the past five decades is essentially utilitarian; it requires that benefits exceed risks to justify federal registration of active ingredients and the products that contain them. By averaging risks across the population while pinpointing benefits, government has justified tens of thousands of separate pesticide-use entitlements. Although these choices may systematically disadvantage groups such as children, farmworkers, or the ill, utilitarians would still defend this regulatory behavior with a conclusion that somehow the greater good has been served.

This interpretation is not meant to diminish the considerable benefits pesticides have provided in the form of disease control and crop protection. I readily admit that DDT alone has saved the lives of millions of children and that insecticides have avoided many billions of dollars in crop losses. But a pattern of thought and analysis has evolved that has systematically neglected the distribution of risk, and this form of thinking is deeply rooted in moral beliefs concerning liberty, utility, and science, which have long been central pillars of most Western political institutions.

Pesticide management, like many other collective choice problems, presents an extraordinarily complex set of prediction and valuation problems. The reasoning that lies behind registrations and tolerances for any single chemical rests on the conclusion that benefits have exceeded costs. Making this judgment, however, normally requires comparing quite disparate types of outcomes—such as health effects and increased crop yields. The only way to make these comparisons is to express the effects in common units of value. For a pesticide that may influence crop production, the incidence of insect-borne disease; jobs; the balance of trade; and the health effects of those who are exposed in food, water, air and the workplace all need to be considered. A probability must be assigned to each possible outcome, which must then be valued. Although the types of outcomes evaluated (as well as the methods of prediction and valuation) have varied wildly among different pesticide decisions, certain patterns are discernible.

Balancing choices were strongly influenced by perceptions of the certainty of anticipated outcomes. Because near-term benefits were believed to be more certain than long-term costs, net welfare or cost-benefit assessment tended to support policies that allow the imposition of significant risks. Accurate understanding of near-term benefits is obviously necessary to justify the costs of bringing new products to the marketplace. Often, these choices are justified by averaging risks, masking variance in risk distribution, and claiming that negative outcomes are both uncertain and distant.

The averaging of risks has been complemented by the opposite analytical strategy with respect to benefits. When Congress or governmental agencies are deliberating over changes to regulations, manufacturers have made the potential costs of risk reduction to individual corporations and specific sectors of the economy very clear. Although it is often considered a separate issue, cost-benefit assessment is fundamentally tied to risk assessment because the probability of future damages must be estimated before these potential social costs can be valued.

A Revolt Against Regulation

A strong and well-organized backlash against precautionary regulation grew quickly during the 1990s. A large coalition—including industry groups protesting the costs of regulation, state and local governments overwhelmed by unfunded federal mandates, and "wise use" advocates claiming government invasion of individual rights—together found widespread Republican Congressional support following the 1994 elections. During the 1980s these groups had been only loosely connected, and they were far less effective in influencing a Democratic Congress. They were active during President Reagan's administration and were especially gratified by the efforts of Vice President Bush, who pressed for regulatory relief. But their true coalition occurred during President Bush's administration—coordinated in part by the White House through the Council on Competitiveness and galvanized when Republicans gained dominance of Congress in 1994.

If there has been any unifying philosophy behind the movement, it has been utilitarianism. The coalition's message is simple: federal regulation has cost far more than the gains it produced. These proponents of regulatory relief emphasize the need for risk and cost-benefit analyses, which they believe should be employed to better determine the net welfare effects of new significant regulations. Their political strategy has become one of broadening the context within which environmental problems were considered to see if funds once spent to lessen the health risks in the environment might provide greater reduction of other types of risks at a lower cost.

The legislative success of environmentalists during the 1970s and 1980s was in part a result of their ability to break environmental problems into discrete (and easily recognizable) compartments such as air and water pollution and pesticides. Yet the utilitarians seized on the artificiality of the problem boundaries, and sought to reconnect the environmental health issues to broader questions of economic welfare using the common metric of social cost. They simply asked: Why bear such high regulatory costs for environmental protection, when greater improvement in welfare could be more efficiently achieved in other arenas? The extraordinary costs and marginal success of the Superfund program—created to clean up orphaned hazardous waste sites—provided an easily ridiculed example of health risk management, and the trading of logging jobs for spotted owls in the Pacific Northwest provided a natural resource management counterpart.

The coalition's demand for synoptic, comprehensive analysis appeared to be simple common sense. But comprehensive analysis alone will hardly compensate for the errors generated by incremental analysis described in this book. These errors have been costly to society in terms of the health and ecological damages incurred, risks imposed, and expenditures paid to avoid or insure against risks. Environmentalists and consumer protection advocates claimed that the utilitarians purpose was less to illuminate and manage the most significant environmental health risks confronting society than to paralyze regulatory machinery through endless analysis.[1]

Spheres of Non-Injurious Freedom

Many have set out to correct the utilitarians' neglect of distributive effects. John Rawls, for example, provided several principles to govern choices which allocate goods and harms. For social institutions to be just, he argued, they must provide (1) the greatest possible system of individual freedom compatible with a similar system of liberty for all, (2) all primary social goods—liberty, opportunity, income and wealth and the bases of self-respect—and distribute them equally in society, unless an unequal distribution is to the advantage of the least favored, and (3) equality of opportunity.[2] To Rawls a just society is one that assigns greater priority to what is "right," not one that only produces the greatest "good."

Rawls's first principle—pursuing the greatest possible individual freedom, compatible with a similar system of liberty for all—provides some important moral guidance for pesticide management. If the imposition of risk without prior and informed consent is a violation of individual rights—constituting a loss of freedom—it seems quite consistent with a liberal view to protect freedom from significant risk. Freedom from risk imposed by others, intentionally or accidentally, should be thought of as another type of individual right deserving legal protection.

In this view, spheres of noninjurious freedom should be maximized and equitably distributed.[3] The central problem posed in this book is that the definition of the boundaries among these spheres—which establish the limits of rights and the allocation of obligations—is organically tied to a highly uncertain, ever-changing, and fractured knowledge base. The most crucial knowledge is understanding when the exercise of one individual's right harms the rights or interests of another. The certainty and significance of the damage will always be contested, as will be the causal link between the purportedly offensive behavior and the hypothesized damage.

In the history of pesticide management, both benefits and harms have been uncertain, and they have fallen on different groups. These conditions foster allegations of injustice and exploitation—claims that may grow from the perception that government regulation has failed to protect the disadvantaged from significant risks or a belief that regulation has imposed excessive costs on business.

Some of the most compelling objections to Rawls's arguments were generated by his second principle, which called for inequalities to be arranged so that they are to everyone's advantage and for those who are least advantaged to benefit most.[4] This requirement applies clearly to scarce resource allocation problems in which some may benefit more than others. But what guidance does it provide for choices, such as those surrounding the use of pesticides, that result in the allocation of both benefits and burdens?

To answer the question in the case of pesticides first requires that the least advantaged be defined perhaps by their heightened susceptibility, their higher level of exposure, or their relative inability to control risk (perhaps for lack of knowledge, wealth, or power). It is clear that the risk-benefit balancing rules in FIFRA and FFDCA have not required protection for those who are most vulnerable. What about the Delaney clause? By requiring zero cancer risk for everyone it might be interpreted as offering protection to the most vulnerable. But, as seen earlier, this clause applies to pesticides under very limited circumstances—only for one type of health effect, cancer, and only when a pesticide concentrates during food processing. This narrow scope prevents the Delaney clause's zero-risk standard from "favoring the least advantaged." A negligible risk standard for all possible types of health effects, however, may offer equal protection of freedom from significant risk imposition. The least advantaged—defined in this case to be children—would be sheltered beneath this shield.

Conditional Precaution

If the most vulnerable deserve health protection, then how should precautionary policy be defined, and what conditions deserve its application? The strictest definition would require that the most exposed and most susceptible individuals be protected with an additional margin of safety to protect against the errors common to risk estimation. In determining maximum allowable contamination levels, risk averaging and the consideration of collective benefits would not be allowed.

This suggestion, however, is conceptually similar to "safety net" policies that ensure equal access to minimum levels of income, health care, and education. Precautionary policy would prevent significant risks from falling on the most vulnerable and would take the form of an acceptable risk ceiling. Practically, this policy would require defining a ceiling on allowable risk from the mixtures of toxic compounds—those likely to produce the same type of toxic effect—that we are likely to encounter in different environments.

Underlying this proposal is a belief that everyone has a fundamental right to be protected from significant and reasonably certain risks imposed by the behavior of others. Definitions of what constitutes a significant risk, and when evidence is reasonably certain, should be articulated as clearly as possible for each environmental health threat confronting society. Precautionary policy is thus conditional, in that it should apply to cases where risks are deemed significant and when certainty is sufficient. Although experts are perhaps best suited to address how statistical certainty surrounding future damage estimates should be expressed, an educated public seems the most appropriate party to define the distinction between socially acceptable and intolerable risks.

There do appear to be defensible exceptions to the precautionary standard. These exceptions occur when (1) the individuals bearing the risks are also the ones who accrue the benefits, (2) these individuals fully understand the estimates of risks and benefits and are aware of alternatives, (3) they are responsible parties, and (4) they consent to the risk imposition. Many decisions to vaccinate appear to meet all of these conditions. Government decisions that permit children to be exposed to significant pesticide risk, however, appear to comply with none of them.

The greatest protest against this principle will come when benefits and risks are both substantial and reasonably certain. Then it should be investigated if strategies might be devised to protect those bearing the risks while safeguarding the benefits. There may be ways, suggested in the previous chapter, to offer special security for children. These should include preventing the use of highly toxic pesticides within common childhood environments such as homes, schools, day care centers, on pets, and on lawns, as well as being vigilant about the presence of these compounds in food and drinking water.

Some justify the imposition of significant risks on others based upon a claim of equality in risk distribution—that is, that everyone is a child at some point; therefore everyone has an equal accumulation of food-borne pesticide risk. Those who take a broad view of the boundaries of individual rights, however, would cry foul. The imposition of significant and relatively certain risks without the knowledge and consent of responsible parties (not children) violates a very fundamental human right. The level of equality in the distribution of the health risk is irrelevant. Instead, the magnitude and certainty of risk for any individual—along with the individual's ability to know, control, and avoid it—determine if and when any exposure has violated a human right beyond an acceptable level.

Looking Back from the Future

During the twenty-first century, debates over environmental health quality will likely occur within a political environment severely strained by increasingly scarce natural resources, regional population pressures, a growing indebtedness of nation states, widening gaps between rich and poor, and increasing competition for more expensive public services. The most dominant environmental force will probably be the growing power of domestic and international markets, which will continually shape political institutions and patterns of consumer demand. The most likely outcome for law is that it will become even more fractured at every level of government from local to global. It will not likely evolve in a systematic pattern that can promote the rational analysis necessary for effective risk assessment and management. Instead, it will probably grow in a way that will promote and protect exchange and markets. The electronic information revolution now well under way guarantees that our database will grow exponentially, but not necessarily in an organized manner. Thus the greater quantity of information will not necessarily translate into a better understanding of how human behavior affects environmental quality or of how contamination affects human health. The pursuit of knowledge through the specialized sciences is also clearly necessary, but it too is insufficient to describe the effects of a technology as complex and diverse as pesticides. Moreover, the fractured law–fractured science entanglement, if not challenged, will continue to be an ideal political strategy that expands the authority of domestic and international markets.

Far in the future, environmental historians may reflect upon how we managed and justified the distribution of risks and benefits associated with pesticides. Few are likely to untangle our current amalgam of risk assessments and cost-benefit analyses. Those who do may be puzzled over the late-twentieth-century obsession with technical forms of analysis that distracted us from seeing the relatively simple moral dimensions of the pesticide problem. Their conclusions regarding the quality of our governance will not likely be tied to the excellence of our science, which by then will look quite primitive, or to the predictive success of our risk estimates. Instead, they will more likely be fascinated by how we decided to allocate risk. Many will look back and wonder about our experimental arrogance and will search for a moral logic more compelling and sensitive to distributive justice than simple utilitarianism.

Abbreviations

BNA	Bureau of National Affairs
CFR	Code of Federal Regulations
DORFA	U.S. Committee on Agriculture, Department of Operations, Research, and Foreign Agriculture
EDF	Environmental Defense Fund
ELR	*Environmental Law Reporter*
EPA	U.S. Environmental Protection Agency
ERC	*Environmental Reporter* Case
FDA	U.S. Food and Drug Administration
FEPCA	Federal Environmental Pesticides Control Act
FFDCA	Federal Food, Drug, and Cosmetic Act
FIFRA	Federal Insecticide, Fungicide, and Rodenticide Act
FR	*Federal Register*
GAO	Government Accounting Office
H. Rep.	Report of the U.S. House of Representatives
HEW	U.S. Department of Health, Education, and Welfare
IARC	International Agency for Research on Cancer
IOM	Institute of Medicine
Mrak Commission	U.S. Dept of Health, Education, and Welfare, *Report of the Secretary's Commission on Pesticides and their relationship to environmental health* (1969)
NAS	National Academy of Sciences
NCI	National Cancer Institute
NIH	National Institutes of Health
NRC	National Research Council
NRDC	National Resources Defense Council
NTIS	National Technical Information Service
NTP	National Toxicology Program
OPP	U.S. Environmental Protection Agency, Office of Pesticide Programs
OSHA	U.S. Occupational Safety and Health Administration
OTA	Office of Technology Assessment
PL	Public law
RFF	Resources for the Future
S. Rep.	Report of the U.S. Senate

SAB	U.S. Environmental Protection Agency, Science Advisory Board
SAP	U.S. Environmental Protection Agency, Scientific Advisory Panel
SEER	U.S. National Cancer Institute, Surveillance, Epidemiology, and End Results Program
UNEP	United Nations Environment Program
USC	*U.S. Code*
USCA	*U.S. Code Annotated*
USGS	U.S. Geological Survey
WHO	World Health Organization

Notes

Chapter 1. The Global Experiment

1. Estimates of yearly pesticide use worldwide range between 4.5 billion and 5 billion pounds. See A. Aspelin, *Pesticide industry sales and usage: 1992 and 1993 market estimates* (Washington, D.C.: EPA, June 1994); see also D. Pimental et al., "Environmental and economic costs of pesticide use," *BioScience* 42 (1992): 750–60; T. Dunlap, *DDT: Scientists, citizens, and public policy* (Princeton, N.J.: Princeton University Press, 1981), 253–54, app. E; EPA, *Pesticide industry sales and usage* (Washington, D.C., Sept. 1995), which states that by 1985 in addition to the nearly 1 billion pounds of pesticides used in agriculture, 1.5 billion pounds of sulfur, wood preservatives, and disinfectants were also used; Mrak Commission, 315, which reports that in 1966 over a billion pounds of pesticides were manufactured in the United States alone, and nearly 18 percent of this total was DDT; and C. Cameron's testimony in House Select Committee to Investigate the Use of Chemicals in Food Products, *Chemicals in food products,* 81st Cong., 1st sess., 1950, 333. Cameron, the scientific director of the American Cancer Society, estimated in 1950 that nearly 80 million pounds of arsenical pesticides—principally in the form of lead and copper arsenate—were applied yearly in the United States.
2. G. Ware, *Pesticides: Theory and application* (New York: W. H. Freeman, 1983).
3. The EPA's requirement that new environmental health and safety data be submitted to justify reregistration of previously registered compounds and products has caused many manufacturers to remove pesticides from the market because data-production costs exceed potential profitability.
4. Aspelin, *Pesticide industry*. Pesticide use in the United States during 1993 totaled approximately 2.2 billion pounds. Aspelin concludes that pesticide usage amounts to approximately 4.4 pounds per person in the United States. In 1993–94, EPA estimated that these included 860 active ingredients registered domestically.
5. U.S. sales are estimated to be $8.5 billion.
6. For example, in 1993 Monsanto purchased Chevron's home and garden Ortho Consumer Products Division for $400 million; American Cyanamid agreed to acquire Shell's agrochemical business; ICI demerged its operations into Zeneca and ICI; Shell divested its 33.3 percent share of Nocil, an Indian company; Roussel-Uclar increased its investment in a subsidiary, Roussel Vietnam, from 11 percent to 60 percent; Nufarm acquired Rhone-Poulenc's phenoxy herbicide plant in the United Kingdom; and American Cyanamid divested its dimethoate business to Wilbur Ellis. See *Agrochemical Service: Update of the companies* (Edinburgh: Wood MacKensie, 1994).
7. A play about the tragedy, *Story of the Gas Tragedy* by Rajiv Saxena, has appeared throughout India. Although it originally portrayed Union Carbide as the chief villain, Saxena recently revised the script to paint a darker picture of the Indian government, which has been slow to disperse settlement funds to victims and has meanwhile accumulated interest payments. Nearly 6,600 death claims have been accepted resulting in the payment of only $2 million as of December 1994. A decade after the release, less than 25 percent of the funds has been dispersed to those claiming injury. See Com-

bined Wire Services, "Victims of Bhopal disaster still suffering a decade later," *Hartford Courant,* 2 Dec. 1994, A15.

8. For a review of these practices, see K. Reilly, "International trade in hazardous pesticides: Risks and remedies," senior essay, Yale University, 1992. See also National Coalition Against the Misuse of Pesticides, *Regulatory and legal chronology of chlordane/heptachlor in the U.S.* (Washington, D.C.: National Coalition Against the Misuse of Pesticides, 1989).
9. See D. Weir and M. Shapiro, *Circle of poison: Pesticides and people in a hungry world* (San Francisco: Institute for Food and Development Policy, 1981). See also United Nations, *Consolidated list of products whose consumption and/or sale have been banned, withdrawn, severely restricted or not approved by governments,* 2d issue (New York: United Nations, 1987).
10. K. Smith, *Pesticides exported from the U.S., 1992–1994* (Los Angeles, Calif.: Foundation for Advancements in Science and Education, 1995).
11. The FDA recognizes this and samples imported foods more heavily than domestically produced foods. See, e.g., GAO, *Pesticides: A comparative study of industrialized nation's regulatory systems* (Washington, D.C., 1993), PEMD-93-17; GAO, *Five Latin American countries' controls over the registration and use of pesticides* (Washington, D.C., 1990), T-RCED-90-57; GAO, *Imported meat and livestock: chemical residue detection and the issue of labeling* (Washington, D.C., 1987), RCED-87-142; GAO, *Pesticides: Better sampling and enforcement needed on imported food* (Washington, D.C., 1986), RCED-86-219; GAO, *Pesticides: Need to enhance FDA's ability to protect the public from illegal residues* (Washington, D.C., 1986), RCED-97-7; GAO, *Agricultural trade: Causes and impacts of increased fruit and vegetable imports* (Washington, D.C., 1986), RCED-88-149BR.
12. A single pesticide may be licensed for use on dozens of crops. Chlorpyriphos—an example considered in chapter 11—was registered for use on nearly sixty crops. See *CFR,* vol. 40, §180 (1995). It is also allowed for use in pet flea collars and is commonly applied as an indoor insecticide to combat fleas and other nuisance insects. EPA requirements that manufacturers submit health and environmental effects data have grown rapidly since 1970. As a result of this effort, we may know more about the health effects of pesticides than many other classes of industrial chemicals.
13. EPA, *OPP annual report* (Washington, D.C., 1994), EPA 735-R-95-001.
14. These statistics are often debated, most likely due to governmental difficulty keeping track of pesticides in the domestic and international marketplace. See, for example, Mrak Commission, 46. In 1969, the Mrak Commission estimated that there were 900 active ingredients in use, formulated into nearly 60,000 separate preparations. There are also an estimated 1,600 "inert" ingredients, which are mixed with the active ingredients; these, however, have rarely been tested for toxicity and have included compounds such as benzene, asbestos, and kerosene.
15. *U.S. Statutes at Large* 86 (21 Oct. 1972), 973.
16. EPA, *OPP Annual Report* (Washington, D.C.: 1994), 17.
17. For example, one regulation limits the amount of the fungicide benomyl to 7 ppm on fresh apples, whereas another limits benomyl to 10 ppm on grapes. See *CFR,* vol. 40, 180.4 (1986); and *CFR,* vol. 21, §193.
18. EPA, *OPP annual report,* 16.
19. D. Pimental and L. Levitan, "Pesticides: Amounts applied and amounts reaching pests," *BioScience* 36, no. 2 (1986): 86–91.

20. NAS, *Alternative agriculture* (Washington, D.C.: National Academy Press, 1989).
21. D. Pimental et al., "Environmental and economic impacts of pesticide use," *BioScience* 41, no. 6 (June 1992).
22. Mrak Commission, 44–46.
23. Pimental et al., *Environmental and economic impacts.*
24. NAS, *Regulating pesticides in food: The Delaney paradox* (Washington, D.C.: National Academy Press, 1987).
25. The term *tolerance* is confusing. It implies that we can tolerate contamination at the maximum allowable level, whereas in reality few tolerances were established to be health protective. It really means the level of contamination that is legally tolerated.
26. Designing electronic computer codes that would allow food intake, residue, and toxicity data—all managed by different agencies—to be linked for the purpose of exposure and risk assessment took two people nearly six months to complete.
27. Cancer risk assessment guidelines had just been adopted by the agency in 1985, with guidance from the White House Office of Science and Technology Policy. See Office of Science and Technology Policy, "Chemical Carcinogens; a review of the science and its associated principles," *FR* 50 (1985): 10371. Even by 1995 there was little consistency among methods used to estimate cancer risks from environmental contaminants across the major regulatory divisions of EPA.
28. The appropriate temporal unit of analysis for cancer risk estimation is the subject of considerable debate. See D. Murdoch, D. Krewski, and J. Wargo, "Cancer risk assessment with intermittent exposure," *J. Risk Anal.* 12, no. 4 (1992): 569–77.
29. J. P. Wargo, "Analytical methodology for estimation of oncogenic risks associated with food use agricultural chemicals," in NAS, *Regulating pesticides in food,* 174–95.
30. These twenty-eight compounds were selected based upon EPA estimates of the quality of animal study and the certainty of effect. Just when quality and certainty become high enough to justify a quantitative indicator of a pesticide's potency (calculated as the slope of the dose-response relation) is a matter for scientific judgment.
31. EPA decides when the strength or "weight" of evidence is sufficient to classify a compound as a possible or probable carcinogen. Once this threshold of quality has been exceeded, the agency will calculate the "cancer potency" of the compound, which is expressed as the slope of the dose-response relation. This variable is also known in the literature on risk assessment as the $q1^*$.
32. *USC* vol. 21, §346b (1984).
33. Although FDA has the authority to administer §409 and to regulate food additives, EPA has the responsibility to regulate all pesticide residues defined as food additives.
34. See NAS, *Regulating pesticides in food,* 19, chap. 6.
35. Processing studies have had the highest "rejection rate" due to the insufficient quality of all types of residue fate studies during the early 1990s. See EPA, *Annual report.*
36. Although the media focused predominantly on Alar, a carcinogenic plant growth regulator commonly used on apples, the NRDC report was a broad attack on the failure of government and industry to account for childhood exposure to pesticides, especially carcinogens and neurotoxins. EPA completed its third peer review of Alar on 26 July 1991 and concluded that both Alar and its metabolite UDMH were probable human carcinogens (Group B). See CBS, "'A' is for apple," *60 Minutes,* 26 Feb. 1989. This was followed by a second program: CBS, "What about apples?" *60 Minutes,* 14 May 1989. See also NRDC, *Intolerable risk: Pesticides in our children's food* (New York: NRDC, 1989).

37. This behavior has been termed an "escape through the *de minimus* window" by Richard Merrill, a scholar on FDA and EPA interpretations of the Delaney clause. See R. Merrill, "FDA's implementation of the Delaney clause: Repudiation of Congressional choice or reasoned adaptation to scientific progress?" *Yale J. on Reg.* 5, no. 1 (1988). See also R. Merrill and M. Taylor, "Saccharin: A case study of government regulation of environmental carcinogens," *Va. J. Nat. Res. Law* 5, no. 1 (1985).
38. This committee was rumored to have been formed over the objections of EPA administrators who actively argued against the appropriation of federal funds for the study of childhood risks from pesticides, fearing the committee would incite conflict over food safety.
39. See chapter 9. The age of key dietary surveys used to estimate pesticide exposure severely limits the quality of exposure forecasts. Data collected in 1977–78 were still used by EPA to estimate exposure in 1995. Advances in food processing, changes in marketing strategies, growing international food markets, and consumer concerns over high-fat diets have all influenced dietary patterns.
40. This appears to be true using either of two measurements of exposure, one expressed as the amount of the compound delivered per unit of body weight, and the other as the amount per unit of surface area. Although the expression of surface-area exposure as milligrams/meters^2/day is preferable as an indicator of organ size, most animal studies of toxicicology were conducted using a dosing regime based upon body weight and expressed in milligrams/kilogram bodyweight/day.
41. Chapter 10 explores how exposure estimates from benomyl vary tremendously depending on the sampling design used to collect residue data.
42. NAS, *Pesticides in the diets of infants and children,* 260–61.
43. Most pediatricians now define childhood as that interval between conception and the time when all organ systems and functions have fully matured, normally by age 18.
44. The biochemical mechanism of tumor induction for most carcinogens is poorly understood. Some toxins are believed to be initiators, inducing cellular damage necessary for later tumor development, whereas other compounds are believed to be promoters that further alter reproductive signals and functions through several stages until uncontrolled reproduction results. Thus exposure to initiators early in life increases the chance of successful promotion and tumor development later in life. Children may therefore be thought of as accumulating carcinogenic risk from pesticide residues in food and drinking water.
45. See chapter 11 for a discussion of childhood exposure to complex mixtures of neurotoxic (cholinesterase-inhibiting) pesticides in the diet; see also NAS, *Pesticides in the diets of infants and children,* 60–61.
46. R. Wiles et al., *Herbicides in drinking water* (Washington, D.C.: Environmental Working Group and Physicians for Social Responsibility, 1995).
47. The debate over endocrine-disruptive chlorinated compounds is discussed further in chapter 8.
48. House Committee on Energy and Commerce, Subcommittee on Health and the Environment, *Health effects of estrogenic pesticides,* 103d Cong., 1st sess., serial 103–87 (21 Oct. 1993). See also T. Colborn, S. vom Saal, and A. M. Soto, "Developmental effects of endocrine disrupting chemicals in wildlife and humans," *Env. Health Pers.* 101, no. 5 (Oct. 1993); L. S. Birnbaum, "Endocrine effects of prenatal exposure to PCBs, dioxins and other xenobiotics: Implications for policy and future research," *Env. Health Pers.* 102, no. 8 (1994): 676–79; D. L. Davis et al., "Medical hypothesis:

Xenoestrogens as preventable causes of breast cancer," *Envir. Health Pers.* 101 (1993): 372–77; and E. Carlsen et al., "Evidence for decreasing quality of semen during past 50 years," *Brit. Med. J.* 304 (1992): 609–13.

Chapter 2. The Urgency of Malaria

1. See, e.g., "Superior new insecticide called DDT," *Sci. Dig.* 15 (Mar. 1944): 93; "DDT considered safe for insecticidal use," *Am. J. Pub. Health* 34 (Dec. 1944): 1312; "DDT may control malaria," *Sci. News Letter* 46 (30 Dec. 1944): 418; C. M. Wheeler, "Control of typhus in Italy 1943–44 by use of DDT," *Am. J. Pub. Health* 36 (1945): 119–29; A. Macchiavello, "Plague control with DDT and 1080: Bubonic plague epidemic, Tumbes, Peru, 1945," *Am. J. Pub. Health* 36 (1946): 842–54; J. P. McEvoy, "DDT magic in Greece," *Reader's Digest* 53 (1948): 115–16; "DDT: Nobel prize in medicine for Dr. Paul Mueller," *Senior Scholastic* 53 (10 Nov. 1948): 9; "DDT saved 5,000,000 in last ten years," *Sci. News Letter* 62 (27 Dec. 1952): 409; "U.S. help to meet DDT shortage," *U.N. Bull.* 11 (1 Aug. 1951): 136.
2. L. J. Bruce-Chwatt, *Essential malariology* (New York: John Wiley, 1993).
3. During this century WHO has consistently estimated that between 1 and 3 million people have died annually from malaria, although others have estimated that as many as 6 million died per year during the 1930s. At the turn of the century, 250 million cases and 2.5 million deaths were estimated to have resulted from malaria. See WHO, *World Health Stat. Q.* (Geneva: WHO, 1988) 41; L. J. Bruce-Chwatt and J. Haworth, "Malaria eradication: Its present status," *Isr. J. Med. Sci.* 1 (1956): 284–89; and W. J. Hayes, *Handbook of pesticide toxicity* (New York: Academic Press, 1991), 11.
4. D. J. Singer and M. Small, *The wages of war 1816–1965: A statistical handbook* (New York: Wiley, 1986). Singer and Small's estimates are based upon battlefield casualties and do not include civilian casualties. The problems of developing comparable and accurate statistics is well recognized by these authors, especially as they attempt to distinguish between civilian and military personnel. This is especially difficult under conditions of civil and guerrilla warfare. This problem is similar to that of estimating the distinction between malaria-induced mortality and morbidity—because malaria may induce death directly, or indirectly by reducing immunity to other illnesses or by simply diminishing the strength of those responsible for subsistence food production or procuring medical supplies. See also J. S. Levy, *War in the modern great power system, 1495–1975* (Lexington: University Press of Kentucky, 1990).
5. WHO, "World malaria situation in 1990," *Bull. WHO* 70, no. 6 (1992): 801–7.
6. IOM, *Malaria: Obstacles and opportunities* (Washington, D.C.: National Academy Press, 1991).
7. W. Takken et al., *Environmental measures for malaria control in Indonesia: an historical review on species sanitation,* Wageningen Agricultural University Papers, 1991, 90–97. L. R. Beck et al., "Remote sensing as a landscape epidemiologic tool to identify villages at high risk for malaria transmission," *Am. J. Trop. Med. & Hyg.* 51, no. 3 (1994): 271–80. See also L. M. Camargo et al., "Unstable hypoendemic malaria in Rondonia: Epidemic outbreaks and work-associated incidence in an agro-industrial rural settlement," *Am. J. Trop. Med. & Hyg.* 51, no. 1 (1994): 16–25; A. A. Gbakima, "Inland valley swamp rice development: Malaria, schistosomiasis, onchocerciasis in south central Sierra Leone," *Pub. Health* 108, no. 2 (1994): 149–57; S. Hewitt et al., "An entomological investigation of the likely impact of cattle ownership on malaria in

an Afghan refugee camp in the North West Frontier Province of Pakistan," *Med. & Vet. Ent.* 8, no. 2 (1994) 160–64; B. M. Khaemba, A. Mutani, and M. K. Bett, "Studies of anopheline mosquitoes transmitting malaria in a newly developed highland urban area: A case study of Moi University and its environs," *East African Med. J.* 71, no. 3 (1994): 159–64.

8. Nearly 2,000 years ago, Marcus Terentius Varro suggested that a malaria-like disease was transmitted by "minute animals, invisible to the eye, which breed in swamps and reach the body by way of nose or mouth and cause diseases which are difficult to get rid of." See J. R. Busvine, *Disease transmission by insects* (Berlin: Springer-Verlag, 1993), 9.
9. Columella, writing in the first century A.D., also recognized an association between marshlands and disease, which he used to suggest sites suitable for country estates: "There should be no marshlands near the buildings . . . for [they] throw off a baneful stench in hot weather and breed insects armed with annoying stings which attack us in dense swarms . . . and which are infected with poison by the mud and decaying filth, from which are often contracted mysterious diseases whose causes are even beyond the understanding of physicians." G. Zattola, *World Health* (Geneva: WHO, 1961).
10. Busvine, *Disease transmission,* 18–27. Busvine provides an excellent summary of the research efforts of Lavaran, Manson, and Ross.
11. Although Ross was largely ignorant of the different species of mosquitoes, he did suspect that one kind was more efficient in promoting the parasite's life cycle than others. Ross experimented on a volunteer who allowed himself to be bitten by several mosquitos that two days before had fed on infected blood. The volunteer remained healthy, mostly because not enough time had elapsed for the parasite to reproduce within the insect. This failure distracted Ross, whose research was further interrupted by a career move that caused him to change the subject of his experiments from humans to birds.
12. IOM, *Malaria.*
13. S. Hewitt et al., "An entomological investigation of the likely impact of cattle ownership on malaria in an Afghan refugee camp in the northwest frontier province of Pakistan," *Med. & Vet. Ent.* 8, vol. 2 (1994): 160–64.
14. IOM, *Malaria.*
15. Grassi and Feletti first found that *P. vivax* caused malaria in 1890.
16. Blackwater fever gets its name from the color of urine, which darkens as the parasite dissolves red blood cells and these cells are discharged through the kidneys. This form of the disease usually occurs in those who have suffered intense, repeated, and untreated infections. It was first discovered by Welch in 1897.
17. *P. malariae* was the first Plasmodium species discovered to cause malaria by Lavaran in 1881.
18. Stephens discovered that *P. ovale* caused malaria in 1922.
19. A. J. Knell, *Malaria* (Amsterdam: Wellcome Trust, 1991). See esp. chap. 2.
20. The origin of malaria in the Americas is unclear. It may have migrated to North America from Asia across the land bridge, or it may have been introduced by European explorers. The disease clearly had an enormous effect upon rates and patterns of settlement, agriculture, and economic development in North America. Although two species of malaria, *Plasmodium vivax* and *Plasmodium malariae,* were introduced by the English when they settled in Jamestown, Virginia, the more deadly *Plasmodium falciparum* was introduced through the African slave trade, most likely from individuals carrying the parasite in their blood (possibly from stowaway mosquitoes).

Malaria became a significant motive for the movement of the Virginia capital to Williamsburg in 1699. The disease became endemic in the southern and western colonies and is known to have been deadly in New England, particularly near Boston, during the early seventeenth century. During the early 1900s nearly 500,000 cases of the disease were reported each year in southern states. See J. Duffy, *Epidemics in colonial America* (Baton Rouge: Louisiana State University Press, 1953). See also IOM, *Malaria,* 39; and W. Biddle, *Field guide to germs* (New York: Henry Holt, 1995).

21. R. Patterson, "Dr. William Gorgas and his war with the mosquito," *Canad. Med. Assoc. J.* 141, no. 6 (15 Sept. 1989): 596–97, 599.
22. For an excellent history of Panama, see G. Mack, *The land divided: A history of the Panama Canal and other isthmian canal projects* (New York: Knopf, 1944).
23. Even the first governor of the Canal Zone, General Davis, chastised Gorgas: "Sanitation! What has that to do with digging a canal? Spending a dollar on sanitation is as good as throwing it into the bay. It is for your good, Gorgas, I say this. You have harped on the mosquito idea until it has become an obsession, and your assistants have caught it too. Do, for goodness sake, get it out of your head." L. J. Warshaw, *Malaria: Biography of a killer* (New York: Reinhart, 1949): 126. The entire book is an extraordinary history, rich in detail, and well corroborated by other histories, though hardly well documented. It is also out of print.
24. "Marshy water, dirty water, stagnant water, might be a public evil—yes; but how could one so accuse the fresh and invigorating rain that had just fallen from the heavens?" See M. Gorgas and B. J. Hendrick, *William Crawford Gorgas: His life and work* (New York: Garden City Publishing, 1924), 176.
25. Ibid., 178.
26. Gorgas was promoted despite the objections of the commission's chair, George Goethals, who complained: "You know Gorgas, that every mosquito you kill costs the United States Government ten dollars." Gorgas responded: "But just think, Colonel Goethals, one of these ten-dollar mosquitoes might bite you, and what a loss that would be to the country." Ibid., 129.
27. W. C. Gorgas, *Sanitation in Panama* (New York: D. Appleton, 1915); see also T. W. Martin, *Doctor William Crawford Gorgas of Alabama and the Panama Canal* (New York: Newcomen Society of England, American Branch, 1947).
28. Having survived the threat of disease while traveling in the first boat to cross the canal, a canoe, he inadvertently paddled into a large blasting operation, even though men on shore furiously tried to warn him away. After stones and dirt rained down upon the canoe, Gorgas remarked: "A rather warm reception." See Gorgas and Hendrick, *Gorgas,* 256.
29. Gorgas went on to become president of the American Medical Association. In 1913 Gorgas redirected his efforts to eradicate pneumonia among black South African mineworkers. He later became surgeon general of the U.S. Army, where he oversaw the health of millions of troops operating in areas infested with insect-borne-disease. King George V of England eventually knighted Gorgas for his service to humankind.
30. Ibid., 137. One final lesson is tied to the extraordinary leadership and organizational skills that Gorgas supplied, perhaps described best by him: "The success of any system of sanitation which is more or less new to any locality will depend a great deal upon the choice of the man who has charge of carrying it into execution. If he believes in it, has tact, is enthusiastic and persevering, it will succeed. If he is discouraged by difficulty and opposition he will fail, even if his system is correct."

31. F. L. Soper and D. B. Wilson, *"Anopheles Gambiae" in Brazil* (New York: Rockefeller Foundation, 1943), 74.
32. Ibid., 74.
33. Ibid., 83.
34. J. R. Davis and R. Garcia, "Malaria mosquito in Brazil," in D. Dahlsten, ed., *Eradication of exotic pests* (New Haven: Yale University Press, 1989), 274–83.
35. Marco Polo is said to have brought pyrethrum home to Europe from the Far East. The heads of the Chrysanthemum flowers were ground and sprinkled in clothing during the Napoleonic wars to control body lice. Early in the twentieth century, the Japanese imported the flower for cultivation and were soon producing several thousand pounds per year. By 1940, the United States was importing nearly 4 million pounds per year. It was available as a spray by 1919, and was well known for its knockdown effect by inducing paralysis. G. McLaughlin, "History of pyrethrum," in J. E. Cassida, ed., *Pyrethrum* (New York: Academic Press, 1973).
36. F. G. Whitfield, "Air transport, insects and disease," *Bull. Ent. Res.* 30, no. 2 (1940): 365–442.
37. Soper and Wilson, *"Anopheles gambiae" in Brazil,* 139.
38. Malaria has had a profound influence upon military operations throughout recorded history. Alexander the Great, who conquered much of the world in the fourth century B.C., eventually succumbed to a fever believed to have been malaria. Throughout its history, Rome has been better protected by the legions of mosquitoes living in surrounding Campagna marshlands than by the organization, resources, and ferocity of its defenders. The armies of Otto the First were brought to a standstill by the disease in 964. The armies of Henry II were incapacitated by fever in 1022. Henry IV, who assaulted Rome on four occasions, learned to withdraw his troops during the malarious summer months. The city's defenders had relatively higher immunity from a lifetime of exposure to the parasite, leaving invaders, particularly those from the northern, less malarious regions, far more vulnerable. The mortality rate among popes of the Roman Catholic Church was so high that by the fourteenth century foreign popes—presumably with less immunity—were not allowed to live in the city. The ecology of malaria dictated land use in the case of at least one Italian settlement, Ostia, which was so malarious that it was transformed into a penal colony. Napoleon was known to have used malaria to his advantage against the British when they attacked his troops in the Walcheren Islands off the southern Netherlands, an area known to be infested with the disease: "We are rejoiced to see that the English themselves are in the morasses of Zeeland. Let them be kept only in check and the bad air and fevers peculiar to the climate will soon destroy their army." His prediction came true as the British became disease-ridden and were forced to retreat. During the Civil War, more than 1.2 million cases of malaria were recorded, resulting in at least 8,000 deaths. See L. J. Bruce-Chewatt, "History of malaria from prehistory to eradication," in W. H. Wernsdorfer and I. McGregor, eds., *Principles and practice of malariology* (Edinburgh: Churchill Livingstone, 1988).
39. IOM, *Malaria.*
40. Ibid., 293.
41. Warshaw, *Malaria,* 167–69.
42. In 1850, the Society of Pharmacy of Paris offered a prize of 4,000 francs to the first chemist to synthesize quinine, yet its artificial production eluded researchers until 1944, when two Harvard chemists, R. Woodward and W. Doering, successfully syn-

thesized the compound in their laboratory. See A. W. Haggis, "Fundamental errors in the early history of cinchona," *Bull. Hist. Med.* 10 (1941): 417; see also Missouri Botanical Garden, *Proceedings of the celebration of the three hundredth anniversary of the first recognized use of cinchon* (St. Louis, Mo., 1931).

43. See Mrak Commission, 45. See also G. Ware, *Pesticides: Theory and application* (New York: W. H. Freeman, 1983), 35. Metals such as lead, arsenic, and mercury have been recognized as poisons and used as pesticides for hundreds of years. The earliest recorded use of arsenic for garden pest control was in China, around A.D. 900. Arsenic in honey was used as ant bait in 1669. Together with lead, arsenic became widely used in agriculture for weed and pest control from 1870 to 1970. Sodium arsenite was used as both an insecticide and a herbicide. Other metal salts gradually came into use during the later nineteenth and early twentieth century, including copper, zinc, chromium, mercury, thallium, and selenium. One of the earliest widely used insecticides was Paris green, or copper acetoarsenite. Before its insecticidal property was recognized in 1867, Paris green was extensively used to color paint green. It was quickly found to be effective in controlling the Colorado potato beetle, and soon was commonly applied as an insecticide to numerous vegetables and fruits. By 1878, more than five hundred tons of Paris green were sold in New York City each year. The use of Paris green aroused considerable protest due to general knowledge of the acute toxicity of arsenic; only six cases of serious poisoning were recorded, however—and one was an overzealous mosquito inspector who, to prove its safety, added too much to his beer in a public demonstration.
44. J. Whorton, *Before "Silent Spring"* (Princeton, N.J.: Princeton University Press, 1974). Whorton's book is an excellent history of the uses of arsenicals in the pre-DDT era. His citations of the claims of the number of sprayed apples that would need to be consumed to cause a toxic effect are precisely the same arguments used over a century later to justify low levels of food additives and contaminants. These arguments rely primarily on the assumption that acute effects are the main concern, and that there is little likelihood of accumulating, chronic effects.
45. McLaughlin, "History of pyrethrum."
46. The market for pyrethrum has been dominated by Kenya since World War II. Before 1962, production came predominantly from farms in excess of 50 acres in size. Following Kenya's independence and the related land redistribution, production shifted to plots less than 1 acre in size, with over 85,000 families organized into cooperatives. See G. Jones, "Pyrethrum production," in Casida, *Pyrethrum.*
47. T. R. Dunlap, *DDT: Scientists, citizens, and public policy* (Princeton, N.J.: Princeton University Press, 1981), 60.
48. E. F. Knipling, "DDT insecticides developed for use by the armed forces," *J. Econ. Ent.* 38 (1945): 201. See also E. F. Knipling, "The greater hazard: Insects or insecticides?" *J. Econ. Ent.* 46 (1953): 1–7.
49. Busvine, *Disease transmission,* 223.
50. P. F. Russell, "Lessons in malariology from World War II," *Am. J. Trop. Med.* 26, no. 5 (1946). See also "Malaria and other insect-borne diseases in the South Pacific campaign: 1942–1945," *Am. J. Trop. Med.* supp. 27 (1947).

Chapter 3. Resistance

1. G. Ware, *Pesticides: Theory and application* (New York: W. H. Freeman, 1993).
2. The United States and other Allied victors of World War I formed the League of

Nations in 1919 to promote world peace. The functions of the league were transferred to the United Nations in 1946, and the efforts of its Malaria Commission provided an important basis for creating the WHO Expert Committee on Malaria. See J. Abbott, *Politics and poverty: A critique of the Food and Agriculture Organization of the United Nations* (New York: Routledge, 1992).

3. WHO Expert Committee on Malaria, "Extract from report on the first session," *Bull. WHO* 1, no. 1 (Geneva: WHO, 1947), 21–28. In one of the rare, early references to the potential ecological damage of pesticides, the committee, almost as a footnote, warned in 1947: "When used from aeroplanes, it may interfere with the normal biological cycles of the treated environment, which may upset the economy of the region, not only from the standpoint of animals but also of plants, both of crops and trees."
4. WHO Expert Committee on Malaria, *Malaria control: Survey and recommendations,* Report of the Second Session of the Expert Committee on Malaria of the Interim Commission, technical report series 1, no. 2 (Geneva: WHO, 1948), 21–28.
5. Ibid., 228. The seeds of a North-South conflict over technology diffusion and control are also evident in the 1948 Expert Committee report. They recognized that DDT was manufactured primarily in the United States, England, and Switzerland, with no production in malarious parts of the world. Worldwide, prices varied between $0.26 to 1.44 per pound. Tropical and subtropical nations felt vulnerable to the pricing decisions of corporations in high-income nations and to the potential disruption of free international trade, and called for the diffusion of production to malarious regions, perhaps coordinating production between WHO and UNICEF, the United Nations International Children's Emergency Fund.
6. Government hesitancy to establish clear "acceptable risk" thresholds is similar in debates concerning cancer risk. EPA has historically avoided adopting a uniform allowable risk level, which can be translated into numbers of expected tumors. In fact, different programs within EPA have informally established different acceptable risk standards. See R. Madison, *Acceptable risk: Differences in EPA programs,* Office of Research and Development (Washington, D.C.: EPA, 1989).
7. WHO Expert Committee on Malaria, *Report of the third session: 1949,* technical report series 8 (Geneva: WHO, 1950).
8. These effects are explored further in chapter 8.
9. WHO, *Malaria eradication: A plea for health* (Geneva: WHO, 1958).
10. The Expert Committee on Malaria met for the sixth time in 1956 to clarify key management concepts that had confused the attempt to reorient their mission from malaria control to global eradication. The committee adopted the following working definitions: "Malaria control implies the reduction of the disease to a prevalence where it is no longer a major public health problem; the concept carries the implication that the programme will be unending, control having to be maintained by continuous, active work. Vector eradication involves the total elimination of all members of the species concerned so that they do not breed when the work is ended. It is therefore a project limited in time. Malaria eradication is the application of the same principle, not to the mosquito but to the malaria parasite . . . the term does not normally imply that vector eradication has been achieved." See WHO Expert Committee on Malaria, *Sixth report,* technical report series 123, no. 5 (Geneva: WHO, 1957).
11. L. J. Bruce-Chwatt, *Essential malariology* (Oxford: Oxford University Press, 1993), 364–65.

12. Ibid., 369. Bruce-Chwatt's estimates are affected by changes in statistical record-keeping, such as the initial availability of data from China in 1977.
13. F. L. Soper and D. B. Wilson, *"Anopheles gambiae" in Brazil* (New York: Rockefeller Foundation, 1943).
14. WHO Expert Committee on Malaria, *Fifth report,* technical report series 80 (Geneva: WHO, 1954).
15. See G. Georgopoulos, "Extension to other insecticides of DDT resistance observed in *Anopheles sacharovi,*" *Bull. WHO* 10 (1954); G. A. Livadas and G. Georgopoulos, "Development of resistance to DDT by *Anopheles sacharovi* in Greece," *Bull. WHO* 8 (1953): 497; and H. Trapido, "Recent experiments on possible resistance to DDT by *Anopheles albimanus* in Panama," *Bull. WHO* 10 (1954).
16. Trapido, "Recent experiments," 23.
17. G. Livadas and D. Athanassatos, "The economic benefits of malaria eradication in Greece," *Rev. Malariol.* 42 (1963): 177–87.
18. J. P. McEvoy, "DDT magic in Greece," *Readers Digest* 53 (1948): 115–16.
19. U.S. scientists found resistance among culicine mosquitoes to chlorinated hydrocarbon insecticides in 1950, and multiple resistance to dieldrin, BHC, and chlordane was found in Mississippi during 1955. This evidence was followed by further cases of resistance to DDT in Java, Saudi Arabia, and Nigeria. The levels of resistance also seemed to correlate with increased disease incidence. Resistance to DDT normally followed several years of spraying, whereas resistance to dieldrin tended to occur more rapidly. WHO Expert Committee, *Sixth report,* 49; WHO, *Vector resistance to pesticides: Fifteenth report of the WHO Expert Committee on Vector Biology and Control,* technical report series 818 (Geneva: WHO, 1992), 2.
20. WHO Expert Committee on Malaria, *Thirteenth report,* technical report series 357 (1967), 35–36.
21. WHO Expert Committee on Malaria, *Sixteenth report,* technical report series 549 (1974).
22. F. H. Collins and S. Paskewitz, "Malaria: Current and future prospects for control," *Ann. Rev. Entomol.* 40 (1995): 195–219. See also K. L. Knight and A. Stone, *A catalogue of the mosquitoes of the world (Diptera culicidae),* 2d ed. (College Park, Md.: Entomological Society of America, 1977).
23. WHO, *Resistance of vectors and reservoirs of disease to pesticides: Tenth report of the Expert Committee on Vector Biology and Control,* technical report series 737 (1986). By 1986, of 50 *Anopheles* species demonstrating resistance, 49 showed resistance to DDT, 24 to organophosphorous compounds, 14 to carbamates, and 10 to pyrethroids; 14 species were found to be resistant to three or four chemical groups. Eleven of the fifty species are important malaria vectors.
24. These species include *Anopheles albimanus, A. culicifacies, A. pseudopunctipennis, A. sacharovi,* and *A. stephensi.*
25. WHO, *A decade of health development in Southeast Asia, 1968–1977,* South Asia series 7 (Geneva: WHO, 1978).
26. Ware, *Pesticides.*
27. Undetected resistance may have evolved much earlier. See J. R. Busvine, *Disease transmission by insects* (Berlin: Springer-Verlag, 1993), 240.
28. HEW, *Report of the Secretary's Commission on Pesticides and their relationship to environmental health* (Washington, D.C., 1969), 47.
29. WHO, *Resistance of vectors: Tenth report,* 312.

30. WHO, *Vector resistance to pesticides,* 26–27.
31. These suggestions have been summarized and adapted from the recommendations of several of the WHO Expert Committees on Vector Biology and Control. See WHO, *Vector resistance to pesticides*; WHO, *Fourteenth report of the WHO Expert Committee on Vector Biology and Control,* technical report series 813 (Geneva: WHO, 1991); WHO, *Thirteenth report of the WHO Expert Committee on Vector Biology and Control,* technical report series 798 (Geneva: WHO, 1989); WHO, *Tenth report of the WHO Expert Committee on Vector Biology and Control,* technical report series 737 (Geneva: WHO, 1986).
32. See WHO, *Vector resistance to pesticides. B.t.,* a larvicidal agent, has the benefit of being derived from an organic source, and it is relatively inexpensive compared with many synthetic pesticides. Because *B.t.* has almost no residual activity, it requires frequent application.
33. These methods may include siting homes and settlements a significant distance from water bodies and marshlands; avoiding clearcutting of tropical and semitropical forests, which tend to increase standing surface water; securing building interiors from access by insects (by enclosing or screening); providing piped underground water supplies in villages; screening water-storage systems such as tanks, jars, and wells; centralizing sewage disposal; ensuring proper solid-waste disposal, particularly the removal or burial of empty water-bearing containers; constructing drainage systems with proper slopes to encourage rapid drainage; clearing drains regularly; clearing brush from shorelines; fluctuating periodically the water level and flow of impoundments.
34. Expeditions were launched to locate cinchona trees in South America even in the early 1700s. Throughout the eighteenth and nineteenth centuries, explorers wandered through Brazil, Peru, Bolivia, and Columbia in search of the trees. Early attempts to transport live seedlings to Europe from South America failed due to unrelated misfortunes ranging from shipwrecks, overturned canoes, mistaken thefts, and plants dying en route. By 1850, Peru, Bolivia, Colombia, and Ecuador had all passed laws claiming national property rights to the plant material, prohibiting its export in any form that might be used to start a plantation elsewhere.
35. A. W. Haggis, "Fundamental errors in the early history of cinchona," *Bull. Hist. Med.* 10, no. 417 (1941): 568. See also Missouri Botanical Garden, *Proceedings of the celebration of the three hundredth anniversary of the first recognized use of cinchona* (St. Louis, Mo.: Missouri Botanical Garden, 1931).
36. W. H. Wernsdorfer, "Epidemiology of drug resistance in malaria," *Acta Tropica* 56 (1994): 143–56.
37. Chloroquine prevents the parasite from polymerizing the heme molecule created as it digests hemoglobin in the trophozoite stage. In its unpolymerized form, the heme molecule is toxic to the parasite. See A. F. Slater and A. Cerami, "Inhibition by chloroquine of a novel haem polymerase enzyme activity in malaria trophozoites," *Nature* 355 (1992): 167–69.
38. WHO Expert Committee on Malaria, *Eighth report,* technical report series 205 (Geneva: WHO, 1961).
39. WHO Expert Committee on Malaria, *Fourteenth report,* 23.
40. This resistance was recorded in Bolivia, Brazil, Ecuador, French Guinea, Guyana, Panama, Peru, Suriname, and Venezuela. In Asia, resistance appeared in China, India, Indonesia, Kampuchea, Lao Peoples's Democratic Republic, Malaysia, Papua New

Guinea, the Philippines, the Solomon Islands, Thailand, Vanuatu, Vietnam, Nepal, Pakistan, and Sri Lanka.

41. WHO, *Resistance of vectors: Tenth report,* 26–31.
42. Wernsdorfer, "Epidemiology of drug resistance," 144; see also E. G. Beausoleil, "Malaria and Drug Resistance," *World Health* (Geneva: WHO, Aug.–Sept. 1986).
43. Correlations in resistance have been also been found between chloroquine and amodiaquine; quinidine and quinine; mefloquine and quinine; chloroquine and pyronaridine; and mefloquine and halofantrine. An inverse relation has been suggested for chloroquine and mefloquine, however; in areas where mefloquine resistance is increasing, chloroquine resistance is reported to be decreasing.
44. D. Payne, "Did medicated salt hasten the spread of chloroquine resistance in *Plasmodium falciparum*?" *Parasitol. Today* 4 (1988): 112–15.
45. C. C. Draper et al., "Serial studies on the evolution of chloroquine resistance in malaria in an area of East Africa receiving intermittent malaria chemosuppression," *Bull. WHO* 63 (1985): 109–18.
46. Wernsdorfer, "Epidemiology of drug resistance," 149–51. Here he cites C. Wongsrichanalai et al., "In vitro sensitivity of *Plasmodium falciparum* isolates in Thailand to quinine and chloroquine, 1984–1990," *Southeast Asian J. Trop. Med. & Publ. Health* 23 (1992): 533–36.
47. The prevalence of malaria in the Amazon region is 38 per 1000, with a population of nearly 15 million. The State of Rondonia has experienced extraordinarily rapid population growth between 1970 and 1990. The highest rates of malaria incidence are reported among those working or living in the rain forest. See L. M. Camargo et al., "Unstable hypoendemic malaria in Rondonia (western Amazon region, Brazil): Epidemic outbreaks and work-associated incidence in an agro-industrial rural settlement," *Am. J. Trop. Med. & Hyg.* 51, no. 1 (1994): 16–25; J. B. Santos, A. Prata, and E. Wanssa, "Mefloquine chemoprophylaxis of malaria in the Brazilian Amazonia," *Revista da Sociedad Brasileira de Medicina Tropical* 26, no. 3 (1993): 157–62; A. C. Marques, "Migrations and the dissemination of malaria in Brazil," *Parasitol. Today* 3 (1986): 166–70; D. R. Sawyer, *Malaria on the Amazon frontier: economical and social aspects of transmission and control,* WHO technical report on malaria control in the Amazon Basin, (Geneva: WHO, 1992), TDR/FIELDMAL/SC/AMAZ/88.3, pp. 7–15.
48. P. H. Brasseur et al., "Patterns of in vitro resistance to chloroquine, quinine, and mefloquine of *Plasmodium falciparum* in Cameroon, 1985–1986," *Am. J. Trop. Med. & Hyg.* 39 (1988): 162–72.
49. U. Brinkmann and A. Brinkmann, "Malaria and health in Africa: The present situation and epidemiological trends," *Trop. Med. & Parasitol.* 42 (1991): 204–13. *P. falciparum* was still quite sensitive to chloroquine on the island of Madegascar, with no instances of R3 resistance recorded by 1993. See D. Fontenille et al., "Malaria transmission and vector biology on Sainte Marie Island, Madagascar," *J. Med. Entomol.* 29, no. 2 (1992): 197; J. Mouchet et al., "Epidemiological stratification of malaria in Madagascar," *Arch. de l'Institut Pasteur de Madagascar* 60, nos. 1–2 (1993): 50.
50. J. G. Breman and C. C. Campbell, "Combating severe malaria in African children," *Bull. World Health Org.* 66, no. 5 (1988): 611–20.
51. B. M. Greenwood, "The impact of malaria chemoprophylaxis on the immune status of Africans," *Bull. WHO* 62, supp. (1984): 69–75.
52. WHO, *Practical chemotherapy of malaria,* technical report series 805 (1990): 81–82.
53. J. C. Beier et al., "Plasmodium falciparum incidence relative to entomologic inocula-

tion rates at a site proposed for testing malaria vaccines in western Kenya," *Am. J. Trop. Med. & Hyg.* 50, no. 5 (1994): 529; see also J. M. Bockarie et al., "Malaria in a rural area of Sierra Leone: pt. 3: Vector ecology and disease transmission," *Ann. Trop. Med. & Parasitol.* 88, no. 3 (1994): 251; K. Gunasekaran et al., "Observations on nocturnal activity and man biting habits of malaria vectors *Anopheles fluviatilis, An. annularis* and *An. culicifacies* in the hill tracts of Koraput District, Orissa, India," *Southeast Asian J. Trop. Med. & Publ. Health* 25, no. 1 (1994): 187; S. Hewitt, M. Kamal, et al., "An entomological investigation of the likely impact of cattle ownership on malaria in an Afghan refugee camp in the northwest frontier province of Pakistan," *Med. & Vet. Entomol.* 8, no. 2 (1994): 160; S. W. Lindsay et al., "Ability of *Anopheles gambiae* mosquitoes to transmit malaria during the dry and wet seasons in an area of irrigated rice cultivation in The Gambia," *J. Trop. Med. & Hyg.* 94, no. 5 (1991): 313. V. Robert and P. Carnevale, "Influence of deltamethrin treatment of bed nets on malaria transmission in the Kou valley, Burkina Faso," *Bull. WHO* 69, no. 6 (1991): 735; T. Smith et al., "Absence of seasonal variation in malaria parasitaemia in an area of intense seasonal transmission," *Acta Tropica* 54, no. 1 (1993): 55.

54. D. A. Warrell, "Treatment and prevention of malaria," in H. M. Giles, and D. Warrell, *Bruce-Chwatt's "Essential Malariology"* (London: Edward Arnold 1993): 191.
55. WHO, *Practical chemotherapy of malaria,* 79–81.
56. G. V. Brown, "Chemoprophylaxis of malaria," *Med. J. Australia* 159 (1993): 187–96.
57. WHO, *Resistance of vectors: Tenth report,* 26.
58. See, for example, "Anti-malaria vaccine successful in animals," *Science Newsletter* 46 (25 Nov. 1944): 345. See also "Vaccine for malaria?" *Newsweek* 24 (27 Nov. 1944): 83.
59. During the blood meal of the female *Anophelene* mosquito, sporozoites enter the human blood. From there, they invade liver cells, beginning an asexual phase of their reproduction, producing thousands of merozoites. Eventually the liver cell ruptures, spilling the merozoites into the bloodstream, where they invade erythrocytes. During a second asexsual stage of development, as many as thirty-six merozoites develop, eventually rupturing the cell, or schizont, and spilling into the blood. This marks the beginning of the illness, including the symptom of fever. Some of the parasites then undergo sexual differention, forming male and female gametocytes. Traveling in the bloodstream, these gametocytes are taken up by a feeding mosquito. Sexual reproduction then occurs in the stomach of the mosquito as exflagellation and zygote formation occur. The resulting ookinete attaches to the wall of the stomach and an oocyst forms within it. Sporozoites form within the oocyst, migrate to the salivary gland, and await the mosquito's next blood meal to complete the cycle of transmission.
60. M. F. Good, "Towards the development of the ideal malaria vaccine: A decade of progress in a difficult field," *Med. J. Australia* 154, no. 4 (1991): 284. For a review of recent efforts in vaccine development, see W. P. Ballou, "Clinical trials of *Plasmodium falciparum* erythrocytic stage vaccines," *Am. J. Trop. Med. & Hyg.* 50, no. 4, supp. (1994): 59; see also J. C. Beier et al., "*Plasmodium falciparum* incidence relative to entomologic inoculation rates at a site proposed for testing malaria vaccines in western Kenya," *Am. J. Trop. Med. & Hyg.* 50, no. 5 (1994): 529; T. R. Jones and S. L. Hoffman, "Malaria vaccine development," *Clin. Microbiol. Rev.* 7, no. 3 (1994): 303; B. L. Pasloske and R. J. Howard, "The promise of asexual malaria vaccine development," *Am. J. Trop. Med. & Hyg.* 50, no. 4, suppl. (1994): 3; A. Saul, "Minimal efficacy requirements for malarial vaccines to significantly lower transmission in epidemic or seasonal malaria," *Acta Tropica* 52, no. 4 (1993): 283.

61. See X. Z. Su et al., "The large diverse gene family var encodes proteins involved in cytoadherance and antigenic variation of *Plasmodium falciparum*-infected erythrocytes," *Cell* 82, no. 1 (14 July 1995): 89–100; D. J. Krogstad et al., "Energy dependence of chloroquine accumulation and chloroquine accumulation and chloroquine efflux in *Plasmodium falciparum,*" *Biochem. Pharm.* 43, no. 1 (9 Jan. 1992): 57–62; T. E. Wellems, A. Wlaker-Jonah, and L. J. Panton, "Genetic mapping of chloroquine-resistance locus on *Plasmodium falciparum* chromosome 7," *Proc. Nat. Acad. Sci.* 88, no. 8 (15 Apr. 1991): 3382–86; N. Angier, "Malaria's genetic game of cloak and dagger," *New York Times,* 22 Aug. 1995, c1, c3.
62. P. L. Alonso et al., "Randomised trial of efficacy of SPf66 vaccine against *Plasmodium falciparum* malaria in children in southern Tanzania," *Lancet* 344, no. 8931 (1994): 1175. See also P. L. Alonso et al., "A trial of the synthetic malaria vaccine SPf66 in Tanzania: Rationale and design," *Vaccine* 12, no. 2 (1994): 181.
63. In this case the protective effect was calculated based upon person-time exposures. The immune response to SPf66 was independent of age. See J. Moscoso et al., "Safety, immunogenicity, and protective effect of the SPf66 malaria synthetic vaccine against *Plasmodium falciparum* infection in a randomized double-blind placebo-controlled field trial in an endemic area of Ecuador," *Vaccine* 12, no. 4 (1994): 337.
64. O. Noya et al., "A population-based clinical trial with the SPf66 synthetic *Plasmodium falciparum* malaria vaccine in Venezuela," *J. Infect. Dis.* 170, no. 2 (1994): 396.
65. M. V. Valero, L. R. Amador, et al., "Vaccination with SPf66, a chemically synthesised vaccine, against Plasmodium falciparum malaria in Colombia," *Lancet* 341, no. 8847 (1993): 705.
66. U. D'Alessandro et al., "Efficacy trial of malaria vaccine SPf66 in Gambian infants," *Lancet* 346, no. 8973 (19 Aug. 1995): 462–67.
67. WHO, "Synopsis of the world malaria situation," *Weekly Epidemiol. Rec.* 22 (1991).
68. The following citations are offered to encourage others to explore the tragic neglect of African children facing malaria: M. K. Aikins, H. Pickering, and B. M. Greenwood, "Attitudes to malaria, traditional practices and bednets (mosquito nets) as vector control measures: A comparative study in five west African countries," *J. Trop. Med. & Hyg.* 97, no. 2 (1994): 81–86; J. Bryce et al., "Evaluation of national malaria control programmes in Africa," *Bull. WHO* 72, no. 3 (1994): 371–81; B. Carme, "*Plasmodium falciparum* malaria in urban zones of high endemic regions in black Africa: Potential seriousness and possible preventive measures," (in French) *Bull. de la Société de Pathologie Exotique* 86, no. 5, pt. 2 (1993): 394–98; B. Carme et al., "Cerebral malaria in African children: Socioeconomic risk factors in Brazzaville, Congo," *Am. J. Trop. Med. & Hyg.* 50, no. 2 (1994): 131–36; B. Hogh et al., "Classification of clinical *falciparum* malaria and its use for the evaluation of chemosuppression in children under six years of age in Liberia, west Africa," *Acta Tropica* 54, no. 2 (1993): 105–15; M. E. Loevinsohn, "Climatic warming and increased malaria incidence in Rwanda," *Lancet* 343, no. 8899 (1994): 714–18; C. Menendez et al., "Malaria chemoprophylaxis, infection of the placenta, and birth weight in Gambian primigravidae," *J. Trop. Med. & Hyg.* 97, no. 4 (1994): 244–48; I. Rambajan, "Highly prevalent falciparum malaria in north west Guyana: Its development history and control problems," *Bull. Pan Am. Health Org.* 28, no. 3 (1994): 193–201; T. Smith et al., "Absence of seasonal variation in malaria parasitaemia in an area of intense seasonal transmission," *Acta Tropica* 54, no. 1 (1993): 55–72; R. W. Snow et al., "Severe childhood malaria in two areas of markedly different falciparum transmission in east Africa," *Acta Tropica* 57,

no. 4 (1994): 289–300; B. Wolde, J. Pickering, and K. Wotton, "Chloroquine chemoprophylaxis in children during peak transmission period in Ethiopia," *J. Trop. Med. & Hyg.* 97, no. 4 (1994): 215–18; C. Ziba et al., "Use of malaria prevention measures in Malawian households," *Trop. Med. & Parasitol.* 45, no. 1 (1994): 70–73.

69. Brinkmann and Brinkmann, "Malaria and health in Africa."
70. K. W. Mott et al., "Parasitic disease and urban development," *Bull. WHO* 68, no. 6 (1990): 691–98.
71. J. A. Najera, "Malaria and the work of WHO," *Bull. WHO* 67, no. 3 (1989): 229–43.

Chapter 4. Beyond Control

1. T. P. Simons, House Committee on Interstate and Foreign Commerce, *Hearings on bills related to insecticides and fungicides,* pt. 1, 61st Cong., 2d sess. (8 Mar. 1910), and pt. 2, 61st Cong., 2d sess. (8 Apr. 1910), 21–22.
2. J. P. Perkins, *Insects, experts, and the insecticide crisis* (New York: Plenum, 1982), 4; A. H. Whitaker, "A history of federal pesticide regulation in the United States to 1947," Ph.D. diss., Emory University, 1974.
3. *U.S. Statutes at Large* 36 (26 Apr. 1910): 331. See also S. Rep. 436, 61st Cong., 2d sess., 1910; and H. Rep. 990, 61st Cong., 2d sess., 1910.
4. "The conferees consider it a matter of fundamental economic as well as social importance to the food industry of this country . . . that the research be pushed through the resources of the Government in order to discover a substitute for lead arsenate as an insecticide for fruits and vegetables. The conferees suggest the desireability of . . . experiments upon the chronic poisonous effects of lead arsenate ingested." Report of the Hunt Committee (1927), cited in T. R. Dunlap, *DDT: Scientists, citizens, and public policy* (Princeton, N.J.: Princeton University Press, 1981), 46; and in J. C. Whorton, *Before "Silent Spring"* (Princeton, N.J.: Princeton University Press, 1974), 186.
5. The duration of a "chronic" toxicity study has changed over time. Studies comparing the effects of arsenic to cryolite were carried out over an eight- to sixteen-week period. Current protocols for a chronic feeding study require a study duration of two years (ibid., 214).
6. The discoverer of the fluoride-based compounds, Simon Markovitch, conducted his own studies of the toxicity of sodium and calcium fluorides and cryolite in rats and found "borderline" mottling of the rats's teeth in 50 percent of rats exposed at the 4 to 7 ppm level. Markovitch concluded that this was equivalent to a fluorine tolerance of 1.4 ppm as a food residue, and that the tolerance could safely be increased by a factor of 10 to allow 0.1 grain per pound. Yet this level offered protection only to 50 percent of animals exposed and offered no safety factor based upon extrapolation from animal studies to humans nor an accounting for variance in the susceptability of some groups such as children. See references to the Smiths' work in Whorton, *Before "Silent Spring,"* 212–47.
7. Ibid., 217–20.
8. Ibid., 225–26.
9. P. A. Dunbar, "FDA looks at pesticides," *Food, Drug, & Cosmetic Law J.* 4 (1949): 234; also cited in Whorton, *Before "Silent Spring,"* 244.
10. H. L. Needleman, "Lead at low dose and the behavior of children," *Neurotoxicol.* 4, no. 3 (1983): 121–33.
11. *FR* 19 (1954): 6738–72; cited in Whorton, *Before "Silent Spring,"* 244.

12. See, for example, A. Downs, "Up and down with ecology: The issue attention cycle," *Public Interest* (summer 1972).
13. House Committee on Interstate and Foreign Commerce, *Hearings on Bills Relating to Insecticides and Fungicides.*
14. House Committee on Agriculture, *Enforcement of the Insecticide Act,* 65th Cong., 2d sess. (8 Jan. 1918); see also House Committee on Agriculture, *Enforcement of the Insecticide Act, Federal Horticultural Board, Eradication of Pink Bollworm of Cotton,* 65th Cong., 3d sess. (8 Jan. 1919).
15. Perkins, *Insects,* 13–14.
16. Dunlap, *DDT,* 254.
17. Testimony by W. G. Reed in House Hearings Before Subcommittee of the Committee on Agriculture, *Federal Insecticide, Fungicide, and Rodenticide Act,* 80th Cong., 1st sess., 11 Apr. 1947, 14. See also Federal Insecticide, Fungicide and Rodenticide Act (FIFRA), *U.S. Statutes at Large* 61 (1947): 163; S. Rep. 199, 80th Cong., 1st sess., 1947; and H. Rep. 313, 80th Cong., 1st sess., 1947.
18. Compounds exchanged within intrastate commerce were exempt from FIFRA requirements, an escape clause that was to become significant if pesticide production took place in states such as California, Texas, and Florida where agriculture was an important component of the economy.
19. Senate, *Requiring that white powder insecticides and fungicides containing poison shall be colored,* 78th Cong., 1st sess., 1943, S. Rept. 240.
20. The warning symbols required by regulations adopted under the authority of the Hazardous Materials Transportation Act or the Occupational Safety and Health Act specify the character of the risk more precisely than required either by the Insecticide Act of 1910 or the Federal Insecticide, Fungicide, and Rodenticide Act of 1947.
21. See, for example, the testimony of J. Conner, General Counsel, National Association of Insecticide and Disinfectant Manufacturers, in House, *Federal Insecticide, Fungicide and Rodenticide Act,* 17.
22. E. P. Laug et al., "Occurrence of DDT in human fat and milk," *A.M.A. Arch. Ind. Hyg. & Occup. Med.* 3 (1951): 245–46.
23. P. Dunbar, Commissioner of FDA, in House Select Committee to Investigate the Use of Chemicals in Food Products, *Chemicals in food products,* 82d Cong., 1st sess., 1950, 36.
24. Ibid., 69–75.
25. Ibid. When questioned about his training on the medical effects of pesticides, Hitchner admitted he had none.
26. W. Hayes, Jr., in House Select Committee to Investigate the Use of Chemicals in Food Products, *Chemicals in food products.*
27. Hayes cited various references during his testimony to support his claim of safety. These included T. M. Stammers and F. G. Whitfield, "Toxicity of DDT to man," *Nature* 157 (1946): 658.
28. J. B. Shepherd et al., "The effect of feeding alfalfa hay containing DDT residue on the DDT content of cow's milk," *J. Dairy Sci.* 23, no. 6 (1949): 549–55.
29. K. T. Hutchinson, asst. secretary of agriculture, to Hon. James J. Delaney, 3 Nov. 1950, in House Select Committee to Investigate the Use of Chemicals in Food Products, *Chemicals in food products,* 4–11.
30. See C. Cameron's testimony in House Select Committee to Investigate the Use of Chemicals in Food Products, *Chemicals in food products,* 332.

31. M. S. Wolff et al., "Blood levels of organochlorine residues and risk of breast cancer," *J. NCI* 86 (1993): 232–34.
32. "DDT more dangerous, fat accumulation hints," *Science News Letter* 62 (20 Sept. 1952): 190. See also "Bird deaths laid to DDT," *Audubon* 53 (1953): 172; G. W. Pearce et al., "Examination of human fat for the presence of DDT," *Science* 116 (1952): 254–56; "DDT eaten in every meal but amount won't harm," *Science News Letter* 66 (1954): 296; G. Knight and C. Richer, "DDT: Miracle or boomerang?" *Science Dig.* 36L (1954): 23–28; "DDT now found in human body," *Consumers Res. Bull.* 32 (1953): 30; "DDT: Danger to soil," *Science Dig.* 35 (1954): 62.
33. *U.S. Statutes at Large* 68 (22 July 1954): 511.
34. *USC,* vol. 28, §346b (1984).
35. NAS, *Regulating pesticides in food: The Delaney paradox* (Washington, D.C.: National Academy Press, 1987), 25.
36. The most obvious deficiency in this approach was the vesting of regulatory responsibility in the same department that had historically viewed its mission as one of promoting increased production. USDA's fascination with pesticides was intense and well documented in their annual *Yearbook of Agriculture,* which records the evolution of major agricultural trends. The yearbook during the 1950s is largely a chronicle celebrating technology that promotes increased crop yields.
37. House Committee on Interstate and Foreign Commerce, *Chemical additives in food,* 84th Cong., 2d sess. (1956), H.R. 4475, H.R. 7605, H.R. 7606, H.R. 8748, H.R. 7607, H.R. 7764, H.R. 8271, H.R. 8275; see also House Interstate and Foreign Commerce Committee, *Food additives: Hearings on bills to amend FFDCA with respect to chemical additives in food,* 85th Cong., 2d sess., 1958.
38. R. A. Merrill, "FDA's implementation of the Delaney clause: Repudiation of Congressional choice or reasoned adaptation to scientific progress?" *Yale J. Reg.* 5, no. 1 (1988); see also R. D. Middlekauff, "The 1950s: The Delaney clause is enacted," *Food Drug & Cosmetic Law J.* 45 (1990): 31–47. The Delaney clause is now included within the Food, Drug, and Cosmetic Act, *USCA* vol. 21, §348(c)(3)(a).
39. House Committee on Interstate and Foreign Commerce, *Food additives,* 85th Cong., 1st and 2d sess. (1957–58). See the testimony of W. C. Martin (NIH) noting that 1,300 substances were in 1957 suspected of being carcinogenic (273); and the testimony of W. C. Hueper (NCI) suggesting it would be wise to be precautionary and that there was no "scientifically valid and practicable method" to determine a safe level of exposure to a carcinogen (389).
40. H. Rep. 2284, 85th Cong., 2d sess., 1958. See also *Congressional Record* 104 (1958): 19358.
41. S. Rep. 2422, 85th Cong., 2d sess., 1958, 11.
42. W. Manchester, *The glory and the dream* (Boston: Little, Brown, 1974), 669–769; see also L. Jenkins, "The role of courts in risk assessment," *Envir. Law Rev.* 16 (Aug. 1986): 187.
43. S. Udall, "The legacy of Rachel Carson," *Saturday Review* (16 May 1964): 23.
44. Monsanto Company, "The desolate year," *Monsanto Magazine* (Oct. 1962).
45. R. White-Stevens, "Rachel Carson dies of cancer: Author of *Silent Spring* was 56," *New York Times,* 15 Apr. 1964, 1.
46. Examples of "protest registrations" have included the use of mercury compounds as algecides in swimming pools; in addition, the use of lindane in home vaporizors was implicated in the death of at least one child. See generally Senate Subcommittee on

Reorganization and International Organizations of the Committee on Government Operations, *Interagency coordination in environmental hazards (pesticides),* 88th Cong. (23 May 1963).

47. State of California, Governor Edmund Brown's Special Committee on Public Policy Regarding Agricultural Chemicals, *Report on agricultural chemicals and recommendations for public policy* (Sacramento, 1960), 974. This report was prepared by a committee chaired by Dr. E. Mrak, who nine years later chaired the HEW committee that recommended creating the EPA.
48. See O. Freeman's testimony in Senate Subcommittee on Reorganization and International Organizations of the Committee on Government Operations, *Interagency coordination in environmental hazards (pesticides),* 94–95.
49. It may be a central tendency of administrative behavior to rationalize past decisions, leading to consistency in patterns of choice, despite new evidence of the harms they may cause.
50. See A. Calabrezze's testimony in Senate Subcommittee on Reorganization and International Organizations of the Committee on Government Operations, *Interagency coordination in environmental hazards (pesticides).*
51. The claim that the nation's food supply was safe was made thirty years later by the president of the NAS Institute of Medicine when he released the report by the Committee on Pesticides in the Diets of Infants and Children. A dominant argument of the committee, however, had been the same as that of Calabrezze—that toxicity and exposure data were so poor that risks could not be forecast accurately.
52. L. Eisley, "Rachel Carson Dies of Cancer: Author of *Silent Spring,*" *New York Times,* 15 Apr. 1964.
53. President's Science Advisory Committee, *Use of pesticides: A report of the President's Science Advisory Committee* (Washington, D.C., 1963).
54. These figures did not account for fungicide applications, nor did they account for the amount of pesticides applied in aerosol sprays, which at that time averaged one can per household; see also President's Science Advisory Committee, *Use of pesticides,* 2.
55. President's Science Advisory Committee, *Use of pesticides.*
56. USDA, Pesticide Regulation Task Force, *Report of the task force on the pesticides regulation division* (Washington, D.C., 1965).

Chapter 5. EPA as the Gatekeeper of Risk

1. PL88-305, *U.S. Statutes at Large* 78 (1964): 190.
2. M. J. Large, "Comments: The Federal Pesticide Control Act of 1972: A compromise approach," *Ecology Law Quart.* 3 (1972): 277–310. Large's history is summarized here.
3. Ibid., 289; 461 F. 2d 293, 4 ERC 1164 (7th Cir. 1972).
4. *EDF v. Hardin,* 428 F. 2d 1093, 1 ERC 1347 (D.C. Cir. 1970). Followed by *EDF v. Ruckelshaus,* 439 F. 2d 584, 2 ERC 1114 (D.C. Cir. 1970).
5. Large, "Comments," 284; 439 F. 2d 595.
6. See D. Hornstein, "Lessons from federal pesticide regulation on the paradigms and politics of environmental law reform," *Yale J. on Reg.* 10 (1993): 431–34. See also M. Lolly, "Comments: Carcinogen Roulette," *Ecology Law Quart.* 49 (1990): 978 n. 22.
7. Large, "Comments," 286–87 nn. 53, 54; 439 F. 2d 595.
8. By 1966, nearly 22 million pounds of aldrin and dieldrin were sold annually. By 1970,

aldrin was registered for use on 80 crops, and dieldrin was allowed for use on 60 crops. See L. McCray, "Mouse livers, cutworms, and public policy: EPA decisionmaking for the pesticides aldrin and dieldrin," in NAS, *Decisionmaking in the EPA: Case studies* (Washington, D.C.: NRC, 1977).

9. Reorganization Order no. 3, §2(a)(1), *U.S. Code Cong. Ad. News* 2996, 91st Cong., 2d sess. (1970).
10. See U.S. Congress, House Committee on Government Operations, Deficiencies in Administration of FIFRA, Hearings May 7 and June 24, 91st Cong., 1st sess., testimony of R. Duggan (1969): 124. See also U.S. Congress, Senate Subcommittee on Agricultural Research and General Legislation of the Senate Committee on Agriculture and Forestry, Hearings on S. 232, S. 272, S. 660, and S. 745, 92d Cong., 1st sess., testimony of W. Ruckelshaus (1971): 296–97.
11. The threat of uncoordinated state laws motivated broad private sector support for federal pesticide regulation throughout most of the past half century.
12. PL92-516, *U.S. Statutes at Large* 86 (21 Oct. 1972): 973 (hereafter FEPCA). See also H. Rep. 92-1540, *FEPCA of 1972 to amend FIFRA to add requirements for pesticide registration and approval by EPA, and enlarge safety criteria and powers of jurisdiction by which EPA may control pesticide production and use,* 92d Cong., 2d sess., 5 Oct. 1972.
13. FEPCA, §2(bb) and §2(j); see also S. Rep. 92-838, FEPCA of 1972, *U.S. Cong. Ad. News 3* (1972): *3997.*
14. *See Environmental Defense Fund v. EPA,* 510 F. 2d 1292, 5 ELR 20243 (D.C. Cir. 1975); and *Chemical Specialities Manufacturers Association v. EPA,* 484 F. Supp. 513, 10 ELR 20, 430 (D.C. Cir. 1980).
15. Large, "Comments," 302 n. 134; FEPCA, §2(e)(4); see also S. Rep. 92-838, 5.
16. A third category of pesticides was proposed in S. 745 for controlling the most toxic compounds by "permit only," to be approved by the EPA administrator. This category was deleted from the final version of the bill that passed the Senate.
17. *U.S. Statutes at Large* 86 (1972): 983; now in *USCA,* vol. 7, §136b (1995).
18. PL94-140, *U.S. Statutes at Large* 89 (1975): 75. An act to extend the FIFRA, as amended and for other purposes.
19. Lolly, "Comments," 981 n. 51; FEPCA, §2(l).
20. FEPCA, §6(c)(3). Suspension normally triggered indemnification, discouraging its use.
21. 469 F. Supp. 892, 13 ERC 1129 (E.D. Mich. 1979).
22. *EDF v. Ruckelshaus,* 439 F. 2d 597, 2 ERC 1121–22 (D.C. Cir. 1970).
23. The court supported EPA's decision only with "great reluctance," and in its judgment of the available evidence did not believe the emergency suspension was justified. The court did not feel, however, that it was appropriate to substitute its judgment for that of EPA. See M. L. Miller, "Federal regulation of pesticides," in J. G. Arbuckle et al., *Environmental law handbook* (Rockville, Md.: Government Institutes, 1991), 347.
24. FEPCA, PL 92-516, *U.S. Statutes at Large* 86 (1972): 979, *USCA,* vol. 7, §137, 92d Cong., 2d sess. *USC,* vol. 7, §136q and §136w.
25. PL 92-516, *U.S. Statutes at Large* 86 (1972): 973. An act to extend FIFRA, and for other purposes.
26. H.R. 92-511, 92d Cong., 1st sess, *Cong. Rec.* 118, daily ed. (9 Nov. 1972): H10,755-57.

27. H. Rep. 100-939, *Federal Insecticide, Fungicide and Rodenticide Act Amendments of 1988,* 100th Cong., 2d sess. (1988): 32.
28. PL 100-532, *USC,* vol. 7, §136a-1(i)(5) (1988). See also Ferguson and Gray, "1988 FIFRA Amendments: A major step in pesticide regulation," *Envt'l Law Rep.* 19 (Feb. 1989): 10070–72, n. 22.
29. See *USC,* vol. 7, §136h. See also House Committee on Agriculture, Hearings on Review of FEPCA, 93d Cong., 1st sess., 1973. This provision led to compensation demands by original registrants if data were later used to support proposals for registration by others. Congress restricted the trade secret protection to data regarding formulas and manufacturing processes, codifying EPA's practices. See *U.S. Statutes at Large* 92 (1978): 812.
30. The number 50,000 is a combination of approximately 35,000 federal registrations and 15,000 state registrations. See House Subcommittee on DORFA, *Regulation of pesticides,* vol. 3, app. to hearings, serial no. 98-22, 98th Cong., 1st sess. (1983): 69. For the 1993 figures, see A. Aspelin, Pesticide industry sales and usage (Washington, D.C.: EPA, June 1994).
31. There are several reasons for this. Testing results may raise questions that require additional testing. Data quality protocols may change, making older data inadequate to support registration. New data requirements may be adopted by EPA. All have the effect of delaying reregistration.
32. Ibid.
33. J. Aidala, "Summary of Hearings of House Subcommittee on Department Operations, Research and Foreign Agriculture Staff Report: 'Regulatory procedures and public health issues in EPA's Office of Pesticide Programs,'" *Cong. Res. Serv.* (13 May 1983): 5.
34. Ibid., 25.
35. National Research Council, *Risk assessment in the federal government: Managing the process* (Washington, D.C.: National Academy Press, 1983).
36. Associate Administrator Todhunter captured the dilemma faced by the Office of Pesticide Programs in the following exerpt from a request for increased internal funding: "During the past 18 months . . . HED [the Hazard Evaluation Division of the Office of Pesticide Programs] has lost a total of 27 scientists with no opportunity to replace them. Further, extramural support [contracts and cooperative agreements] is expected to decrease by one million dollars in FY 1984. This decrease will most likely result in a reduction in the level of extramural scientific support for our ambitious goals for the establishment of a high volume [of] registration standards, and increased levels of registration actions in FY 1984. . . . The proposed increase to scientific staff would also substantially improve confidence in OPP's scientific reviews." J. Todhunter to Dr. Hernandez, acting administrator of EPA, memorandum, 14 Mar. 1983.
37. J. Miller, "GAO blasts EPA on monitoring systems," *Gov. Comp. News,* 25 Nov. 1991.
38. NAS, *Regulating pesticides in food: The Delaney paradox* (Washington, D.C.: National Academy Press, 1987).
39. *CFR,* vol. 40, §154.7 (1995).
40. Special Review was formerly known as "Rebuttable Presumption Against Registration (RPAR)," and it is applied to compounds undergoing reregistration as well. See §6 of FIFRA. See, e.g., "Ethylene bisdithiocarbamates," *FR* 57 (1992): 7484.
41. *USCA,* vol. 7, §136a(d)(1)(C) (1994); *U.S. Statutes at Large* 86 (1972): 981.

42. M. L. Miller, Federal regulation of pesticides, (1991); See also H. Rep. 100-939, *Federal Insecticide, Fungicide and Rodenticide Act Amendments of 1988,* PL 100-532 (1988): 28. This report provides a concise summary of key legislative activity throughout the 1988 "FIFRA Lite" amendments.
43. FIFRA, 18.
44. GAO, *Pesticides: Thirty years after "Silent Spring" many long-standing concerns remain* (Washington, D.C.), GAO/T-RCED-92-77 (1992): 6.
45. E. Johnson, House Subcommittee on DORFA, *Regulation of pesticides,* vol. 3, app. to hearings, serial no. 98-22, 98th Cong. (1984).
46. House Subcommittee on DORFA, *Regulation of pesticides,* vol. 3, 41–53. This is also known as the "Benbrook Report" after its primary author, Charles Benbrook, chief of the DORFA staff. Benbrook went on to become the executive director of the NAS Board on Agriculture, where he launched several important committees, including those producing the following NAS publications: *Regulating pesticides in food: The Delaney paradox* (1987); *Alternative agriculture* (1989); and *Pesticides in the diets of infants and children* (1993).
47. House Subcommittee on DORFA, *Regulation of pesticides,* 3:59.
48. Ibid., 48.
49. Ibid., 59.
50. J. Aidala, "Summary of Hearings of House Subcommittee on DORFA Staff Report."
51. EPA, *Status report: Bush Task Force inititatives* (Washington, D.C., 1 Oct. 1982).
52. EPA made the toxicity testing protocols voluntary.
53. EPA, *Briefing booklet for the White House Office of Policy Development briefing on FIFRA and TOSCA* (Washington, D.C., 1983).
54. J. Todhunter, "Balancing agriculture's needs with environmental needs," speech before 1982 Beltwide Cotton Production-Mechanization Conference, Las Vegas, Nev., 6 Jan. 1992.
55. See H. Rep. 100-939, 100th Cong., 2d sess., 1988, 29.
56. *USCA,* vol. 7, §136a-1(d)(5)(B)(ii).
57. House Committee on Government Operations, Subcommittee on Environment, Energy, and Natural Resources, *Pesticides: Thirty years after "Silent Spring"—Status of EPA's review of older pesticides,* 102d Cong., 2d sess., 23 July 1992.
58. GAO, *Pesticides: Reregistration delays jeopardize success of proposed policy reforms,* (Washington, D.C., 29 Oct. 1993), T-RCED-94-48.
59. GAO, *Pesticides: Pesticide re-registration may not be completed until 2006* (21 May 1993), RCED-93–94.
60. M. Kraft and N. Vig, "Environmental policy from the seventies to the nineties: Continuity and change," in N. Vig and M. Kraft, *Environmental policy in the 1990s* (Washington, D.C.: Congressional Quarterly Press, 1994).
61. Ibid., 17–19.
62. See chapter 12, nn. 38–40.
63. The history of the reregistration program appears to demonstrate that EPA has demanded more rigorous standards of proof to support claims of risk than to support claims of safety. The reasons for this seem several. First, claims of risk are used to adjust or remove entitlements and their associated benefits. Second, the presumption of safety was nurtured by USDA's previous declarations that if used according to label instructions, the products posed little threat. If EPA becomes aware of new evidence of significant risk, such as the cancer-inducing potential of a pesticide, it reverses the

presumption of safety to be one of significant risk. The burden of proof then lies with the manufacturer to offer convincing evidence that the registration should be sustained.

Chapter 6. Risk Assessment and Tolerance Setting

1. L. Fisher, EPA Office of Pesticides and Toxic Substances, to Sen. Edward M. Kennedy, 30 Mar. 1992, in response to letter from Sen. Kennedy on 19 Feb. 1992. I am not suggesting Fisher's responsibility for these effects. In this case the definition of safety was derived from animal studies that demonstrated a "no observed adverse effect" level. This level was divided by a safety factor, often 100, to arrive at an acceptable daily intake ("reference dose") for the average human.
2. In testimony before the Senate Labor and Human Resources Committee, Fisher also stated that she was unaware of any instance in which EPA set or adjusted a pesticide tolerances to ensure the health protection of infants or children.
3. See chapters 4 and 5 for a discussion of the registration and reregistration process.
4. The tolerance-setting responsibility of EPA under FFDCA contrasts with its registration responsibility under FIFRA described in the previous two chapters. Whereas registration specifies allowable uses, required labeling, and methods of application, tolerances specify maximum allowable residue levels in various foods.
5. Mrak Commission, vol. 7.
6. *USC,* vol. 7, §136(a) (1978).
7. Ibid.
8. It became USDA policy in 1964 not to register any new pesticide until it had received a tolerance from FDA under the Miller Amendment, it was exempted from the need to obtain a tolerance, or it was demonstrated that no residue would result from use of the compound. Manufacturers petitioned FDA for a tolerance while making an application for registration to USDA. This of course changed in 1970 when EPA was created and became responsible both for registration and tolerance-setting. See D. F. Rohrman, "The law of pesticides," *J. Pub. Law* (1968).
9. *U.S. Statutes at Large* 68 (22 July 1954): 511.
10. *USC,* vol. 21, §346(b) (1984).
11. NAS, *Regulating pesticides in food: The Delaney paradox* (Washington, D.C.: National Academy Press, 1987), 25.
12. §409 as amended reads: "(3) No such regulation shall issue if a fair evaluation of the data before the Secretary—(A) fails to establish that the proposed use of the food additive, under the conditions of use to be specified in the regulation, will be safe: Provided, That no additive shall be deemed to be safe if it is found to induce cancer when ingested by man or animal, or if it is found, after tests which are appropriate for the evaluation of the safety of food additives, to induce cancer in man and animal." *USC,* vol. 21, §348(c)(3)(A).
13. See *USC* vol. 21, §342(a)(2)(C) (1984): "where a pesticide chemical has been used in or on a raw agricultural commodity in conformity with an exemption granted or a tolerance prescribed under section 408 and such raw agricultural commodity has been subjected to processing such as canning, cooking, freezing, dehydrating, or milling, the residue of such pesticide chemical remaining in or on such processed food shall . . . not be deemed unsafe if such residue in or on the raw agricultural commodity has been removed to the extent possible in good manufacturing practice and the concen-

tration of such residue in the processed food when ready to eat is not greater than the tolerance prescribed for the raw agricultural commodity."

14. NAS, *Pesticides in the diets of infants and children* (Washington, D.C.: National Academy Press, 1993), 258.
15. NAS, *Regulating pesticides in food,* 28.
16. See R. Jackson's testimony presented at Senate Committee on Labor and Human Resources, *Safety of pesticides in food act of 1991,* 102d Cong., 1st sess., 10 July 1991.
17. EPA claims that this has not occurred, but if not, it is very difficult to explain the absence of §409 tolerances relative to §408 tolerances. See NAS, *Regulating pesticides in food,* 258.
18. The history of attempted escapes from the Delaney prohibition by FDA is presented well by Richard Merrill, perhaps the most knowledgable scholar on the Delaney Clause. Much of this brief history is derived from his article "FDA's implementation of the Delaney clause: Repudiation of Congressional choice or reasoned adaptation to scientific progress?" *Yale J. on Reg.* 5 (1988): 1. See also R. A. Merrill and M. Schewel, "FDA regulation of environmental contamination in food," *Virg. L. Rev.* 66, no. 8 (1980).
19. R. Merrill and M. Taylor, "Saccharin: A case study of government regulation of environmental carcinogens," *Va. J. of Nat. Res. Law* 5, no. 1 (1985).
20. EPA, "Guidelines for carcinogen risk assessment," *FR,* 51 (1986): 33992.
21. EPA, *List of suspected or demonstrated pesticide carcinogens* (Washington, D.C., 1985).
22. *USC,* vol. 21, §321(s) (1982).
23. DES was first synthesized in 1938 by Edward Dodds, a British physician, and his colleagues. The drug was often prescribed to women believed to have insufficient estrogen during pregnancies. See Y. Brackbill and H. Berendes, "Dangers of Diethylstylbestrol: Review of a 1953 paper," *Lancet* 2 (1978): 520; A. Herbst, H. Ulfelder, and D. Poskanzer, "Adenocarcinoma of the vagina," *New Eng. J. of Med.* 284 (1971): 878–81. These articles are cited in the review by T. Colborn, D. Dumanoski, and J. P. Myers, *Our stolen future: How we are threatening our fertility, intelligence, and survival—a scientific detective story* (New York: Penguin, 1996).
24. Drug Amendments of 1962, *USC,* vol. 21, §348(c)(3)(A) and §376(b)(5)(B) (1982).
25. Merrill, "FDA," 27–28, n. 149.
26. Ibid., 26, n. 146; see also R. M. Cooper, "Stretching Delaney till it breaks," *Regulation* (Nov.–Dec. 1985): 11.
27. Ibid., 29–32; "Saccharin and its salts: Removal from Generally Regarded As Safe List; provisional regulation prescribing conditions of safe use," *FR* 36 (1971): 12109.
28. "Chemical compounds in food producing animals: Criteria and procedures for evaluating assays for carcinogenic residues," *FR* 44 (1979): 17070.
29. "Saccharin and its salts: Proposed rule making," *FR* 42 (1977): 19996.
30. PL 95–203, *U.S. Statutes at Large* 91 (1977): 1451.
31. See NAS, *Saccharin: Technical assessment of risks and benefits,* pt. 1 (Washington, D.C.: National Academy Press, 1978); and NAS, *Food safety policy: Scientific and societal considerations* (Washington, D.C.: National Academy Press, 1979).
32. Merrill and Taylor, "Saccharin."
33. Merrill, "FDA," 38–41; *Scott v. FDA,* 728 F. 2d 322 (6th Cir. 1984).
34. "Acrylonitrile copolymers intended for use in contact with food: notice of proposed rule making," *FR* 39 (1974): 38907.

35. "Indirect additives: Polymers, acrylonitrile copolymers used to fabricate beverage containers," *FR* 49 (1984): 36635.
36. Merrill, "FDA," 34 n. 183; *Monsanto Co. v. Kennedy,* 613 F. 2d 947 (D.C. Cir. 1979).
37. Ibid., 7.
38. FDA, "Listing of D&C Orange No. 17 for use in externally applied drugs and cosmetics," *FR* 51 (1986): 28331, 28342; and FDA, "Listing of D&C Red No. 19 for use in externally applied drugs and cosmetics," *FR* 51 (1986): 28331, 28346, 28359.
39. Merrill, "FDA," 44–45 nn. 239–40. 636 F. 2d 323 (D.C. Cir. 1979).
40. FDA, *FR* 51 (7 Aug. 1986): 28331.
41. *Public Citizen v. Young,* 831 F. 2d 1108, 1111 (D.C. Cir. 1987).
42. Although it is difficult to estimate this number, in 1996 there are approximately 325 active ingredients in food-use pesticides, which are combined with nearly 1,800 inactive ingredients. The inactive compounds remain unregulated as residues under §408, or as food additives under §409. EPA has broken the food supply into roughly 675 different types of food on which residues might appear. Thus there is a potential for an extraordinarily large number of separate pesticide-food combinations, each of which requires separate scrutiny to determine what contamination level should be considered safe.
43. NAS, *Toxicity testing: Strategies to determine needs and priorities* (Washington, D.C.: National Academy Press, 1984). These 8,000-odd food additives constitute roughly 18 percent of all commercial chemicals to which humans are exposed. Among the 34,000 chemicals which are considered to be toxic by the National Institute for Occupational Safety and Health, more than 2,300 had carcinogenicity studies that remained unevaluated in 1980. See Merrill, "FDA," 17.
44. B. Ames, "Letters: Cancer and diet," *Science* 224 (1984): 658.
45. EPA, *List of suspected or demonstrated pesticide carcinogens* (Washington, D.C., 1985, 1989, 1993).
46. See NAS, *Regulating pesticides in food.*
47. *CFR,* vol. 40, §180 (1987).
48. NAS, *Regulating pesticides in food,* 19, table 1–1.
49. Reviews for numerous fungicides—several of which have some carcinogenic potential in animals—have often taken well over a decade to complete. Once a thorough review is conducted, the hearing and appeal procedures often can delay a final decision for three to five years.
50. Pesticide exposure is calculated as the product of residue and food intake estimates. Food intake has been surveyed by USDA for over forty years, normally at ten-year intervals. People of different ages and ethnic backgrounds are surveyed in different seasons and regions of the country. Those surveyed keep a diary of the types and amounts of food they consume and the results are standardized into grams of food per kilogram of body weight per day. These surveys have been used by EPA to estimate pesticide exposure since the early 1980s, but not without problem. USDA keeps track of approximately 5,000 different types of food, ranging from pizza to chicken pie. EPA must then attempt to break the food "as reported eaten" into raw food forms, for which pesticide residue limits have been set. There are approximately 675 of these food forms. Pizza would be broken into water, wheat flour, tomato paste, tomato sauce, cheese, etc., by percentage of total weight. Notice that even the components of the prepared meal are often processed foods such as tomato paste. Further consolidation of the food forms is necessary to estimate exposure to pesticides, because tolerances are

set for only approximately 375 different types of food. This file is occasionally updated by EPA, but given the explosion in types of processed food, particularly complex, ready-to-eat meals, the potential for error in estimating intake, and therefore pesticide residues, is significant.

51. *FR* 50 (1985): 10371. Office of Science and Technology Policy, *Chemical Carcinogens: A review of the science and its associated principles* (14 Mar. 1985). Five categories of carcinogens have been established: (1) Class A: Human Carcinogen: Epidemiological evidence is sufficient to demonstrate a convincing relation between exposure and cancer in humans. (2) Class B: Probable Human Carcinogen: Two subcategories have been established: (B1): Sufficient evidence of carcinogenicity from animal studies with limited evidence from epidemiological studies. (B2): Sufficient evidence of cancinogenicity from animal studies with inadequate or no epidemiological data. (3) Class C: Possible Human Carcinogen: Limited evidence of carcinogenicity from animal studies with no human evidence. (4) Class D: Not Classifiable as to Human Carcinogencity: Inadequate or no animal and human data for carcinogenicity. (5) Class E: Evidence of Non-Carcinogenicity for Humans: No evidence of carcinogenic activity in at least two adequate animal and epidemiological studies. Three criteria influence the classification of a compound's carcinogenicity. The first is the type of study (animal or human), the second is the strength of causal relationship demonstrated by the study, and the third is the quality of study. For example, an animal study may demonstrate a strong dose-related linkage between exposure and malignant tumors, but the poor quality of study design, or implementation, may preclude its classification as a B2 carcinogen.
52. NAS, *Regulation of pesticides in food,* 67.
53. Ibid., 68, table 3–9.
54. "Regulation of pesticides in food: Addressing the Delaney Parodox policy statement," *FR* 53 (1988): 41104.
55. These included trifluralin (spearmint and peppermint oils), benomyl (raisins and tomato products), phosmet (cottonseed oil), mancozeb (raisins, and brans of barley, oats, rye and wheat), dicofol (dried tea), DDVP (packaged and bagged nonperishable processed foods and dried figs), and chlordimeform (dried prunes). All of these chemicals are classified as animal carcinogens by EPA. See *FR* 54 (30 June 1989): 27700.
56. *FR* 55 (25 Apr. 1990): 17560.
57. *FR* 55 (18 May 1992): 20481.
58. *Les v. Reilly,* 968 F. 2d 985 (9th Cir. 1992), cert. denied, 113 S.Ct. 1361 (1993).
59. *FR* 59 (30 Mar. 1994): 14980. Updated list of pesticides and uses potentially affected by the Delaney Clause of the Federal Food, Drug, and Cosmetic Act.
60. S. Breyer, *Breaking the vicious circle* (Cambridge, Mass.: Harvard University Press, 1994).
61. Executive Order 12866, for example, created a threshold of significance which must be exceeded before risk and cost-benefit analyses are required. "Significant regulatory actions" are defined primarily in terms of the estimated costs of regulation. "Significance" is not defined (although it could and perhaps should be) based upon the damages to be avoided through the act of regulation, i.e., the benefit of regulation that is tied to the magnitude of risk.
62. NAS, *Risk assessment in the federal government: Managing the process,* (Washington, D.C., National Academy Press, 1983). This committee sharply criticized the variability of risk assessment standards within federal agencies in this report. Now, more than

a decade later, it is clear that great variability in methods and standards remain. The NAS committee responsible for the report suggested that federal agencies adopt an approach which clearly segregates the process of "risk assessment" from that of "risk management." This separation, the committee argued, would "protect" science from politics. Since this report, most federal agencies have formally distinguished the "risk assessment" process, which they believe to be scientific, from the "risk management" process, which they believe to be political. This was a well-meaning attempt to protect the quality of science in risk assessments from hidden political biases and value judgments believed to be restricted to "risk management."

63. Disclosure has been a serious problem in two areas. Before *Ruckelshaus v. Monsanto,* public access was restricted from data on chemical toxicity. Second, public access to EPA's dietary exposure databases and computer programs was restricted even in 1995. See T. McGarity, *Reinventing rationality: The role of regulatory analysis in the federal bureacracy* (New York: Cambridge University Press, 1991); T. McGarity and S. A. Shapiro, "The trade secret status of health and safety testing information: Reforming agency disclosure policies," *Harvard Law Rev.* 93 (1980): 837; and *Ruckelshaus v. Monsanto Co.,* 467 U.S. 986, 104 S. Ct. 2862, 81 L. Fd. 2d 815 (1984), which found that the disclosure requirements of FIFRA's §10 were reasonable, despite the registrant's property interests in data. See also J. E. Bonine and T. O. McGarity, *The law of environmental protection,* 2d ed. (St. Paul, Minn.: West, 1994), 675.
64. See EPA Science Advisory Board, *Reducing risk: Setting priorities and strategies for environmental protection* (Washington, D.C., 1990); See also EPA, *Unfinished business: A comparative assessment of environmental problems* (Washington, D.C., 1992).
65. NAS, *Carcinogens and anticarcinogens in the human diet* (Washington, D.C.: National Academy Press, 1996). By contrast risk communication has become the dominant form of managing that is recognizable by the public. Hazard labeling of alcohol, tobacco, fat, and calories provide examples.

Chapter 7. The Human Ecology of Pesticide Residues

1. EPA, *Pesticide industry sales and usage: 1992 and 1993 market estimates* (Washington, D.C., June 1994).
2. Sales figures may be poor reflections of use patterns, because there is no guarantee where or when pesticides purchased will be used.
3. An excellent critique of available use surveys was prepared by R. C. Edwards and his colleagues in 1993. They concluded, "Current pesticide use surveys . . . are not adequately identifying and reporting active ingredients for pesticides that cause human health and/or environmental concerns. . . . A comprehensive assessment of recent pesticide use patterns is difficult with currently available data, particularly for fruit and vegetable crops." See R. C. Edwards et al., *Toward a complete and consistent data set: An assessment of pesticide use surveys for ten commercial crops, 1975–1991* (Purdue University, 1993). One of the more thorough attempts to estimate trends in use has been conducted by RFF. They found a pattern of decreasing volume of use between 1979 and 1992 for fungicides, herbicides, and insecticides on major food crops. Fungicide use declined 12.5 percent, while herbicide use declined 19 percent, and insecticide use declined by 51 percent. These estimates are for eighteen major field and vegetable crops only and do not include fruits, nuts, minor use crops, or other nonagricultural uses, which are significant. Part of their explanation was that

a federal agricultural subsidy program, known as Payment in Kind, heavily compensated farmers during the 1980s for not planting crops in an effort to ensure stable commodity prices. See L. Gianessi and J. Anderson, *Pesticide use trends in U.S. Agriculture, 1979–1992,* discussion paper PS-93-1, National Center for Food and Agricultural Policy (Washington, D.C.: National Center for Food and Agricultural Policy, 1993).

4. EPA has not released use data by pesticide since it was created, preferring instead to issue reports of data aggregated by major pesticide class. EPA estimates vary from those prepared by RFF, with the largest difference being for 1982 insecticide use which EPA reports to be 122 million pounds greater than RFF. See EPA, *Pesticide industry sales and usage: 1992 and 1993,* for a comparison.
5. These estimates were summarized in National Research Council, *Alternative agriculture* (Washington, D.C.: National Academy Press, 1988), 46.
6. L. Gianessi, "The quixotic quest for chemical-free farming," *Issues in Sci. & Tech.* (Fall 1993): 32.
7. L. Gianessi and C. A. Puffer, "Regulatory policy, new technology, and mother nature," *Resources* (Fall 1989).
8. Care should be taken not to correlate the weight of a pesticide with its potential to induce health or ecological damage. The most potent pesticides may require the lowest application concentrations.
9. G. Ware, *Pesticides: Theory and application* (New York: W. H. Freeman, 1983).
10. R. Wiles et al., *Tapwater blues: Herbicides in drinking water* (Washington, D.C.: Environmental Working Group and Physicians for Social Responsibility, 1994).
11. R. L. Metcalf, "An increasing public concern," *EPA Journal* 10, no. 5 (1984): 30–31.
12. L. Gianessi, *U.S. pesticide use trends, 1966–1989: Report to EPA Office of Policy Analysis from Resource for the Future* (Washington, D.C.: EPA, 1992).
13. P. Schuck, *Agent Orange on trial: Mass toxic disasters in the courts* (Cambridge, Mass.: Belknap, 1987). See also M. Fingerhut et al., "Cancer mortality in workers exposed to 2,3,7,8-TCDD," *New Eng. J. Med.* 324(1991): 212.
14. For a review of the exposure to dioxins—especially 2,3,7,8-TCDD—and associated health risks, see L. C. Dickson and S. C. Buzik, "Health risks of 'dioxins': A review of environmental and toxicological considerations," *Vet. & Human Toxicol.* 35, no. 1 (1993): 68; O. Dohr et al., "Modulation of growth factor expression by 2,3,7,8-tetrachlorodibenzo-p-dioxin," *Exper. & Clin. Immunogenetics* 11, nos. 2–3 (1994): 142; M. P. Holsapple et al., "A review of 2,3,7,8-tetrachlorodibenzo-p-dioxin-induced changes in immunocompetence: 1991 update," *Toxicol.* 69, no. 3 (1991): 219; J. E. Huff et al., "Long-term carcinogenesis studies on 2,3,7,8-tetrachlorodibenzo-p-dioxin and hexachlorodibenzo-p-dioxins," *Cell Bio. & Toxicol.* 7, no. 1 (1991): 67; R. Neubert et al., "Effects of small doses of dioxins on the immune system of marmosets and rats," *Annals N.Y. Acad. of Sci.* 685 (1993): 662; A. Schecter et al., "Dioxins in U.S. food and estimated daily intake," *Chemosphere* 29, nos. 9–11 (1994): 2261; B. G. Svensson et al., "Parameters of immunological competence in subjects with high consumption of fish contaminated with persistent organochlorine compounds," *Internat. Arch. Occup. & Envir. Health* 65, no. 6 (1994): 351; L. Tollefson, "Use of epidemiology data to assess the cancer risk of 2,3,7,8-tetrachlorodibenzo-p-dioxin," *Reg. Toxicol. & Pharm.* 13, no. 2 (1991): 150; C. C. Travis and H. A. Hattemer-Frey, "Human exposure to dioxin," *Sci. of the Total Envir.* 104, nos. 1–2 (1991): 97.

15. "Intent to cancel or restrict registrations of pesticide products containing toxaphene," *FR* 47 (29 Nov. 1982): 53784.
16. Instead of searching every food sample for every type of pesticide, the government has tried to be more strategic, often concentrating sampling efforts where they suspect violative residue levels. But this approach can have little influence in reducing risks, because FDA monitoring efforts have been directed toward testing foods that make up only a small proportion of the average diet.
17. S. J. Eisenreich, B. B. Looney, and J. D. Thornton, "Airborne organic contaminants in the Great Lakes ecosystem," *Env. Sci. Tech.* 15 (1981): 30–38.
18. R. D. Arthur, J. D. Cain, and B. F. Barrentine, "Atmospheric levels of pesticides in the Mississippi Delta," *Bull. Env. Cont. Tox.* 15 (1976): 129–134.
19. "How to use DDT in your home," *House Beautiful* 87 (Dec. 1945): 124.
20. R. Carson, *Silent spring* (Boston: Houghton Mifflin, 1962): 178–81.
21. S. Tanabe et al., "Global pollution of marine mammals by PCBs, DDTs, and HCHs (BHCs)," *Chemosphere* 12 (1983): 1269–75.
22. E. Atlas and C. S. Giam, "Global transport of organic pollutants: Ambient concentrations in the remote marine atmosphere," *Science* 211 (1980): 163–65.
23. D. E. Glotfelty, J. N. Seiber, and L. A. Liljedahl, "Pesticides in fog," *Nature* 325 (1987): 602–5.
24. Government and Regulatory Affairs Committee, *SETAC: The Society of Environmental Toxicology and Chemistry News* 11, no. 4 (1991): 9.
25. T. J. Murphy, Atmospheric inputs of chlorinated hydrocarbons to the Great Lakes, in J. O. Nriagu and M. S. Simmons, eds., *Toxic contaminants in the Great Lakes* (New York: Wiley and Sons, 1984): 53–76.
26. See T. L. Wu, "Atrazine in estuarine water and the aerial deposition of atrazine into Rhode River, Maryland," *Water Air Soil Pollution* 15 (1981): 173–84. See also W. E. Pereira and F. D. Hostettler, "Nonpoint source contamination of the Mississippi River and its tributaries by herbicides," *Envir. Sci. & Tech.* 27, no. 8 (1993): 1542–52. For an excellent summary of the literature on herbicide and fungicide exposure, see C. R. Clement and T. Colborn, *Herbicides and fungicides: A perspective on potential human exposure* (1991).
27. C. G. Wright and R. B. Leidy, "Chlordane and heptachlor in the ambient air of houses treated for termites," *Bull. Env. Cont. Tox.* 26 (1982) 617–23.
28. "DDT wall and drawer paper kills insects," *Better Homes and Gardens* 24 (May 1946): 123.
29. C. S. Giam et al., "Phthalate esters, PCB and DDT residues in the Gulf of Mexico atmosphere," *Atmos. Envir.* 14 (1980): 65–69.
30. T. F. Bidleman and R. Leonard, "Aerial transport of pesticides over the Northern Indian ocean and adjacent seas," *Atmos. Envir.* 16 (1982): 1099–107; T. F. Bidleman et al., "Atmospheric transport of organochlorines in the North Atlantic gyre," *J. Mar. Res.* 39 (1981): 443–64.
31. S. Tanabe, H. Hidaka, and R. Tatsukawa, "PCB's and chlorinated hydrocarbon pesticides in Antarctic atmosphere and hydrosphere," Chemosphere 12 (1983): 277–88.
32. One of the best overviews of the literature on pesticide residues in diverse environmental media is provided by Robert Spear. See his "Recognized and possible exposure to pesticides," in W. Hayes, ed., *Handbook of pesticide toxicology* (New York: Academic Press, 1991): 245–74.
33. Ibid., 263.

34. A. W. Taylor et al., "Volatilization of dieldrin and heptachlor residues from field vegetation," *J. Agric. Food Chem.* 25 (1977):542.
35. G. S. Hartley, "Evaporation of pesticides," *Adv. Chem. Ser.* 86 (1969): 115–34.
36. This rate of "wicking" is affected by the concentration of the pesticide in the soil, the chemical and physical properties of the soil, its vapor pressure, and the rate of water evaporation.
37. Spear, "Recognized and possible exposure."
38. S. Tanabe, H. Iwata, and R. Tatsukawa, "Global contamination by persistent organochlorines and their ecotoxicological impact on marine mammals," *Sci. Total Envir.* 154 (1994): 163–77.
39. G. Ware et al., "Pesticide drift," *J. Econ. Entom.* 62 (1969): 840–46.
40. See L. A. Barrie et al., "Arctic contaminants: Sources, occurrence, and pathways," *Sci. Total Environ.* 122 (1992): 1–74; W. E. Cotham and T. F. Bidleman, "Estimating the deposition of organic contaminants to the Arctic," *Chemosphere* 22 (1991): 165–88; G. Forget, "Pesticides and the third world," *J. Toxicol. Envir. Health* 32 (1991): 11–31; H. Iwata et al., "Distribution of persistent organochlorines in the oceanic air and surface seawater and the role of ocean on their global transport and fate," *Envir. Sci. Technol.* 27 (1993): 1080–98; H. Iwata et al., "Geographical distributions of persistent organochlorines in air, water, and sediments from Asia and Oceania, and their implications for global redistribution from lower latitudes," *Envir. Pollut.* 85 (1994): 15–33; M. Kawano et al., "Bioconcentration and residue patterns of chlordane compounds in marine mammals: Invertebrates, fish, mammals, and seabirds," *Envir. Sci. Technol.* 22 (1988): 792–97; D. Martineau et al., "Levels of organochlorine chemicals in tissues of beluga whales (Delphinapterus leucas) from the St. Lawrence estuary, Quebec, Canada," *Arch. Environ. Toxicol.* 16 (1987): 137–47; D. C. G. Muir et al., "Artic marine ecosystem contamination," *Sci. Total Environ.* 122 (1992a): 75–134; R. J. Norstrom et al., "Organochlorine contaminants in Arctic marine food chains: Identification, geographical distribution, and temporal trends in polar bears," *Envir. Sci. Technol.* 22 (1988): 1063–71; D. Peakall et al., *Animal Biomarkers as Pollution Indicators* (London: Chapman and Hall, 1992), 291; A. Ramesh et al., "Characteristic trends of persistent organochlorine contamination in wildlife from a tropical agricultural watershed, South India," *Arch. Envir. Contam. Toxicol.* 23 (1992): 26–36; S. Tanabe et al., "Global distribution and atmospheric transport of chlorinated hydrocarbons: HCH (BHC) isomers and DDT compounds in the western Pacific, eastern Indian and Antartic oceans," *J. Oceanogr. Soc. Jpn.* 38 (1982): 137–48; S. Tanabe et al., "Specific pattern of persistent organochlorine residues in human breast milk from South India," *J. Agric. Food Chem.* 38 (1990): 899–903; S. Tanabe et al., "Fate of HCH (BHC) in tropical paddy fields: Application test in South India," *Int. J. Environ. Anal. Chem.* 45 (1991b): 45–53; R. Tatsukawa et al., "Global transport of organochlorine insecticides: An 11-year case study (1975–1985) of HCHs and DDTs in the open ocean atmosphere and hydrosphere," in D. A. Kurtz, ed., *Long range transport of pesticides* (Chelsea, Mich.: Lewis, 1990): 127–41; D. J. Thomas et al., "Arctic terrestrial ecosystem contamination," *Sci. Total Envir.* 122 (1992): 135–64.
41. Pesticides also vary according to the degree of specificity of their toxic effect among species. The mechanism of toxicity may be limited to an individual species or it may extend across genera. Organophosphate and carbamate insecticides, for example, are toxic to insects because they inhibit an enzyme that is crucial for normal nervous system function. This enzyme is called "cholinesterase" (ChE), and its inhibition disrupts the transmission of electronic signals across the gaps among nerve cells. Yet these

insecticides cause similar effects in birds and humans. DDT, by contrast, is a contact poison to many insects, whereas its immediate effect on humans is insignificant. Thus understanding the "specificity" of any insecticide is critical to forecasting its ecological or human health effects.

42. G. M. Woodwell, C. F. Wurster, and P. A. Isaacson, "DDT residues in an East Coast estuary: A case of biological concentration of a persistent insecticide," *Science* 156 (1967): 821–24.
43. "It is a bird of spectacular appearance and beguiling habits, building its floating nests in shallow lakes. . . . It is called the 'swan grebe' with reason, for it glides with scarcely a ripple across the lake surface, the body riding low, white neck and shining black head held high. The newly hatched chick is clothed in soft gray down; in only a few hours it takes to the water and rides on the back of the father or mother, nestled under the parental wing covers." Carson, *Silent spring,* 47.
44. DDD was known in 1948 to destroy parts of the adrenal cortex, although this effect was initially believed to be confined to dogs. Later evidence suggested a similar effect in humans, and the compound was prescribed to suppress adrenal function in humans for the treatment of breast and prostate cancers. See also A. A. Nelson and G. Woodard, "Severe adrenal cortical atrophy, cytotoxic and hepatic damage produced in dogs by feeding 2,2,-bis(para-chlorophenyl)-(DDD or TDE)," *Arch. Pathol.* 48 (1948): 387–94; and B. Zimmerman et al., "Effects of DDT on human adrenal: Attempt to use adrenal destructive agent in treatment of disseminated mammary and prostatic cancer," *Cancer* 9 (1956) 940–48.
45. R. J. Barker, "Notes on some ecological effects of DDT sprayed on elms," *J. Wildlife Mgt.* 22 (1958): 269–74.
46. Z. E. Wallace et al., "Cushing's syndrome due to adrenocortical hyperplasia," *New Eng. J. Med.* 265 (1961) 1088–93.
47. "DDT news: Kills gypsy-moth caterpillar," *Time* 44, no. 72 (31 July 1945).
48. S. Dreistadt and D. Weber, "Gypsy moth in the Northeast and Great Lakes states," in D. L. Dahlsten, R. Garcia, and H. Lorraine, eds., *Eradication of exotic pests* (New Haven: Yale University Press, 1990).
49. Carson, *Silent spring,* 158–61. See also A. Worrell, "Pests, pesticides, and people," *Am. Forests* 66 (1960): 39–81.
50. C. E. Lundholm, "Influence of chlorinated hydrocarbons, Hg2+ and methyl-Hg+ on steroid hormone receptors from eggshell gland mucosa of domestic fowls and ducks," *Arch. of Tox.* 65, no. 3 (1991): 220–27.
51. D. A. Ratcliff, "Broken eggs in peregrine eyries," *Br. Birds* 51 (1958): 23–26; see also D. A. Ratcliff, "Decrease in eggshell weight in certain birds of prey," *Nature* 215 (1967): 208–10.
52. D. W. Anderson and J. J. Hickey, "Eggshell changes in certain North American birds," *Proc. Inter. Ornithol. Cong.* 15 (1970): 514–40.
53. R. A. Faber and J. J. Hickey found that the common merganser and the red-breasted merganser experienced eggshell thinning caused by PCBs, but not DDE. Chickens remain immune to the thinning effects of DDT or DDE, and the Bengalese finch experiences eggshell thickening in response to DDT exposure. See their "Eggshell thinning, chlorinated hydrocarbons, and mercury in inland aquatic bird eggs, 1960 and 1970," *Pest. Monitor* 7 (1973): 27–36.
54. K. Cooper, "Effects of pesticides on wildlife," in W. Hayes, ed., *Handbook of pesticide toxicity* (New York: Academic Press, 1991), 483.

55. T. J. Peterle, "DDT in Antarctic snow," *Nature* 224 (1969): 620.
56. J. H. Tatton, "Organochlorine pesticides in Antarctica," *Nature* 215 (1967): 346–48.
57. S. Tanabe, H. Tanaka, and R. Tatsukawa, "Polychlorobiphenyl, o-DDT, and hexachlorocyclohexane isomers in the western North Pacific ecosystem," *Arch. Environ. Contam. Toxicol.* 13 (1984): 731–38.
58. Tanabe et al., "Global contamination," 171–72.
59. Most deaths from pesticides followed direct accidental ingestion, and many of these victims have been children. Comparing the relative toxicity of oral to dermal exposure for 67 compounds in 1969, T. B. Gaines and W. Hayes found that 64 compounds were more toxic if exposure was from ingestion rather than skin contact. The average ratio of ingestion/dermal toxicity was 4:2. See W. Hayes, "Dosage and other factors influencing toxicity," in Hayes, *Handbook of pesticide toxicology*. See also W. Hayes, "Toxicity of pesticides to man: Risks from present levels," *Proc. R. Soci. London,* ser. b, 167 (1967): 101–127; T. B. Gaines, "Acute toxicity of pesticides," *Toxicol. Appl. Pharm.* 14 (1969): 515–34.
60. Respirators fitted with filters provide an excellent opportunity to measure inhalation exposure and to then test various physiological characteristics that may be evidence of a toxic response.
61. G. E. Quinby and A. B. Lemmon, "Parathion residues as a cause of poisoning in crop workers," *J. Am. Med. Assoc.* 166 (1958): 740–46.
62. R. C. Spear et al., "Fieldworkers' response to weathered residues of parathion," *J. Occup. Med.* 19 (1977): 406–10. See also R. C. Spear, W. J. Popendorf, and T. H. Milby, "Worker poisonings due to paraoxon residues," *J. Occup. Med.* 19 (1977): 411–14.
63. GAO, *Lawn care pesticides, risks remain uncertain while prohibited safety claims continue* (Washington, D.C., 1990), GAO/RCED-90-134, pp. 1–24.
64. For a wonderful reflection on the ecology, risks, and aesthetics of lawn care, see F. H. Bormann, D. Balmori, and G. Geballe, *Reinventing the American lawn* (New Haven: Yale University Press, 1993).
65. Mrak Commission.
66. Ibid., 121.
67. J. C. Modin, "Chlorinated hydrocarbon pesticides in California bays and estuaries," *Pest. Mon. J.* 3 (1969): 1–7, cited in Mrak Commission, 129.
68. G. E. Smith and G. G. Isom, "Investigation of effects of large-scale applications of 2,4-D on aquatic fauna and water quality," *Pest. Mon. J.* 1 (1967): 16–21.
69. Mrak Commission, 116.
70. A 150-foot separation may offer little protection from a water soluble contaminant, spilled or applied to sandy soils near a shallow well. Some pesticides have migrated several miles underground.
71. Monsanto Company, *Well water sampling report* (St. Louis, Mo.: Monsanto, 1987).
72. NAS, *Pesticides in the diets of infants and children* (Washington, D.C.: National Academy Press, 1993), 238–41. See also EPA, National Survey of Pesticides in Drinking Water Wells: Phase I, NTIS doc. no. PB-91-125765 (Washington, D.C., 1990). Although this study gives some indication of the scale of contamination of groundwater supplies, the sample sizes were so small that they provide a poor basis for estimating exposure through tap water.
73. USGS, "National water summary 1986: Hydrologic events and groundwater quality," USGS Water Supply Paper 2325 (Washington, D.C., 1988). See also USGS, "National

water summary 1986: Hydrologic events and water supply and use," USGS Water Supply Paper 2350 (Washington, D.C., 1988).

74. Alachlor and metolachlor are closely related in chemical structure, as are atrazine, simazine, and cyanazine, all chlorinated triazines.
75. D. B. Baker and R. P. Richards, "Herbicide concentration patterns in rivers draining intensively cultivated farmlands of northwestern Ohio," in D. Weigmann, ed., *Pesticides in terrestrial and aquatic environments,* Virginia Water Resources Research Center (Virginia Polytechnic Institute and State University, 1989).
76. E. M. Thurman et al., "Herbicides in surface waters of the midwestern U.S.: The effect of spring flush," *Env. Sci. Tech.* 25 (1991): 1794–96.
77. State of California Department of Fish and Game, "Hazard assessment of the rice herbicides molinate and thiobencarb to aquatic organisms in the Sacramento River system," Administrative Report 90-1, CDFG (Sacramento, 1990).
78. *CFR,* vol. 40, §141.50(a) (1991).
79. This summary was derived from an extraordinary literature review prepared by R. Levine, "Recognized and possible effects of pesticides in humans," in W. Hayes, ed., *Handbook of pesticide toxicity* (New York: Academic Press, 1991), 275–360.
80. "Aldicarb food poisoning from contaminated melons," *California, Mortality and Morbidity Weekly Report* 35 (1986): 254–58.
81. The rate of dissipation commonly follows first-order kinetics, in which the logarithm of concentration is linearly related to time, eventually approaching, but in theory never reaching, zero. NAS, *Pesticides in the diets of infants and children,* 207; citing G. Zweig, "The vanishing zero: The evolution of pesticide analyses," *Essays in Toxicol.* 2 (1970): 155.
82. The mixture is placed on absorbant paper, and components are separated and quantified with solvents.
83. L. C. Mitchell, "A new indicator for the detection of chlorinated pesticides on the paper chromatograph," *Assoc. Off. Agric. Chem. J.* 35 (1952): 928; cited in J. D. Rosen and F. M. Gretch, "Analytical chemistry of pesticides: Evolution and impact," in G. J. Marco, R. M. Hollingsworth, ad W. Durham, eds., *"Silent Spring" revisited* (Washington, D.C.: American Chemical Society, 1987).
84. P. A. Mills, J. H. Onley, and R. A. Gaither, "Rapid method for chlorinated pesticide residues in non-fatty foods," *Assoc. Off. Agric. Chem. J.* 46, no. 2 (1963): 186–91.
85. This brief description of detection technology is summarized from Rosen and Gretch, "Analytical chemistry of pesticides," 127–41.
86. As EPA continues its reregistration of individual pesticides, it commonly adjusts tolerances to reduce health risks. These adjustments are still made at the discretion of the agency, which uses a risk-benefit balancing criterion contained in FIFRA. See chapters 5 and 6 for a more complete discussion.
87. The single residue methods must now be submitted by the manufacturer to EPA before a new pesticide may be registered, to demonstrate that the compound may be detected in crops, soil, and water where it is likely to reside.
88. The Fish and Wildlife Service within the Department of the Interior monitors pesticide residues in fish and wildlife, primarily for the purposes of managing rare, threatened, or endangered species.
89. NAS, *Pesticides in the diets of infants and children,* 224–27.
90. In dairy products, chlorinated compounds were the most commonly detected pesticide. DDE, a metabolite of DDT, was detected in 49.3 percent of milk samples, dieldrin in

22.5 percent, BHC in 16.4 percent, DDT in 10.8 percent, heptachlor epoxide in 7.7 percent, PCB (Aroclors) in 4.1 percent, lindane in 2 percent, endrin in 0.4 percent, and aldrin in 0.3 percent. Chlorinated compounds were also frequently found in eggs. Nearly 41 percent of those tested contained DDE, 15.5 percent had dieldrin, 14.7 percent had DDT, 9.6 percent had PCB, and 1.5 percent had heptachlor epoxide. Imported egg products had higher levels of DDT, heptachlor epoxide, lindane and BHC, although sample sizes for imported products were far lower than for domestic foods. Nearly 70 percent of 15,200 samples of beef and 89 percent of over 11,000 samples of poultry contained pesticide residues. The accumulation of chlorinated compounds in meat, poultry, and fish is demonstrated by the finding that DDT and/or metabolites were found in 69.8 percent of meat and 89.1 percent of poultry; dieldrin in 50.7 percent of meat and 72.5 percent of poultry; HCB in 34.5 percent of meat and 15.9 percent of poultry; and heptachlor+heptachlor epoxide in 20.7 of meat and 16.5 percent of poultry tested. The average level of detection of these compounds was less than 1 ppm, but their prevalence and accumulative potential in human tissue make even these levels of interest. Nearly 70 percent of all fish and shellfish sampled had residues of DDE, 46 percent had detectable residues of dieldrin, 44 percent had DDT residues, and 42 percent TDE residues. Nearly 19 percent of animal feed tested during the period of study contained residues of DDE, 17.7 percent contained DDT; 10.8 percent had diedrin, and 10.3 percent had TDE. Similarly, residues in hay were dominated by DDT, DDE, BHC, and dieldrin. See R. E. Duggan et al., *Pesticide residue levels in foods in the U.S. from July 1, 1969, to June 30, 1976* (Washington, D.C.: FDA, 1983).

91. E. L. Gunderson, "FDA total diet study, April 1982–April 1984, dietary intakes of pesticides, selected elements, and other chemicals," *J. Assoc. Off. Anal. Chem.* 71, no. 9 (1988): 1200–1209.
92. Each year, FDA collects food samples, tests them in laboratories for pesticide residues, and publishes the results. See N. J. Yess, E. L. Gunderson, and R. R. Roy, "U.S. Food and Drug Administration monitoring of pesticide residues in infant foods and adult foods eaten by infants/children," *Jour. A.O.A.C. Int'l.* 76, no. 3 (1993): 492–507; FDA, "Food and Drug Administration Pesticide Program-Residue monitoring, 1991," *J. Assoc. Off. Anal. Chem.* 75 (1992): 136A–158A; FDA, "Food and Drug Administration Pesticide Program-Residue monitoring, 1990," *J. Assoc. Off. Anal. Chem.* 74 (1991): 121A–142A; FDA, "Food and Drug Administration Pesticide Program-Residue monitoring, 1989," *J. Assoc. Off. Anal. Chem.* 73 (1990): 127A–146A; FDA, "Food and Drug Administration Pesticide Program-Residue monitoring, 1988," *J. Assoc. Off. Anal. Chem.* 72 (1989): 133A–152A; FDA, "Food and Drug Administration Pesticide Program-Residue monitoring, 1987," *J. Assoc. Off. Anal. Chem.* 71 (1988): 156A–174A.
93. NAS, *Pesticides in the diets of infants and children,* 218.
94. See Yess et al., "U.S. Food and Drug Administration monitoring of pesticide residues," 492: "Most infant foods and adult foods eaten by infants/children that were analyzed for pesticide residues under FDA's 3 approaches to pesticide residue monitoring had residue levels well below EPA tolerances or FDA action levels. In relatively few instances, under regulatory monitoring, residues were found for which there was no tolerance for that particular commodity. Overall the findings corroborate results presented in earlier reports that indicate the safety of the food supply relative to pesticide residues."
95. R. Wiles and C. Campbell, *Pesticides in children's food,* Environmental Working Group Report (Washington, D.C., 1993).

96. Ibid.
97. Public Voice for Food and Health Policy, *Agrichemicals in America: Farmer's reliance on pesticides and fertilizers, a study of trends over 25 years* (Washington, D.C.: Public Voice for Food and Health Policy, 1993), 89–92.
98. Although FDA does not keep track of cases where more than one pesticide residue appears on a single sample, they do keep track of sample numbers, which makes it possible to construct a database of all individual sample results. This database may then be sorted by sample number to identify those samples contaminated by more than one pesticide. This method still presumes that all residue spikes were reported, which is not likely for residues detected well below the legally allowable limit.
99. "Instead of vigilence, U.S. inspection of imported food reminds some of the old fable where one monkey covers up his ears, another his mouth, another his eyes. You hear no evil, you speak no evil, you see no pesticide residues." Sen. P. Leahy, opening statement, Senate Committee on Agriculture, Nutrition, and Forestry, *Hearing on the circle of poison: Impact on American consumers,* 102d Cong., 1st sess., 20 Sept. 1991, 3.
100. USDA Economic Research Service, *Foreign agricultural trade of the United States,* 1990 supp. (Washington, D.C., 1992).
101. GAO, *Pesticides: U.S. and Mexican Fruit and Vegetable Pesticide Programs Differ,* (Washington, D.C., 1993), T-RCED-93-9.
102. GAO, *Better regulation of pesticide exports and pesticide residues in imported food is essential* (Washington, D.C., 1979), CED 79-43.
103. C. Smith and S. L. Beckmann, "Export of pesticides from U.S. ports in 1990: Focus on restricted pesticide exports," in Senate Committee on Agriculture, Nutrition and Forestry, *Hearing on the circle of poison,* 2.
104. The term "restricted use" has special legal significance in the United States because it allows registration only under specified conditions, due to high levels of toxicity or potential for environmental contamination. Pesticides classified as restricted use were under consideration for cancellation of registration. Special use restrictions such as applicator training; protective equipment such as respirators, gloves, or rain gear; and special application methods may be required for use of these compounds in the United States. To be effective, this form of "risk management" requires government vigilence, an understanding by farmers of the risks and methods of their management, and the availability of control equipment.
105. L. W. Goodman, "Foreign toxins: Multinational corporations and pesticides in Mexican agriculture," in C. S. Pearson, ed., *Multinational corporations, environment, and the third world* (Durham, N.C.: Duke University Press, 1987): 98.
106. United Nations Food and Agriculture Organization, *FAO Trade Yearbook 1989* 42 (1990): 311–13.
107. A. Wright, "Third world pesticide production," *Global Pesticide Campaigner* 1 (June 1991): 3.
108. GAO, *Agricultural trade: Causes and impacts of increased fruit and vegetable imports* (Washington, D.C., 10 May 1986), GAO/RCED-88-149BR.
109. GAO, *Pesticides: Limited testing finds few exported unregistered pesticide violations on imported food* (Washington, D.C., 1993), RCED-94-1.
110. GAO, *Five Latin American countries' controls over the registration and use of pesticides* (Washington, D.C., 1990), T-RCED-90-57.
111. F. Powledge, "The farm bill: Toxic shame," *Amicus J.* 13, no. 1 (1991): 40.

112. GAO, *Pesticides: Better sampling and enforcement needed on imported food* (Washington, D.C., 1986), RCED-86-219.
113. GAO, *Pesticides: Need to enhance FDA's ability to protect the public from illegal residues* (Washington, D.C., 1986R), CED-97-7.
114. GAO, *Imported meat and livestock: Chemical residue detection and the issue of labeling* (Washington, D.C., 1987), RCED-87-142.
115. GAO, *Pesticides: A comparative study of industrialized nation's regulatory systems* (Washington, D.C., 1993), PEMD-93-17.
116. GAO, *Five Latin American countries' controls over the registration and use of pesticides* (Washington, D.C., 1990), T-RCED-90-57.
117. One of the sole methods of human excretion of DDT and its metabolites such as DDE is through breast milk. In India, the interior of homes and other buildings are still routinely sprayed with DDT, and it is also allowed for use for crop protection in many tropical nations. In a recent survey of cereals, spices, milk, butter, and edible oils, 85 percent were contaminated with DDT residues. The levels found in fruits and vegetables were quite small (.06 ppm or less) whereas levels in vegetable oils and dairy products ranged between 0.59 and 4.8 ppm. See B. S. Kaphalia et al., "Organochlorine pesticide residues in different Indian cereals, pulses, spices, vegetables, fruits, milk, butter, Deshi ghee, and edible oils," *J. Assoc. Off. Anal. Chem.* 73, no. 4 (July–Aug. 1990): 509–12.
118. R. R. Ofner and A. O. Calvery, "Determination of DDT and its metabolite in biological materials by use of the Schecter-Haller method," *J. Pharm. Exp. Tehr.* 85 (1945): 363–70.
119. F. W. Kutz et al., "Effects of reducing DDT usage on total DDT storage in humans," *Pest. Mon. J.* 11 (1977): 61–63.
120. These compounds induce cytochrome P-450 activity and hydroxylate testosterone. See J. Haake et al., "The effects of organochlorine pesticides as inducers of testosterone and benzo[a]pyrene hydroxylases," *Gen. Pharmac.* 18, no. 2 (1987).
121. E. P. Laug, F. M. Kunze, and C. S. Prickett, "Occurrence of DDT in human fat and milk," *AMA Arch. Ind. Hyg. & Occup. Med.* 3 (Mar. 1951): 245–46. The work of Laug and his colleagues at FDA was known to Rachel Carson and cited in *Silent Spring* in 1962. This article is also cited by T. Dunlap in his *DDT* (Princeton, N.J.: Princeton University Press, 1981).
122. G. Woodard, R. R. Ofner, and C. M. Montgomery, "Accumulation of DDT in body fat and its appearance in the milk of dogs," *Science* 102 (1945): 177–78.
123. Carson, *Silent spring,* 23.
124. See generally A. A. Jensen et al., *Contaminants in human milk* (Boca Raton, Fla.: CRC Press, 1990), 228; A. Nair and M. K. Pillai, "Trends in ambient levels of DDT and HCH residues in humans and the environment of Delhi, India," *Sci. Total Envir.* 121 (1992): 145; P. E. Spicer and R. K. Kereu, "Organochlorine insecticide residues in human breast milk: a survey of lactating mothers from a remote area in Papua, New Guinea," *Bull. Envir. Contam. & Toxicol.* 50, no. 4 (1993): 540; M. F. Stevens et al., "Organochlorine pesticides in Western Australian nursing mothers," *Med. J. Australia* 158, no. 4 (1993): 238; M. S. Wolff, "Occupationally derived chemicals in breast milk," *Am. J. Ind. Med.* 4 (1983): 259–82; M. Wolff et al., "Blood levels of organochlorine residues and risk of breast cancer," *J. NCI* 85, no. 8 (1993): 648–52.
125. NAS, *Pesticides in the diets of infants and children,* 239–44.

126. D. R. Mattison, "Pesticide concentrations in Arkansas breast milk," *J. Arkansas Med. Soc.* 88, no. 11 (1992): 553–57.
127. S. Slorach and R. Vaz, *Assessment of human exposure to selected organochlorine compounds through biological monitoring,* UNEP/WHO/Swedish National Food Administration (Geneva, 1983).
128. H. Bouwman et al., "Transfer of DDT used in malaria control to infants via breast milk," *Bull. WHO* 70, no. 2 (1992): 241. See also H. Bouwman et al., "Levels of DDT and metabolites in breast milk from Kwa-Zulu mothers after DDT application for malaria control," *Bull. WHO* 68, no. 6 (1990): 761.
129. H. Bouwman, "Malaria control and levels of DDT in serum of two populations in KwaZulu," *J. Toxicol. & Envir. Health* 33 (1991): 141–55; see also Bouwman et al., "Transfer of DDT," 241–50.
130. Cited in Bouwman et al., "Transfer of DDT," 241–50.
131. NAS, *Pesticides in the diets of infants and children,* 243.

Chapter 8. The Susceptibility of Children

1. P. A. von Hohenheim-Paracelsus, as quoted in M. O. Amdur, J. Doull, and C. Klaassen, eds., *Casarett and Doull's toxicology: The basic science of poisons,* 4th ed. (New York: McGraw Hill, 1991), 5.
2. It also underlies the science of pharmacology, which explores both effective and safe dosages of drugs.
3. The association between exposure to a toxic agent and an adverse effect is referred to as a dose-response relation. It is generally recognized through formal and controlled scientific experiments designed both to identify types of hazards associated with a technology or chemical agent and to identify how variation in levels of contamination lead to variation in adverse effects. To provide knowledge useful for regulating the quality of environmental health, many questions surrounding the dose-response function should be addressed, including: What are the types of damages, or damage endpoints, for which we should test? Is there a threshold of exposure that must be exceeded before an adverse effect results? Does the severity of the effect increase in a linear relation to an increase in dose? Is there age-related variance in susceptibility to the toxin? Does the controlled test provide a reasonable basis for extrapolating the results to the real world—e.g., which species of laboratory animal are likely to provide the most accurate basis for predicting human responses to toxins? Should the maximum tolerated dose be administered, or should it more closely approximate expected human exposure? All of these questions are important for a clear understanding of any dose-response relation, which serves as an indicator of relative toxicity among chemicals and provides a crucial foundation for regulating human exposures.
4. NAS, *Pesticides in the diets of infants and children* (Washington, D.C.: National Academy Press, 1993). Much of this section is derived from the literature that supported this report.
5. Ibid., 28.
6. Ibid., derived from J. Karlberg, "On the construction of the infancy-childhood-puberty growth standard," *Acta Paediatrica Scandinavia* suppl. 356 (1989): 26–37.
7. NAS, *Pesticides in the diets of infants and children,* 24.
8. S. M. Cohen and L. B. Ellwein, "Genetic errors, cell proliferation, and carcinogene-

sis," *Cancer Res.* 51 (1991): 6493–505; I. B. Weinstein, "Mitogenesis is only one factor in carcinogenesis," *Science* 251 (1991): 387–88.

9. J. E. Norman, "Breast cancer in women irradiated early in life," in V. R. Hunt et al., *Environmental factors in human growth and development,* Banbury report no. 11, Cold Spring Harbor Laboratories (Cold Spring Harbor, N.Y.: Cold Spring Harbor Labs, 1982); see also A. B. Miller et al., "Mortality from breast cancer after irradiation during fluoroscopic examinations in patients being treated for tuberculosis," *New Eng. J. Med.* 321 (1989): 1285–89. Unfortunately, much of the remaining data correlating higher levels of cell proliferation with higher cancer incidence come from animals, two examples being N-nitrosomethylurea and vinyl chloride.
10. See V. N. Anisimov, *Role of age in host sensitivity to carcinogens,* IARC Scientific Publication 51 (Lyons: IARC, 1983): 99–112; R. T. Drew et al., "The effect of age and exposure duration on cancer induction by a known carcinogen in rats, mice and hamsters," *Toxicol. Appl. Pharm.* 68 (1983): 120–30; V. N. Anisimov, "Modifying effects of again on N-methyl-N-nitrosourea-induced carcinogenesis in female rats," *Exp. Pathol.* 19 (1981): 81–90; V. N. Anisimov, "Modifying effects of agin on N-methyl-N-nitrosourea-induced carcinogenesis in female rats," *Exp. Pathol.* 19 (1981): 81–90.
11. D. J. Murdoch, D. Krewski, and J. Wargo, "Cancer risk assessment with intermittent exposure," *Risk Anal.* 12, no. 4 (1992): 569–77.
12. The rates of increase may also be relevant to childhood risks from pesticides. Extracellular water volume and body surface area appear to be linearly related to metabolic rates throughout childhood. By contrast, total body fat increases quickly during infancy and during adolescence, especially for females.
13. E. J. Calabrese, *Age and susceptibility to toxic substances* (New York: Wiley and Sons, 1986).
14. Ibid., 153.
15. Ibid., 154.
16. NTP, *NTP Technical report on the perinatal toxicology and carcinogenesis studies of ethylene thiourea (CAS no. 96-45-7) in F3441N Rats and B6C3F1 mice (feed studies),* NTP TR 388 (Research Triangle Park, N.C.: NTP, 1992).
17. J. Jeyaratnam, "Health problems of pesticide usage in the Third World," *Br. J. Ind. Med.* 42 (1985): 505–6. R. Levine, "Recognized and possible effects of pesticides in humans," in W. Hayes, ed., *Handbook of pesticide toxicity* (New York: Academic Press, 1991): 275.
18. J. R. Davis et al., "Family pesticide use and childhood brain cancer," *Arch. of Env. Contam. and Toxicol.* 21, no. 1 (1993): 87–92. See also P. Kristensen et al., "Cancer in offspring of parents engaged in agricultural activities in Norway: Incidence and risk factors in the farm environment," *Int. J. of Cancer* 65, no. 1 (1996): 39–50.
19. FFDCA, *USC,* vol. 21, 321. See Drug Amendments Act of 1962. The law requires substantial evidence to support claims of effectiveness: "'Substantial effectiveness' means evidence consisting of adequate and well controlled investigations, . . . by experts qualified . . . to evaluate the effectiveness of the drug . . . , on the basis of which it could fairly and responsibly be concluded . . . that the drug will have the effect it purports or is represented to have."
20. The absence of any requirement for drug safety testing in humans is curious, in that the act of licensing a drug results in a grander-scale uncontrolled human experiment as a result of efficacy testing. Although dosage is controlled in efficacy trials, long-term studies of chronic adverse health effects are not normally conducted. See A. K.

Done, S. N. Cohen, and L. Strebel, "Pediatric clinical pharmacology and the 'therapeutic orphan,'" *Ann. Rev. Pharmacol. Toxicol.* 17 (1977): 561–73. See also NAS, *Pesticides in the diets of infants and children,* 54.

21. W. J. Hayes, "Studies in humans," in Hayes, *Handbook of pesticide toxicology,* 216–20.
22. D. L. Glaubiger et al., "The relative tolerance of children and adults to anti-cancer drugs," *Frontiers of Rad. Ther. & Oncol.* 16 (1981): 42–49.
23. G. Powis and M. P. Hacker, *The toxicity of anticancer drugs* (New York: Pergamon Press, 1991). These values were calculated on a milligram per square meter basis; they would have demonstrated more significant variation had they been calculated on a milligram per kilogram body weight per day basis.
24. W. G. Woods, M. O'Leary, and M. E. Nesbit, "Life-threatening neuropathy and hepatotoxicity in infants during induction therapy for acute lymphoblastic leukemia," *J. Pediatr.* 98 (1981): 642–45.
25. R. E. McKinney et al., "A multicenter trial of oral zidovudine in children with advanced human immunodeficiency virus disease," *New Eng. J. Med.* 324 (1991): 1018–25.
26. Exposure to lead from the environment comes from numerous sources. An important source of lead toxicity in children is paint, which has severely contaminated over 20 million buildings in the United States. In 1990, 4.4 million children lived in housing built before 1950, and therefore known to be contaminated by lead paint, which was not phased out from sales until 1980. Leaded gasoline, still in the process of being phased out in the United States, has resulted in broadscale contamination of soil, plants, and humans. Drinking water is supplied by pipes with lead fittings and lead solder to millions of children. Contaminated dust and soil may be eaten or inhaled, especially by infants and young children. Lead may also contaminate food, both from the process of bioaccumulation and from packaging, especially from lead solder in food cans and poorly glazed cookware dishes. In a 1990 report to Congress on lead poisoning, it was estimated that between 3 and 4 million children have blood lead levels greater than 15 μg/dl, a level that has been demonstrated to induce behavioral effects in children. See A. F. Crocetti et al., "Determination of numbers of lead-exposed U.S. children by areas of the U.S.: An integrated summary of a report to the U.S. Congress on childhood lead poisoning," *Envir. Health Perspect.* 89 (1990): 109–20; see also P. Mushak, "Defining lead as the premier environmental health issue for children in America: Criteria and their quantitative application," *Envir. Res.* 59, no. 2 (1992): 281–309.
27. N. L. Day and G. A. Richardson, "Prenatal alcohol exposure: A continuum of effects," *Semin. Perinatol.* 15, vol. 4 (1991): 271–79.
28. R. A. Goyer, "Transplacental transport of lead," *Envir. Health Persp.* 89 (1990): 101–5.
29. J. Perino and C. B. Erinhart, "The relation of subclinical lead level to cognitive and sensorimotor impariment in black preschoolers," *J. Learn. Dis.* 7 (1974): 26–30. H. L. Needleman et al., "Deficits in psychologic and classroom performance of children with elevated dentine lead levels," *New Eng. J. Med.* 300 (1979): 689–95; J. W. Graef, "Management of low level lead exposure," in M. L. Needleman, ed., *Low level lead exposure: The chemical duplication of current research* (New York: Raven Press, 1980);
30. D. P. Alfano and T. L. Petit, "Postnatal lead exposure and the cholinergic system," *Physiol. Behav.* 34 (1985): 449–55; D. Bellinger et al., "Low level lead exposure and

infant development in the first year," *Neurobehav. Toxicol. Teratol.* 8 (1986): 151–61; D. Bellinger et al., "Longitudinal analyses of prenatal and postnatal lead exposure and early cognitive development," *New Eng. J. Med.* 316 (1987): 1037–43; K. Dietrich et al., "Low level fetal lead exposure effect on neurobehavioral development in early infancy," *Pediatrics* 80 (1987): 721–30; M. R. Moore et al., "Prospective study of the neurological effects of lead in children," *Neurobehav. Toxicol. Teratol.* 4, no. 6 (1982): 739–43; M. R. Moore, M. J. McIntosh, and I. W. Bushnell, "The neurotoxicology of lead," *Neurotoxicol.* 7, no. 2 (1986): 541–56; H. L. Needleman, "Lead at low dose and the behavior of children," *Neurotoxicol.* 4, no. 3 (1983): 121–33; D. Otto and L. Reiter, "Developmental changes in slow cortical potentials of young children with elevated lead body burden, neurophysiological considerations," *Ann. N.Y. Acad. Sci.* 425 (1984): 377–83; D. Otto et al., "Five year follow-up study of children with low-to-moderate lead absorption: Electrophysiological evaluation," *Envir. Res.* 38, no. 1 (1985): 168–86; H. A. Ruff and P. E. Bijur, "The effects of low to moderate lead levels on neurobehavioral functions of children," *J. Dev. Behav. Pediatr.* 10, no. 2 (1989): 103–9; G. Winneke et al., "Comparing the effects of perinatal and later childhood lead exposure on neuropsychological outcome," *Envir. Res.* 38, no. 1 (1985): 155–67.

31. NAS, *Pesticides in the diets of infants and children,* 63.
32. E. K. Silbergeld, "Mechanisms of lead neurotoxicity, or looking beyond the lamppost," *FASEB (Federation of American Society of Experimental Biologists) J.* 6, no. 13 (1992): 3201–6.
33. These include erythrocyte acetylcholinesterase, plasma pseudocholinesterase, and arylesterase. See D. J. Ecobichon and D. S. Stephens, "Perinatal development of human blood esterases," *Clin. Pharmacol. Therap.* 14 (1973): 41–47.
34. In 1986 Congress passed the Federal Emergency Planning and Community Right-to-Know Act, requiring that EPA create an inventory of more than 300 toxic chemicals commonly released to the environment. This inventory has become known as the Toxic Release Inventory (TRI). EPA, *The Toxics Release Inventory: A national perspective,* EPA Report 560/4-89-006 (Washington, D.C., 1987); see also Office of Technology Assessment, *Neurotoxicity: Identifying and controlling poisons of the nervous system* (Washington, D.C., 1991).
35. A. K. Done, "Developmental pharmacology," *Clin. Pharmacol. Therap.* 5 (1964): 432–79.
36. E. L. Goldenthal, "A compilation of LD50 values in newborn and adult animals," *Toxicol. Appl. Pharmacol.* 18 (1971): 185–207.
37. F. C. Lu, D. C. Jessup, and A. Lavallee, "Toxicology of pesticides in young versus adult rats," *Food Cosmet. Toxicol.* 3 (1965): 591–96.
38. G. M. Benke and S. D. Murphy, "The influence of age on the toxicity and metabolism of methyl paration and parathion in male and female rats," *Toxicol. Appl. Pharmacol.* 31 (1975): 269.
39. NAS, *Pesticides in the diets of infants and children,* 53.
40. J. Gershanik et al., "The gasping syndrome and benzyl alcohol poisoning," *New Eng. J. Med.* 307 (1982): 1384–88.
41. E. E. Tyrala et al., "Clinical toxicology of hexachlorophene in newborn infants," *J. Pediatr.* 91 (1977): 481–86.
42. A. Warner, "Drug use in the neonate," *Clin. Chem.* 32 (1986): 721–27.
43. L. Ries et al., *SEER cancer statistics review, 1973–1991,* HEW, NCI, NIH pub. no. 94-2789 (Bethesda, Md.: NIH, 1994).

44. These regions include: Connecticut, metropolitan Atlanta, Hawaii, Iowa, metropolitan Detroit, New Mexico, San Francisco–Oakland SMSA (Standard Metropolititan Statistical Area), Utah, and Seattle–Puget Sound.
45. Ries et al., *SEER,* 41.
46. These statistics are age-adjusted incidence statistics for the SEER survey regions. See Ries et al. *SEER,* 428.
47. These statistics are age-adjusted U.S. mortality statistics. See Ries et al., *SEER,* 429.
48. These estimates were made by R. Doll and R. Peto, "The causes of cancer: Quantitative estimates of avoidable risks of cancer in the United States today," *J. Natl. Cancer Inst.* 66 (1981): 1191–308; and adjusted by B. N. Ames, L. S. Gold, and W. C. Willett in "The causes and prevention of cancer," *Proc. Natl. Acad. Sci.* 92, no. 12 (1995), to lie between 20 and 40 percent.
49. B. E. Henderson, R. K. Ross and M. C. Pike, "Toward the primary prevention of cancer," *Science* 254 (1991): 1131–38; J. R. Harris et al., "Medical Progress-Breast cancer," *New Eng. J. Med.* 327 (1992): 319–28; and G. A. Colditz et al., "Type of postmenopausal hormone use and risk of breast cancer," *Cancer Causes & Control* 3 (1992): 433–39.
50. IARC, Schistosomes, liver flukes and helicobacter pylori (Lyons: IARC, 1994); cited in Ames et al., "Causes and prevention of cancer."
51. M. M. Braun et al., "Genetic component of lung cancer: Cohort study of twins," *Lancet* 344 (1994): 440–43.
52. Damaged DNA may cause a mutation, a permanent transmissible change in genetic material, loss of genetic material from a chromosome, or the translocation of a fragment from one chromosome to another.
53. Ames et al., "Causes and prevention of cancer," 5258.
54. B. Ames, M. Profet, and L. S. Gold, *Proc. Natl. Acad. Sci.* 87 (1990): 7777–81.
55. L. S. Gold et al., "Interspecies extrapolation in carcinogenesis: Prediction between rats and mice," *Envir. Health Persp.* 81 (1989): 211–19; and L. S. Gold et al., "The fifth plot of the carcinogenic potency data base," *Envir. Health Persp.* 100 (1993): 65–135.
56. The effect of employing different types of contamination estimates is demonstrated in chapter 10: one approach may make risks disappear, whereas another can raise alarm.
57. L. S. Gold et al., "A carcinogenic potency database of the standardized results of animal bioassays," *Envir. Health Pers.* 58:(1984) 9–319; see also L. S. Gold et al., "Third chronological supplement to the Carcinogenic Potency Database: Standardized results of animal bioassays published through December 1986 and by the National Toxicology Program through June 1987," *Envir. Health Perspect.* 84 (1990): 215–86. Relative potency was judged using an estimate known as the "TD50," that dose that cuts the number of tumor-free individuals in half at the end of the dosing period. Using the TD50 as an estimate of carcinogenic potency derived from the number of tumors produced at specific dose levels for each toxic compound, they found that 40 percent of the estimates from compared studies lie within a factor of two, 80 percent lie within a factor of five, and 85 percent lie within a factor of ten. Where discordant results were detected, they found that the compounds were only weak carcinogens, active only at the maximum tolerated dose.
58. There were 41 comparisons among rat studies, 25 comparisons among mice, and 4 in hamsters. Nearly 75 percent involved only two comparisons, whereas 25 percent involved three or four experiments.

59. L. S. Gold et al., "Reproducibility of results in 'near-replicate' carcinogenesis bioassays," *J. Nat. Cancer Inst.* 6, no. 11 (1987): 49–58.
60. E. E. McConnell, "The maximum tolerated dose: The debate," *J. Am. Coll. Toxicol.* 8 (1989): 1115–20.
61. If fifty individuals are tested for each sex-species-dose, and the incidence of tumors in controls was 10 percent (5/50), then a minimum statistically significant response at the MTD would be 20 percent (10/50). See NAS, *Issues in risk assessment* (Washington, D.C.: National Academy Press, 1993), 49.
62. W. Hayes, "Dosage and other factors influencing toxicity," in Hayes, *Handbook of pesticide toxicology.*
63. J. M. Sontag, N. P. Page, and U. Saffiotti, "Guidelines for carcionogen bioassay in small rodents," *Carcinog. Tech. Rep.*, ser. 1, HEW Pub. (NIH 76-801) (Bethesda, Md.: National Cancer Institute, 1976).
64. The MTD is by definition the inverse of a compound's potency in inducing a toxic response—the higher the MTD, the lower the potency. Another indicator of toxic potency is the LD50, or that single dose that is expected to kill half of exposed animals. The TD50 is an inverse measure of carcinogenic potency: it is defined as the dose that if administered chronically for the normal life span of the species will halve the probability of remaining tumorless throughout that period. See R. Peto et al., "The TD50: A proposed general convention for the numerical description of the carcinogenic potency of chemicals in chronic-exposure animal experiments," *Envir. Health Persp.* 58 (1984): 1–8. In addition to the TD50, the potency of a carcinogen may be estimated based upon the slope of the dose response curve, in the dose region of the curve that best represents estimated human exposure. This slope is commonly referred to as "q1," and is the coefficient of the linear term of the multistage model of carcinogenesis. The 95 percent upper bound confidence limit of this slope is known as the q1*. The simple linear equation [q1*]*[dose] yields an estimate of expected cancer risk at that dose. See NAS, *Issues in risk assessment,* 22–24.
65. Congress, Office of Technology Assessment, *Identifying and regulating carcinogens,* OTA-BP-H-42 (Washington, D.C., Nov. 1989).
66. See comments by M. Davis in NAS, *Issues in risk assessment,* 88.
67. Counterclaims have been made by J. K. Haseman, who examined 52 separate studies and found that in over two-thirds if one-half the MTD had been used instead of the MTD, no carcinogenic effect would have been found. See J. K. Haseman, "Issues in carcinogenicity testing: Dose selection," *Tox. App. Pharm.* 5 (1985): 371–86.
68. NAS, *Pesticides in the diets of infants and children,* 18.
69. S. Preston-Martin et al., "Epidemiologic evidence for the increased cell proliferation model of carcinogenesis," in B. E. Butterworth et al., eds., *Chemically induced cell proliferation: Implications for risk assessment* (New York: Wiley-Liss, 1991), 64.
70. Ibid., 64.
71. P. Armitage and R. Doll, "The age distribution of cancer and a multistage theory of cancer," *Brit. J. Cancer* 8 (1954): 1–12; P. Armitage, "Multistage models of carcinogenesis," *Envir. Health Persp.* 63 (1985): 195–201; T. J. Slaga et al., "Studies on the mechanism of skin tumor promotion: Evidence for several stages in promotion," *Proc. NAS* 77 (1980): 3659–63; H. Pitot, L. Barsness, and T. Kitagawa, "Stages in the process of hepatocarcinogenesis in rat liver," *Carcinogen.* 2 (1978): 433–42.
72. S. Moolgavkar and D. Venzon, "Two-event models for carcinogenesis: Incidence curves for childhood and adult tumors," *Math. Biosci.* 47 (1979): 55–57; see also S.

Moolgavkar and A. Knudson, "Mutation and cancer: A model for human carcingenesis," *J. Nat. Cancer Inst.* 64 (1981): 977–89; S. Moolgavkar, A. Dewanji, and D. Venzon, "A stochastic two-stage model for cancer risk assessment: vol. 1: The hazard function and the probability of tumor," *Risk Anal.* 8 (1988): 383–93; and EPA, "Guidelines for carcinogen risk assessment," *FR* 51, 33992 (1986)

73. They distinguish among normal cells, two types of damaged cells, and two types of mutated cells. The models permit normal cells to divide or die at various birth and death rates, and the entire carcinogenic process is assumed to have two stages. In the first stage, a damaged cell is formed following exposure to some xenobiotic agent through the formation of DNA adducts, single strand breaks, chromosomal translocation, or some other genetic perturbation. This damage is assumed to be in only one strand of DNA and to be reparable at some variable rate. If it is not repaired, when the cell divides, the DNA damage is assumed to become fixed in one of the daughter cells, creating the first type of mutated cell. The other daughter cell is assumed derived from the DNA strand without the aberration, and is therefore normal. In the second stage, the entire process of damage, repair, birth, and death is repeated. For those cells experiencing the second round of mutation, the result is a malignant cell that is expected to progress to malignancy. See C. J. Portier and A. Kopp-Schneider, "A multistage model of carcinogenesis incorporating DNA damage and repair," *Risk Anal.* 11, no. 3 (1991): 535–43.
74. D. J. Murdoch, D. Krewski, and J. Wargo, "Cancer risk assessment with intermittent exposure," *Risk Anal.* 12, no. 4 (1992): 569–77.
75. D. Krewski, J. P. Wargo, and R. Rizek, "Risks of dietary exposure to pesticides in infants and children," in R. Kroes, ed, *Monitoring dietary intakes* (Berlin: Springer Verlag, 1991), 75–89; and J. P. Wargo, "Analytical methodology for estimating oncogenic risks of human exposure to agricultural chemicals in food crops," in NAS, *Regulating pesticides in food,* 174–95.
76. Within the multistage model advanced by Armitage and Doll, early life exposures are more effective in producing tumors than later exposures, if the first stage (initiation) is dose-dependent. The effects of intermittent or time-dependent exposure on cancer risk have been explored by many scientists. See K. Crump and R. Howe, "The multistage model with a time-dependent dose pattern: Applications to carcinogenic risk assessment," *Risk Anal.* 4 (1984): 163–76; T. C. Thorslund, C. Brown, and G. Charnley, "Biologically motivated cancer risk models," *Risk Anal.* 7 (1987): 109–19; R. Kodell, D. Gaylor, and J. Chen, "Using average lifetime dose rate for intermittent exposures to carcinogens," *Risk Anal.* 7 (1987): 339–45; P. F. Morrison, "Effects of time-variant exposure on toxic substance response," *Envir. Health Pers.* 76 (1987): 133–40; D. J. Murdoch and D. Krewski, "Carcinogenic risk assessment with time-dependent exposure patterns," *Risk Anal.* 8 (1988): 521–30; D. Krewski and D. J. Murdoch, "Cancer modeling with intermittent exposures," in S. H. Moolgavkar, ed., *Scientific issues in quantitative cancer risk assessment* (New York: Birkhauser Boston, 1990), 196–214.
77. G. Pietra, H. Rappaport, and P. Shubik, "The effects of carcinogenic chemicals in newborn mice," *Cancer* 14 (1961): 308–17; see also B. Toth, "A critical review of experiments in chemical carciogenesis using newborn animals," *Cancer Res.* 28 (1968): 727–38.
78. S. D. Vesselinovitch, N. Mihailovich, and G. Pietra, "The prenatal exposure of mice to urethan and the consequent development of tumors in various tissues," *Cancer Res.*

27 (1967) :2333–37; S. D. Vesselinovitch et al., "Broad spectrum carcinogenicity of ethylnitrosourea in newborn and infant mice," *Proc. Am. Assoc. Cancer. Res.* 12 (1971): 56; and S. D. Vesselinovitch et al., "The effect of age, fractionation, and dose on radiation carcinogenesis in various tissues of mice," *Cancer Res.* 31 (1971): 2133–42.

79. M. Naito, Y. Naito, and A. Ito, "Effect of age at treatment on the incidence and location of neurogenic tumors induced in Wistar rats by a single dose of N-ethyl-N-nitrourea," *Jpn. J. Cancer Res.* 72 (1981): 569–77.
80. B. Toth, "A critical review of experiments in chemical carcinogenesis using newborn animals," *Cancer Res.* 28 (1968): 727–38.
81. NAS, *Pesticides in the diets of infants and children,* 70–76.
82. S. D. Vesselinovitch et al., "Conditions modifying development of tumors in mice at various sits by Benzo(a)pyrene," *Cancer Res.* 35 (1975): 2948–53.
83. S. D. Vesselinovitch et al., "Neoplastic response of mouse tissues during perinatal age periods and its significance in chemical carcinogenesis," National Cancer Institute Monograph 51 (1976): 247.
84. R. T. Drew et al., "The effect of age and exposure duration on cancer induction by a known carcinogen in rats, mice and hamsters," *Tox. Appl. Pharm.* 68 (1983): 120–30.
85. V. M. Craddock, "Cell proliferation and experimental liver cancer," in H. M. Cameron, D. A. Linsell, and G. P. Warick, eds., *Liver Cell Cancer* (Amsterdam: Elsevier, 1976), 152–201.
86. M. C. Dryoff et al., "Correlation of O4-ethyldeoxythymidine accumulation, hepatic initiation and hepatocellular carcinoma induction in rats continuously administered diethylnitrosamine," *Carcinogen.* 7, no. 2 (1986): 241–46.
87. W. Lijinsky and R. M. Kovatch, "The effect of age on susceptibility of rats to carcinogenesis by two nitrosamines," *Jpn. J. Cancer Res.* 77 (1986): 1222–26.
88. P. C. Chan and T. L. Dao, "Effects of dietary fat on age-dependent sensitivity to mammary carcinogenesis," *Cancer Letters* 18 (1983): 245–49; citing J. Russo et al., "Effect of dietary fat and BHT on rat mammary gland differentiation and susceptibility to carcinogenesis," *Proc. Am. Assoc. Cancer Res.* 22 (1981): 113.
89. P. J. Landrigan, "Commentary: environmental disease—a preventable epidemic," *Am. J. Pub. Health* 82 (1992): 941–43.
90. Descriptive epidemiology may simply provide statistics on the frequency and distribution of disease. Longitudinal epidemiology examines trends in variables over time. Ecologic epidemiology explores relationships between disease incidence and exposures. Case-control studies compare disease incidence in affected and unaffected populations and attempt to identify causal variables. Prospective studies identify populations of interest and monitor patterns of behavior, exposure, and disease incidence over an appropriate time period for the risk of interest.
91. Whereas vinyl chloride produces a rare angiosarcoma of the liver, organophosphate insecticide poisoning produces symptoms that are quite flu-like: nausea, vomiting, fever, chills, and if severe, convulsions and death. Moderate poisoning may therefore easily be mistaken for a variety of illnesses by the unsuspecting clinician. See I. F. H. Purchase, J. Stafford, and F. M. Paddle, "Vinyl chloride: A cancer case study," in D. B. Clayson, D. Krewski, and I. Munro, *Toxicological risk assessment,* vol. 2, *General criteria and case studies* (Boca Raton, Fla.: CRC Press, 1985): 167–94. C. Maltoni et al., "Carcinogenicity bioassays of vinyl chloride monomer: A model of risk assessment on an experimental basis," *Envir. Health Persp.* 41 (1981): 3–29.
92. See, generally, D. Krewski, "Risk and risk management: Issues and approaches," in

R. S. McColl, ed., *Environmental health risks: Assessment and management* (Waterloo, Ont.: University of Waterloo Press, 1987): 29–51.

93. OPP, Economic Analysis Branch, Biological and Economic Analysis Division, *Pesticide industry sales and usage, 1988 market estimates* (Washington, D.C., 1989).
94. J. R. Davis, R. C. Brownson, and R. Garcia, "Family pesticide use in the home, garden, orchard, and yard," *Arch. Envir. Contam. Tox.* 22, no. 3 (1992): 260–66.
95. R. A. Lowengart et al., "Childhood leukemia and parents' occupational and home exposures," *J. Nat'l Can. Inst.* 79, no. 1 (1987): 39–46.
96. X. O. Shu et al., "A population-based case-control study of childhood leukemia in Shanghai," *Cancer* 62, no. 3 (1988): 635–44.
97. J. D. Buckley et al., "Occupational exposures of parents of children with acute nonlymphocytic leukemia: A report from the Children's Cancer Study Group," *Cancer Res.* 49, no. 14 (1989): 4030–37.
98. E. A. Holly et al., "Ewing's bone sarcoma, paternal occupational exposure, and other factors," *Am. J. Epidem.* 135, no. 2 (1992): 122–29.
99. J. R. Davis et al., "Family pesticide use and childhood brain cancer," *Arch. Envir. Cont. Tox.* 24 (1993): 87–92.
100. E. Gold et al., "Risk factors for brain tumors in children," *Am. J. Epidem.* 109, no. 3 (1979): 309–19.
101. T. H. Sinks, "N-nitroso compounds, pesticides, and parental exposures in the workplace as risk factors for childhood brain cancer: A case-control study," *Diss. Abstr. Int'l* 46, no. 6 (1985): 1888b.
102. E. S. Hansen, H. Hasle, and F. Lander, "A cohort study on cancer incidence among Danish gardeners," *Am. J. Ind. Med.* 21, no. 5 (1992): 651–60.
103. L. O'Leary et al., "Parental occupational exposures and risk of childhood cancer: a review," *Am. J. Ind. Med.* 20 (1992): 651–60.

Chapter 9. The Diet of a Child

1. These figures do not include decomposition products or manufacturing contaminants. See NAS, *Toxicity testing: Strategies to determine needs and priorities* (Washington, D.C.: National Academy Press, 1984), 5.
2. Research Triangle Institute, *Interim report no. 1: The construction of a raw agricultural commodity consumption data base,* RTI Project no. 252U (Research Triangle Park, N.C.: Research Triangle Institute, 1983): 2123–27; see also D. S. Saunders and B. J. Petersen, *An introduction to the Tolerance Assessment System Hazard Evaluation Division, Office of Pesticide Programs* (Washington, D.C.: EPA, 1987).
3. Nearly 5,000 different types of foods were reported eaten by people surveyed in the 1977–78 National Food Consumption Survey. See USDA, Consumer Nutrition Division, *Nationwide food consumption survey: Food intakes, individuals in 48 states, year 1977–78,* report no. I-1 (Hyattsville, Md.: Consumer Nutrition Division, Human Nutrition Information Service, 1983).
4. *CFR,* vol. 40, §180.4 (1986).
5. House of Representatives, Subcommittee on Department Operations, Research, and Foreign Agriculture of the Committee on Agriculture, *Regulation of pesticides: Appendix to hearings,* Serial 98–22, 98th Cong., 1983, 86. This is also known as the "Benbrook Report" after Charles Benbrook, its primary author.
6. The Dietary Risk Evaluation System is a computer program that EPA uses to estimate human exposure and risk from pesticides in the nation's food supply. This pro-

gram estimates food intake of roughly 700 raw and processed food forms by breaking complex foods such as pizza into component forms such as tomato paste, wheat flour, milk fat, etc.

7. Two NAS committees, that which produced *Regulating pesticides in food: The Delaney paradox* (Washington, D.C.: National Academy Press, 1987) and another, which created *Pesticides in the diets of infants and children* (Washington, D.C.: National Academy Press, 1993), recommended that EPA refine their computer programs to account for variance in food intake, residue, and toxicity data. These suggestions were largely ignored until 1994 when the Office of Prevention, Pesticides, and Toxic Substances came under the direction of Lynn Goldman.
8. NAS, *Regulating pesticides in food.*
9. Stewart Dary and Ross Brennan, two graduate students at the Thayer School of Engineering at Dartmouth College between 1984 and 1986, made important contributions to our understanding of the variance in patterns of food intake, and its implications for childhood exposure to pesticides. See S. C. Dary, "Uncertainty in exposure modelling: Use of a microcomputer system to examine the latest pesticide data," Master's thesis (Hanover, N.H., Dartmouth College, 1986); and R. A. Brennan, "The development and use of a microcomputer system for analysis of pesticide data," Master's thesis (Hanover, N.H., Dartmouth College, 1986).
10. J. P. Wargo, "Methodology for estimation of oncogenic risks associated with food use agricultural chemicals," in NAS, *Regulating pesticides in food.*
11. D. Krewski, J. Wargo, and R. Risek, "Risks of dietary exposure to pesticides in infants and children," in I. MacDonald, ed., *Monitoring dietary intake* (New York: Springer-Verlag, 1991), 75–89.
12. These data sets were evaluated according to criteria including: (1) representativeness and projectability; (2) survey type: record, recall, frequency; (3) survey duration; (4) seasonality; (5) diversity of populations sampled; (6) sample size; (7) longtitudinal character of survey; and (8) inclusion of demographic parameters.
13. USDA, Consumer Nutrition Division, *Nationwide Food Consumption Survey, 1977–78.* The USDA has conducted national food consumption surveys periodically since 1942. Until 1965 these surveys collected information on food consumption at the household level. Since then, three studies (1965–66, 1977–78, and 1987–88) have been conducted that collected food-intake information for both households and individuals within households.
14. In 1985, USDA initiated the Continuing Survey of Food Intakes by Individuals (CSFII), which includes food consumption for individuals only. The CSFII was initiated to provide timely data on U.S. diets and the diets of special population groups of concern, and to sample diets periodically over a year as a basis for estimating "usual" intake. The 1985 CSFII was designed to sample women 19–50 years, their children 1–5 years, and men 19–50 years; the 1986 CSFII was designed to sample just women and children in these age groups. For the purposes of this study, data from the two years were pooled for children and women, whereas reports for males were not considered. For the CSFII, USDA used a multistaged stratified-area-probability sample design such as that described for the NFCS above. Six nonconsecutive days of dietary intake data were obtained using a 24-hour recall for each day of dietary data. Data were obtained once every two months, and different days of the week were sampled each month. The first interview was administered in person by a trained interviewer, and the remainder were conducted by telephone. Additional information collected

included height and weight, age, gender, ethnicity, and income. Individual food consumption data was collected on three consecutive days using a single 24-hour recall followed by two days of diet records. The food-intake data included descriptions and quantities of all foods and beverages consumed at home and away from home. See the following reports published by USDA's Nutrition Monitoring Division, Human Nutrition Information Service: *Nationwide food consumption survey: Continuing survey of food intakes of individuals, women 19–50 years and their children 1–5 years, 1-day, 1985,* report no. 85-1 (Hyattsville, Md., 1985); *Nationwide food consumption survey: Continuing survey of food intakes of individuals, low-income women 19–50 years and their children 1–5 years, 1-day, 1985,* report no. 85-22 (Hyattsville, Md., 1986): *Nationwide food consumption survey: Continuing survey of food intakes of individuals, men 19–50 years, 1-day, 1985,* report no. 85-3 (Hyattsville, Md., 1986); *Nationwide food consumption survey: Continuing survey of food intakes of individuals, low-income women 19–50 years and their children 1–5 years, 1-day, 1986,* report no. 86-2 (Hyattsville, Md., 1987); *Nationwide food consumption survey: Continuing survey of food intakes of individuals, women 19–50 years and their children 1–5 years, 1-day, 1986,* report no. 86-1 (Hyattsville, Md., 1987); *Nationwide food consumption survey: Continuing survey of food intakes of individuals, women 19–50 years and their children 1–5 years, 4 days,* 1985, report no. 85-4 (Hyattsville, Md., 1987); *Nationwide food consumption survey: Continuing survey of food intakes of individuals, low-income women 19–50 years and their children 1–5 years, 4 days, 1985,* report no. 85-5 (Hyattsville, Md., 1988).

15. GAO, *Nutrition Monitoring: Mismanagement of nutrition survey has resulted in questionable data* (Washington, D.C., 1991), GAO/RCED91-117. USDA convened an independent panel of statistical experts to review the sampling design and results, and this panel "did not recommend use of the data" unless users employed the greatest caution because the data may be biased estimates of the nation's dietary intake.
16. E. L. Gunderson, "FDA total diet study, April 1982–April 1984: Dietary intakes of pesticides, selected elements, and other chemicals," *J. Assoc. Off. Anal. Chem.* 71 (1988): 1200–1209.
17. See B. N. Ames, R. Magaw, and L. S. Gold, "Ranking possible carcinogenic hazards," *Science* 236 (1987): 271–80.
18. FDA, *FDA pesticide program: Residues in foods* 1987, vol. 71 (Washington, D.C., 1988): 156A–174A. For the Total Diet Studies, foods are collected from retail markets in urban areas across the United States and then analyzed for contaminants of interest, including pesticides. Foods are collected four times per year, once from four geographical areas of the United States. For each collection, identical foods are purchased from grocery stores in three cities within a geographical area. These three subsamples (one from each of the three cities) are combined to form a sample for analysis.
19. S. M. Nusser, A. L. Carriquiry, and W. A. Coward, "A semiparametric transformation approach to estimating usual daily intake distributions," University of Iowa, 1986.
20. J. W. Marr and J. A. Heady, "Within- and between-person variation in dietary surveys: Number of days needed to classify individuals," *Human Nutr. & Appl. Nutr.* 40 (1986): 347–64; see also B. P. Basiotis, S. O. Welsh, F. J. Cronin, J. L. Kelsay, and W. Mertz, "Number of days of food intake records required to estimate individual and group nutrient intakes with defined confidence," *J. Nutr.* 117 (1987): 1638–41.
21. *CFR,* vol. 40, 180.4 (1986), and *CFR,* vol. 21, §190 (1986).

22. S. B. White, "The construction of a raw agricultural commodity database," Research Triangle Inst. Project 252U (Research Triangle Park, N.C.: Research Triangle Institute, 1983): 2123–27. Consumption estimates are expressed in grams of commodity per kilogram of body weight per day, and they have been adjusted by standard error estimates so that they represent the outer bound of the 95 percent confidence interval for the population mean. One may only conclude that if a similarly sized sample were surveyed from the same population at the same time, then the estimate of the population mean consumption level would not exceed the original 95 percent outer bound estimate. One cannot conclude that 95 percent of the individual consumption reports will fall below this level. That outer bound would likely be far higher, and could be predicted using standard deviation data for each commodity. Even given a sample size of over 30,000 people, a normal distribution can not be assumed, suggesting that a substantial percentage of the population will consume higher levels than the average estimates used in this study for individual foods.
23. See G. H. Beaton, J. Milner, P. Corey, M. Cousins, E. Stewart, M. de Ramons, D. Hewitt, P. V. Grambsch, N. Kassim, and J. A. Little, "Sources of variance in 24-hour dietary recall data: Implications for nutrition study design and interpretation," *Am. J. Clin. Nutr.* 32 (1979): 2546; G. H. Beaton, J. Milner, V. McGuire, T. E. Feather, and J. A. Little, "Source of variance in 24-hour dietary recall data: Implications for nutrition study design and interpretation, carbohydrate sources, vitamins, and minerals," *Am. J. Clin. Nutr.* 37 (1983): 986; Marr and Heady, "Within- and between-person variation in dietary surveys," 347; D. McGee, G. Rhoads, J. Hankin, K. Yano, and J. Tillotson, "Within-person variability of nutrient intake in a group of Hawaiian men of Japanese ancestry," *Am. J. Clin. Nutr.* 36 (1982): 657; S. M. Nusser, A. L. Carriquiry, and W. A. Fuller, "A semiparametric transformation approach to estimating usual daily intake distributions," University of Iowa, 1991; C. T. Sempos, N. E. Johnson, E. L. Smith, and C. Gilligan, "A two-year dietary survey of middle-aged women: Repeated dietary records as a measure of usual intake," *J. Am. Diet. Assoc.* 84 (1984): 1008.
24. Intake data were aggregated across meals for each day, and then it was averaged over all days reported by each individual. Foods were then arrayed from most consumed to least consumed; water was excluded, however, due to problems in the manner that EPA estimated the percentage of foods constituted by water. See A. B. Ershow and K. P. Cantor, *Total water and tapwater intake in the U.S.: Population-based estimates of quantities and sources* (Bethesda, Md.: Federation of American Societies for Experimental Biology, 1989).
25. A private baby-food manufacturer conducted infant nutrition surveys to monitor infant feeding practices, nutrient intake, nutrition contribution and changes since 1972. Participants were selected from a national birth list of infants 2 to 12 months of age. Parents of infants were contacted using a mail questionnaire. The sample was selected to provide a balance of age and geographic area (Northeast, North Central, South, West) so that approximately 90 infants were sampled for each one-month age class. That is, approximately 90 two-month-olds, three-month-olds, four-month-olds, etc., were sampled, and for each age class the 90 samples were obtained across the four geographic regions. Data were pooled for the two survey years, 1983 and 1986, for the purposes of this analysis. This resulted in a sample size of 904 infants. This group was analyzed together and was broken into quarters of the first year of life. Parents completed a four-day food diary for the infants that included breast milk and water intake,

and they were remunerated for participation. Telephone follow-up was used as necessary to complete or clarify information provided. Although the study was designed to provide a balanced survey, it does not provide a statistically projectable sample. Still, the results are important simply as verification of the intake levels found within the USDA National Food Consumption Survey. At the time of this study, the Gerber survey appeared to provide the best alternative to the NFCS to estimate infant food intake. See G. A. Purvis and S. J. Bartholmey, "Infant feeding practices: Commercially prepared baby foods," in R. C. Tsang and B. L. Nichols, eds., *Nutrition during infancy* (Philadelphia: Hanley and Belfus, 1988): 399–417.

26. R. Wiles, B. Cohen, C. Campbell, and S. Elderkin, *Tapwater blues: Herbicides in drinking water* (Washington, D.C.: Environmental Working Group and Physicians for Social Responsibility, 1994); see also G. Hallberg, "Pesticide pollution of groundwater in the humid United States," *Agr. Ecosys. & Envir.* 26 (1989): 299–367.
27. This question is addressed in chapter 11.

Chapter 10. Averaging Games

1. In 1995 EPA initiated a rare joint review of triazine herbicides, including atrazine, cyanazine, and simizine, known to have widely contaminated Midwestern aquifers.
2. See, for example, debates concerning the relative toxicity of alachlor and metolachlor for use on corn and soybeans in 1986.
3. L. Fisher's testimony before Senate Labor and Human Resources Committee, *Hearings on S. 1074* (10 July 1991), S. Hrg. 102-252.
4. See NAS, *Toxicity testing: Strategies to determine needs and priorities* (Washington, D.C.: National Academy Press, 1984). See also NAS, *Pesticides in the diets of infants and children* (Washington, D.C.: National Academy Press, 1993).
5. In 1990, EPA refused to revoke food additive tolerances for benomyl on raisins and tomato products as requested by the State of California, the Natural Resources Defense Council, Public Citizen, the AFL-CIO, and several individuals in 1989. This refusal prompted a lawsuit, *Les v. Reilly,* which was decided by the U.S. Court of Appeals, Ninth Circuit, and which set aside the agency's order. EPA proposed the revocation of these food-additive tolerances. See 968 F. 2d 985 (9th Cir. 1992), cert. denied, 113 S. Ct. 1361 (1993). See also "Revocation of food additive regulations for benomyl, mancozeb, phosmet, and trifluralin," *FR* 58 (14 July 1993): 37862; *FR* 58 (1993): 29318; and *FR* 48 (1993): 48456; all concerning tolerance revocation.
6. "DuPont, plaintiffs settle benlate case, $4.25 million payment is fraction of claims," Bureau of National Affairs, Daily Report for Executives, 13 Aug. 1993, A155. In 1993, nearly 400 lawsuits were then pending against du Pont concerning benomyl, many in Florida and Hawaii. P. Power, "Jury says du Pont must pay," *Tampa Tribune,* 8 June 1996, 1.
7. For a summary, see J. Quest, "Third peer review of benomyl-MBC," EPA memorandum, to J. Mitchell 14 Feb. 1989; J. Quest to H. Jacoby, "Peer review of benomyl and MBC," 31 Mar. 1986. See also B. E. Wiechman et al., *Long-term feeding study with methyl 1-(butylcarbamoyl)-2-benzimidazole-carbamate (INT 1991, benomyl, benlate) in mice,* report no. 20-82, by Haskell Laboratory for E. I. du Pont De Nemours Co., Inc., EPA MRID 00096514 (Wilmington, Del.: 26 Jan. 1982); C. K. Wood, P. W. Schneider, and H. J. Trochimowicz, *Long-term feeding study with 2-benzimidazole-carbamic acid, methyl ester (MBC, INE-965) in mice,* report no. 70-82 by Haskell Lab-

oratory for E. I. du Pont De Nemours Co., EPA MRID 00154676 (Wilmington, Del.: 26 Jan. 1982); R. Beems, H. Til, and C. Vander Heijden, *Carcinogenicity study with carbendazim (99% MBC) in mice,* summary report no. R4936 of the Central Institute for Nutrition and Food Research, EPA MRID 00153420 (Washington, D.C., 1976); H. Sherman, *Long-term feeding study in rats with 1-butyl-carbamyl-2-benzimidaazole-carbamic acid, methyl ester INT 1991, benlate(R), benomyl,* Haskell Laboratory Report no. 232-69, submitted by E. I. du Pont de Nemours Co., CDL: 050427-Q, EPA MRID 00097284 (Wilmington, Del., 1969).

8. The metabolite is methyl-2-benzimidazole carbamate.
9. Based upon these data, EPA has estimated the human oncogenic potency, or q1*, to be 0.0042 tumors/(mg chemical/kgbw/day). To estimate cancer risk from this fungicide, EPA multiplies this figure by anticipated levels of benomyl exposure. See EPA, *List of suspected carcinogens* (Washington, D.C., 1995). See also *FR* 58 (1993): 48456, *FR* 58 (1993): 57862, and *FR* 58 (1993): 29318.
10. W. G. Ellis et al., "Relationship of periventricular overgrowth to hydrocephalus in brains of fetal rats exposed to benomyl." *Teratogen., Carcinogen., & Mutagen.* 8, no. 6 (1988): 377.
11. W. G. Ellis et al., "Benomyl-induced craniocerebral anomalies in fetuses of adequately nourished and protein-deprived rats," *Teratogen., Carcinogen., & Mutagen.* 7, no. 4 (1987): 357.
12. T. B. Barnes et al., "Reproductive toxicity of methyl-1-(butylcarbamoyl)-2-benzimidazole carbamate (benomyl) in male Wistar rats," *Toxicol.* 28, nos. 1–2 (1983): 103. See also S. D. Carter and J. W. Laskey, "Effect of benomyl on reproduction in the male rat," *Tox. Let.* 11, nos. 1–2 (1982): 87.
13. S. D. Carter et al., "Effect of benomyl on the reproductive development of male rats," *J. Toxicol. & Envir. Health* 13, no. 1 (1984): 53.
14. J. M. Goldman et al., "Effects of the benomyl metabolite, carbendazim, on the hypothalamic-pituitary reproductive axis in the male rat," *Toxicol.* 57, no. 2 (1989): 173.
15. R. A. Hess et al., "The fungicide benomyl (methyl 1-(butylcarbamoyl)-2-benzimidazolecarbamate) causes testicular dysfunction by inducing the sloughing of germ cells and occlusion of efferent ductules," *Fund. & Appl. Toxicol.* 17, no. 4 (1991): 733. See also M. Nakai, R. A. Hess, et al., "Acute and long-term effects of a single dose of the fungicide carbendazim (methyl 2-benzimidazole carbamate) on the male reproductive system in the rat," *J. Androl.* 13, no. 6 (1992): 507.
16. J. B. Mailhes and M. J. Aardema, "Benomyl-induced aneuploidy in mouse oocytes," *Mutagen.* 7, no. 4 (1992): 303.
17. *CFR,* vol. 40, §180.4 (1986), and *CFR,* vol. 21, 193.
18. See chapter 6. The food intake data are derived from the 1985–86 USDA Nationwide Food Consumption Survey. See also D. Krewski, J. Wargo, and R. Risek, "Risks of dietary exposure to pesticides in infants and children," in I. MacDonald, ed., *Monitoring dietary intake* (New York: Springer-Verlag, 1991): 75–89; J. P. Wargo, "Methodology for estimation of oncogenic risks associated with food use agricultural chemicals," in NAS, *Regulating pesticides in food: The Delaney paradox* (Washington, D.C.: National Academy Press, 1987); and NAS, *Pesticides in the diets of infants and children,* esp. chap. 5.
19. The residue data were collected by FDA between 1988 and 1989 using two different designs: one known as surveillance sampling, and the other known as compliance sampling. These sampling designs are applied to domestic and imported food. For the

purpose of this study, both domestic and imported surveillance sample results were employed in the exposure analyses. Compliance data, however, were deleted because they are more likely to reflect anomalies of high residue levels than "usual" patterns of contamination.

20. See, generally, the discussion of pesticide residues in foods presented in chapter 7. For a critique of FDA's pesticide residue testing program, see GAO, *Pesticides: Need to enhance FDA's ability to protect the public from illegal residues,* RCED-97-7 (Washington, D.C., 1986).
21. There is a noticeable absence of FDA attention to processed foods, especially fruit juices. Apple juice, for example, was not tested for benomyl residues during the two-year survey period; orange juice was tested only once; and grape juice was not tested. The absence of these data make exposure assessment for children difficult. If juices are omitted from analyses, estimates will likely underestimate exposure, particularly for children, a population that consumes large quantities of juice compared with adults. I chose to assume that 100 percent of residues transfer from raw food to juice, unless data exists to indicate otherwise. This assumption will likely provide an overestimate of exposure. The toxicological significance of the overestimate may easily be tested through sensitivity techniques that assume alternative levels of raw-to-processed residue transfer (e.g., 0 percent, 50 percent, and 100 percent). The absence of pesticide tolerances for processed foods is also obvious in *CFR.* Similarly, the absence of current data reflecting the effects of modern food-processing techniques on pesticide residues makes exposure estimation difficult. Given these conditions, it seems both prudent and reasonable to assume 100 percent transfer of residues.
22. Most tolerances have been set with a concern for economic benefits, balanced against health considerations, and therefore do not necessarily reflect a safe level of exposure. Residue levels beneath the tolerance could therefore pose a human health hazard.
23. Food intake data used in these analyses were derived from the 1985–86 USDA continuing survey of children ages 1–5 and women, ages 19–50, described in chapter 9. The chief limitations of these data are associated with (1) their age, (2) the small sample sizes (n = 170 for one-year-olds, n = 195 for two-year-olds, n = 225 for three-year-olds, n = 191 for four-year-olds, and n = 209 for five-year-olds), (3) a significant decline in response rate for the survey over the six-day period, and (4) the fact that the sampling design was not structured to ensure that each age class sample was statistically representative of the national population. By contrast, the strength of these data lie in their youth relative to the 1977–78 data. Also, because these data were collected on six days, at two-month intervals over an entire year, the sample may be more representative of actual intra-individual variation in intake than the sample taken over three consecutive days in 1977–78. Person-day intake values were averaged over the days of consumption reported for each individual (ranging between one and six days each), thereby producing an average daily intake amount for each individual. These values were not adjusted by sample weights because the weights provided by EPA were designed to adjust for nonrepresentativeness of women in the sample rather than for nonrepresentativeness of the children. Average daily intake is expressed in all cases in grams of food per kilogram of body weight per day. Dietary intake reports were broken into forms of food for each individual, and intake was then averaged across the number of days during which the child reported consumption. Because 170 one-year-olds were surveyed, 170 average daily intake values were recorded for each of the foods. In this case only 14 foods are considered due to the absence of sufficient

residue data for other crops or foods that may legally contain residues of benomyl. Individual food-intake values were then multiplied by food-specific residue values to produce person-specific exposure values. For each food with residue data judged to be sufficient, mean contamination levels were calculated. Exposure values were then summed across foods for each person to produce person-day exposure estimates.

24. Averaging residue levels for each food assured that few of the children—roughly 10 percent—remain unexposed in the estimate. Only children who did not report eating the foods were assumed to have no pesticide exposure.
25. See J. W. Marr and J. A. Heady, "Within- and between-person variation in dietary surveys: Number of days needed to classify individuals," *Human Nutr. & Appl. Nutr.* 40 (1986): 347–64.
26. Using full residue distributions to estimate childhood exposure is more important if the adverse health effect of concern is acute. An approach is suggested in the following chapter.
27. Normally, cancer risk estimates presume a lifetime level of exposure. In this case I am estimating exposures only during a single year. Simply multiplying the exposure estimate by the potency factor will produce a lifetime risk, and will falsely assume that the individual ate a similar set of foods with a similar set of residue levels for a lifetime. This is unreasonable, to say the least. One solution is simply to estimate the increment in cancer risk accrued during the period of exposure, one year, which may be done by dividing the product of (exposure * potency factor) by 75—the average life span. This produces an estimate of an accumulation of risk during a single year. If this approach is used, cancer risks for benomyl fall between zero and and roughly 5 × 10(−7). Because there is no standard for allowable cancer risk accumulation during any year, it could be calculated by dividing the "unwritten" EPA lifetime standard of one in one million excess cancer risk by the average number of years in life. This simply amortizes allowable cancer risk equally across an average lifetime and allows accumulation of 1.43 × 10(−8), roughly an order of magnitude lower than the risk that appears to be accumulating from annual childhood exposure to benomyl. This approach of considering risk accumulation year-by-year does not account for the more rapid rate of cell division in children, which would increase the rate of childhood risk accrual. See D. J. Murdoch, D. Krewski, and J. Wargo, "Cancer risk assessment with intermittent exposure," *Risk Anal.* 12, no. 4 (1992): 569–77. As mentioned above, FDA has little incentive to track down and verify residue spikes or "detections" beneath the legal tolerance. All of the FDA surveillance data sets contain high proportions of data labeled as "non-detections," where no residue is reported. How should the "non-detection" data be interpreted? One approach is to assume that these cases really represent no residue. Alternatively, because FDA has reported the average detection limit for benomyl as 30 ppb, one could assume that the "zeros" may be as high as 30 ppb. Under this assumption a distribution of residues lying between zero and 30 ppb might be assumed. If one assumes that all of the reported zeros in the data set are actually at the detection limit, the distributions shift only modestly to the right, with the most significant distinction occurring in the very low exposure range (less than 1 microgram/kg bodyweight/day). This effect appears to have greatest importance if the compound in question is highly toxic or if the contaminated food constitutes a high proportion of the diet. The estimates also have not been adjusted by the percentage of crop treated by the pesticide. EPA customarily adjusts field trial residue data by "percent crop treated" statistics, thereby reducing estimated exposures. They will also multiply the number

of cases where no residue was detected by the "percent crop treated" statistic and assume that residues in those cases exist at the limit of quantitation whereas the remainder of "non-detects" lie at zero.

28. J. Eickhoff, B. Petersen, and C. Chaisson, *Anticipated residues of benomyl in food crops and potential dietary exposure and risk assessment,* Technical Assessment Systems, Inc., data submitted to EPA from du Pont as FIFRA data requirement (Washington, D.C.: EPA, 1989).
29. *CFR,* vol. 40, §180.294 (1986).
30. Comparison across these cases is difficult because sampling was conducted on different foods, using different sampling designs. Residues on raw apples, oranges and grapes were assumed to transfer completely to apple, orange, and grape juice for all cases except for NFPA residues (where the only positives included apples and apple juice) and the tolerance analysis, where tolerance-level residues were assumed.
31. It is more likely that this interpretation could significantly influence exposure estimates if (1) the food is highly consumed by the population of interest, (2) the chemical of concern is highly toxic, (3) a high proportion of the residue sample results lie below the limit of quantitation, or (4) positive results are reasonably near the detection limit.

Chapter 11. The Complex Mixture Problem

1. Several others provided valuable advice and guidance, including Philip Landrigan, chair of the Department of Community Medicine at Mt. Sinai School of Medicine, New York City; Donald Mattison, dean of the School of Public Health, University of Pittsburgh; Michael Gallo, Robert Wood Johnson School of Medicine, New Brunswick, N.J.; Sheryl Bartlett of Health and Welfare Canada; and Richard Thomas at the NRC's Board on Environmental Studies and Toxicology. The results of this collaboration are summarized in NAS, *Pesticides in the diets of infants and children* (Washington, D.C.: National Academy Press, 1993).
2. L. J. Fuortes, A. D. Ayebo, and B. C. Kross, "Cholinesterase-inhibiting insecticide toxicity," *Amer. Fam. Phys.* 47, no. 7 (1993): 1613–20.
3. J. Tafuri and J. Roberts, "Organophosphate poisoning," *Ann. Emerg. Med.* 16 (1987): 193–202.
4. M. A. Gallo and N. J. Lawryk, "Organic phosphorus pesticides," in W. Hayes, ed., *Handbook of pesticide toxicology* (New York: Academic Press, 1991).
5. F. R. Sidell and J. Borak, "Chemical warfare agents: pt. 2: Nerve agents," *Ann. Emerg. Med.* 21, no. 7 (July 1992): 865–71. See also A. P. Watson, T. D. Jones, and J. D. Adams, "Relative potency estimates of acceptable residues and reentry intervals after nerve agent release," *Ecotoxicol. Envir. Safety* 23, no. 3 (June 1992): 328–42; N. B. Munro et al., "Treating exposure to chemical warfare agents: Implications for health care providers and community emergency planning," *Envir. Health Pers.* 89 (Nov. 1990): 205–15.
6. T. Litovitz et al., "Annual report of the American Association of Poison Control Centers national data collection system," *Am. J. Emerg. Med.* 9 (1990): 461–509.
7. S. Brown, R. Ames, and D. Mengle, "Occupational illnesss from cholinesterase-inhibiting pesticides among agricultural applicators in California, 1982–85," *Arch. Envir. Health* 44 (1989): 34–39.
8. D. J. Ecobichon, "Toxic effects of pesticides," in M. O. Amdur, J. Doull, C. S. Klaassen,

eds., *Casarett and Doull's Toxicology: The basic science of poisons,* 4th ed. (New York: McGraw Hill, 1991).

9. Z. Weizman and S. Sofer, "Acute pancreatitis in children with anticholinesterase insecticide intoxication," *Pediatrics* 90, no. 2, pt. 1 (Aug. 1992): 204–6.
10. E. Savage and T. Keefe, "Chronic neurological sequelae of acute organophosphate pesticide poisoning," *Arch. Envir. Health* 43, no. 1 (Jan. 1988): 38–45.
11. D. Ratner, B. Oren, and K. Vigder, "Chronic dietary anti-cholinesterase poisoning," *Israel J. Med. Sci.* 19 (1983): 810–14.
12. D. Slamenova et al., "Decemtione (Imidan)-induced single-strand breaks to human DNA, mutations at the hgprt locus of V79 cells, and morphological transformations of embryo cells," *Envir. Mol. Mutagen.* 20, no. 1 (1992): 73–78.
13. H. Soreq and H. Zakut found an inheritable amplification of a ChE gene that encoded defective BChE in a family that had experienced prolonged exposure to methyl parathion. They suggest that because cholinesterases are important to the development of many types of cells, amplification and overexpression of their corresponding genes could affect fertility and be related to the progression of various tumor types. See H. Soreq and H. Zakut, "Amplification of butyrylcholinesterase and acetylcholinesterase genes in normal and tumor tissues: Putative relationship to organophosphorous poisoning," *Pharm. Res.* 7, no. 1 (1990): 1; S. H. Zahm and A. Blair, "Pesticides and non-Hodgkin's lymphoma," *Cancer Res.* 52, no. 19 suppl. (1 Oct. 1992): 5485s–88s.
14. NRC, *Drinking water and health,* vol. 9, *Selected issues in risk assessment, Safe Drinking Water Committee, BEST/Commission on Life Sciences* (Washington, D.C.: National Academy Press, 1989).
15. M. B. Abou-Donia and D. M. Lapadula, "Mechanisms of organophosphorus ester-induced delayed neurotoxicity: Type I and type II," *Ann. Rev. Pharm. & Toxicol.* 30 (1990): 405; see also J. L. de Bleecker, J. L. de Reuck, and J. L. Willems, "Neurological aspects of organophosphate poisoning," *Clin. Neurol. Neurosurg.* 94, no. 2 (1992): 93–103; see also M. C. Cherniak, "Toxicological screening for organophosphorus-induced delayed neurotoxicity: Complications in toxicity testing," *Neurotoxicol.* 9 (1988): 249–72; M. D. Whorton and D. L. Obrinsky, "Persistence of symptoms after mild to moderate acute organophosphate poisoning among 19 farm field workers," *J. Toxicol. Envir. Health* 11 (1983): 347–54; S. Vasilesque, K. V. Rao, and N. Mihailovich, "Neoplastic response of mouse tissues during perinatal age periods and its significance in chemical carcinogenesis," in J. M. Rice, ed., *Perinatal Carcinogenesis,* NCI Monograph no. 51, DHEW publ. no. (NIH) 79-1622 (Washington, D.C., 1979).
16. J. de Bleecker et al., "Prolonged toxicity with intermediate syndrome after combined parathion and methyl parathion poisoning," *J. Toxicol. Clin. Toxicol.* 30, no. 3 (1992): 333–45, discussion 347–49; see also J. de Bleecker, K. Van Den Neucker, and J. Willems, "The intermediate syndrome in organophosphate poisoning: Presentation of a case and review of the literature," *J. Toxicol. Clin. Toxicol.* 30, no. 3 (1992): 321–29, discussion 331–32; N. Senanayake and L. Karalliedde, "Neurotoxic effects of organophosphorous insecticides: An intermediate syndrome," *New Eng. J. Med.* 316 (1987): 761–63.
17. Abnormal brain bioelectrical activity in rats was reported following repeated exposures to chlorphenvinphos (CVP), an organophosphate pesticide. Even after plasma and erythrocyte cholinesterase levels had rebounded to normal levels, abnormal EEG's remained. See S. Gralewicz et al., "Changes in brain biolelectrical activity (EEG) after

repetitive exposure to an organo-phosphate anticholinesterase," *Polish J. of Occup. Med.* 4, no. 2 (1991): 183–89.

18. H. Sinczuk-Walczak, "Status of the nervous system of workers engaged in the production of organophosphate pesticides," *Med. Prac.* 41, no. 4 (1990): 238–45.
19. F. Yelamos et al., "Acute organophosphate insecticide poisonings in the province of Almeria: A study of 187 cases," *Medicina Clinica* 98, no. 18 (9 May 1992): 681–84.
20. N. M. Maizlish et al., "A behavioral evaluation of pest control workers in short-term, low-level exposure to the organophosphate diazinon," *Am. J. Indust. Med.* 12, no. 2 (1987): 153–72.
21. O. Devinsky, J. Kernan, and D. M. Bear, "Aggressive behavior following exposure to cholinesterase inhibitors," *J. Neuropsych. Clin. Neurosci.* 4, no. 2 (1992): 189–94.
22. S. D. Murphy, "Toxic effects of pesticides," in Doull, Klaassen, and Amdur, *Casarett and Doull's Toxicology.*
23. NRC, *Drinking water and health,* vol. 9.
24. E. L. Baker, M. Warren, and M. Zack, "Epidemic malathion poisoning in Pakistan malaria workers," *Lancet* 1, no. 8054 (1978): 31–34.
25. J. P. Frawley et al., "Marked potentiation in mammalian toxicity from simultaneous administration of two anticholinesterase compounds," *J. Pharm. Exp. Ther.* 121 (1957): 96–106. H. Moeller and J. A. Rider, however, tested human response to a mixture of EPN with malathion at maximum, legally allowable levels and found no depression of plasma or erythrocyte cholinesterase. See H. C. Moeller and J. A. Rider, "Cholinesterase depression by EPN and malathion," *Pharmacol.* 2 (1960): 84; J. E. Cassida, R. L. Baron, M. Eto, and J. L. Engel, "Potentiation and neurotoxicity induced by certain organophosphates," *Biochem. Pharmacol.* 12 (1963): 73–83. See also S. D. Cohen and S. D. Murphy, "Malathion potentiation and inhibition of hydrolysis of varioius carboxylic esters by triorthotoyl phosphate (TOTP) in mice," *Biochem. Pharmacol.* 20 (1971): 575–87; S. D. Murphy, R. L. Anderson, and K. P. DuBois, "Potentiation of the toxicity of malathion by triorthotolyl phosphate," *Proc. Soc. Exp. Biol. Med.* 100 (1959): 483–87; and K. P. DuBois, "Potentiation of the toxicity of organophosphorus compounds," *Adv. Pest. Control Res.* 4 (1961): 117–51.
26. NRC, *Drinking water and health,* vol. 9, 152–53.
27. D. D. McCollister et al. calculated the ratio of expected to observed LD50s for 50–50 mixtures of Ronnel, an organophosphate insecticide, with each of 10 other organophosphorus insecticides. Six of the pairs yielded [LD50 (other) / LD50 (Ronnel)] ratios greater than 1. The LD50—a measure of acute toxicity—is the dosage that caused 50 of the exposed animals to die prematurely. See D. D. McCollister, F. Oyen, and V. K. Rowe, "Toxicological studies of o,o-dimethyl-O-(2,4,5-trichlorophenyl) phosphorothionate (Ronnel) in laboratory animals," *J. Agric. Food Chem.* 7 (1959): 689.
28. NAS, *Complex mixtures* (Washington, D.C.: National Academy Press, 1989).
29. H. Veldstra, "Synergism and potentiation with special reference to the combination of structural analogues," *Pharm. Rev.* 8 (1956): 339–87.
30. B. W. Wilson and J. D. Henderson, "Blood esterase determinations as markers of exposure," *Rev. Envir. Contam. Toxicol.* 128 (1992): 55–69.
31. M. A. O'Malley and S. A. McCurdy, "Subacute poisoning with phosalone, an organophosphate insecticide," Worker Health and Safety Branch, *West J. Med.* 153, no. 6 (Dec. 1990): 619–24; see also D. B. Rama and K. Jaga, "Pesticide exposure and

cholinesterase levels among farm workers in the Republic of South Africa," *Sci. Total Envir.* 122, no. 3 (29 July 1992): 315–19.

32. C. Karr et al., "Organophosphate pesticide exposure in a group of Washington state orchard applicators," *Envir. Res.* 59, no. 1 (Oct. 1992): 229–37; see also A. A. Kamal et al., "Serum choline esterase and liver function among a group of organophosphorus pesticides sprayers in Egypt," *J. Toxicol. Clin. Exp.* 10, nos. 7–8 (Nov.–Dec. 1990): 427–35; and T. Ueda et al., "Health effects and personal protection in pest control workers using organophosphorus insecticides," *Nippon Koshu Eisei-Zasshi* 39, no. 3 (Mar. 1992): 147–52.
33. F. Lander and S. Lings, "Variation in plasma cholinesterase activity among greenhouse workers, fruitgrowers, and slaughtermen," *Br. J. Ind. Med.* 48, no. 3 (Mar. 1991): 164–66. See also F. Lander and K. Hinke, "Indoor application of anti-cholinesterase agents and the influence of personal protection on uptake," *Arch. Envir. Contam. Toxicol.* 22, no. 2 (Feb. 1992): 163–66; F. Lander et al., "Anti-cholinesterase agents uptake during cultivation of greenhouse flowers," *Arch. Envir. Contam. Toxicol.* 22, no. 2 (Feb. 1992): 159–62.
34. R. McConnell et al., "Monitoring organophosphate insecticide-exposed workers for cholinesterase depression: New technology for office or field use," *J. Occup. Med.* 34, no. 1 (Jan. 1992): 34–37.
35. S. B. Markowitz, "Poisoning of an urban family due to misapplication of household organophosphate and carbamate pesticides," *J. Toxicol. Clin. Toxicol.* 30, no. 2 (1992): 295–303. See also S. R. Muldoon and M. J. Hodgson, "Risk factors for nonoccupational organophosphate pesticide poisoning," *J. Occup. Med.* 34, no. 1 (1992): 38–41.
36. M. B. Abou-Donia and D. M. Lapadula, "Mechanisms of organophosphorus ester-induced delayed neurotoxicity: Type I and type II," *Ann. Rev. Pharm. & Toxicol.* 30 (1990): 405. See also O. J. Kasilo, T. Hobane, and C. F. Nhachi, "Organophosphate poisoning in urban Zimbabwe," *J. Appl. Toxicol.* 11, no. 4 (Aug. 1991): 269–72; F. Yelamos et al., "Acute organophosphate insecticide poisonings," 681–84.
37. C. N. Pope and T. K. Chakraborti found that plasma ChE inhibition may be a useful quantitative index for the degree of brain cholinesterase inhibition in neonatal and adult rats following organophosphate exposure. They found no significant differences between brain and plasma cholinesterase inhibition across administered doses of three common organophosphate pesticides, methyl parathion, parathion, and chlorpyrifos. See C. N. Pope and T. K. Chakraborti, "Dose-related inhibition of brain and plasma cholinesterase in neonatal and adult rats following sublethal organophosphate exposures," *Toxicol.* 73, no. 1 (1992): 35–43; see also EPA, SAB/SAP Joint Study Group on Cholinesterase, *Review of cholinesterase inhibition and its effects* (Washington, D.C., 1990); and, EPA Risk Assessment Forum Technical Panel, *Cholinesterase inhibition as an indicator of adverse toxicological effect* (Washington, D.C., 1988).
38. C. N. Pope et al., "Long-term neurochemical and behavioral effects induced by acute chlorpyrifos treatment," *Pharmacol. Biochem. Behav.* 42, no. 2 (June 1992): 251–56.
39. NAS, *Pesticides in the diets of infants and children,* 61.
40. E. Mutch et al. examined human interindividual variation for five enzymes involved in organophosphate toxicity: red blood cell acetylcholinesterase, lymphocyte neuropathy target esterase (NTE), serum cholinesterase (ChE), serum paraoxonase, and serum arylesterase. They found that AChE and arylesterase were normally distributed, whereas NTE, ChE, and paraoxonase were significantly nonnormal, and they concluded that this variation could have important effects on susceptibility to

organophosphate toxicity. See E. Mutch, P. G. Blain, and F. M. Williams, "Interindividual variations in enzymes controlling organophosphate toxicity in man," *Hum. Exp. Tox.* 11, no. 2 (Mar. 1992): 109–16.

41. Tominage and Imai report a family in which seven members had the cholinesterase deficiency gene E1s. Cited in M. J. McQueen, "Clinical and analytical considerations in the utilization of cholinesterase measurements," *Clinica Chimca Acta* 237, nos. 1–2 (1995): 91–105. Gallo and Lawryk, "Organic phosphorus pesticides," 994.
42. B. H. Fuller and G. M. Berger, "Automation of serum cholinesterase assay—paediatric and adult reference ranges," *S. Afr. Med. J.* 78, no. 10 (17 Nov. 1990): 577–80; published erratum appears in *S. Afr. Med. J.* 79, no. 2 (19 Jan. 1990): 110.
43. A. A. Kamal et al., "Serum choline esterase and liver function among a group of organophosphorus pesticides sprayers in Egypt," *J. Toxicol. Clin. Exp.* 10, nos. 7–8 (Nov.–Dec. 1990): 427–35.
44. LD50s were lower for dialifor, methamidiphos, and famphur. See T. B. Gaines and R. E. Linder, "Acute toxicity of pesticides in adult and weanling rats," *Fund. Appl. Tox.* 7 (1986): 299–308.
45. See F. C. Lu, D. C. Jessup, and A. Lavallee, "Toxicity of pesticides in young versus adult rats," *Food Cosmet. Tox.* 3 (1965): 591–96; see also J. Brodeur and K. P. DuBois, "Comparison of acute toxicity of anticholinesterase insecticides to weanling and adult male rats," *Proc. Soc. Exp. Biol. Med.* 114 (1963): 509–11.
46. Pope and Chakraborti, "Dose-related inhibition."
47. M. E. Stanton and L. P. Spear, "Conference on the qualitative and quantitative comparability of human and animal developmental neurotoxicity," *Neurotox. Ter.* 12 (1990): 261–67. There is some conflicting evidence demonstrating that adult rats and hens are more susceptible to OPIDN than their young. The embryonic chick has been a model for developmental biology and has been shown to be a good model for identifying the neurotoxic effects of environmental pollutants, including organophosphate insecticides. Moretto et al. found that 70 percent inhibition in peripheral nerve NTE resulted in axonal degeneration and paralysis, whereas similar NTE inhibition in chicks did not produce changes in retrograde axonnal transport nor OPIDN. See A. Moretto et al., "Age sensitivity to organophosphate-induced delayed polyneuropathy: Biochemical and toxicological studies in developing chicks," *Biochem. Pharmacol.* 41, no. 10 (15 May 1991): 1497–504; see NTE recovery rates in peripheral nerves were found to be more rapid in chicks than in hens (half-life of three versus five days). M. Peraica et al. (1993) concluded that chicks resistance to OPIDN might be due to a less effective initiation, or a more efficient repair mechanism, or both. See M. Peraica et al., "Organophosphate polyneuropathy in chicks," *Biochem. Pharmacol.* 45, no. 1 (Jan. 1993): 131; and M. Lotti, "The pathogenesis of organophosphate polyneuropathy," *Crit. Rev. Toxicol.* 21, no. 6 (1991): 465–87.
48. EPA, "Guidelines for health risk assessment of chemical mixtures," *FR* 51 (1987): 34014.
49. EPA, "2,3,5,6-Tetrachloro-2,5-cyclohexadiene-1,4-dione: Proposed significant new use of a chemical substance," *FR* 58 (1993): 27980; see also EPA, "Standards of performances for new stationary sources: Municipal waste combustors," *FR* 56 (1991): 5488; EPA, "Burning of hazardous waste in boilers and industrial furnaces," *FR* 56 (1991): 7134; EPA, "Water programs: Guidelines establishing test procedures for the analysis of pollutants," *FR* 56 (1991): 5090; EPA, "Hazardous waste management system: Identification and listing of hazardous waste, proposed exclusion," *FR* 55 (1990):

13556; EPA, "Risk assessment forum report on toxicity equivalency factors for chlorinated dibenzo-p-dioxins and dibenzofurans," *FR* 54 (1989): 46767; EPA, "Emission guidelines for municipal waste combustors," *FR* 54 (1989): 52209; EPA, *Risk assessment forum: Interim procedures for estimating risks associated with exposures to mixtures of chlorinated dibenzo-p-dioxins and dibenzofurans* (Washington, D.C., 1987).

50. EPA, "Hazard ranking scheme," *FR* 55 (1990): 51532.
51. Methods to estimate risks from complex mixtures have also been reported in the following literature: S. Safe, "Toxicology, structure function relationship, and human and environmental health impacts of polychlorinated biphenyls: Progress and problems," *Envir. Health Pers.* 100 (Apr. 1993): 259–68; S. Safe, "Polychlorinated biphenyls (PCBs), dibenzo-p-dioxins (PCDDs), dibenzofurans (PCDFs), and related compounds: Environmental and mechanistic considerations which support the development of toxic equivalency factors (TEFs)," *Crit. Rev. Toxicol.* 21, no. 1 (1990): 51–88; D. Krewski and R. D. Thomas, "Carcinogenic mixtures," *Risk Anal.* 12, no. 1 (1992): 105–13; B. D. Beck et al., "Utilization of quantitative structure-activity relationships (QSARs) in risk assessment: Alkylphenols," *Reg. Toxicol. & Pharm.* 14, no. 3 (1991): 273–85; B. A. Schwetz and R. S. Yang, "Approaches used by the U.S. National Toxicology Program in assessing the toxicity of chemical mixtures," IARC Scientific Publications 104 (Lyons, 1990): 113–20; E. A. Blinova, "Analysis of methodological approach to the evaluation of toxicity of complex mixtures of relatively constant composition used in industry," *Gigiena Truda i Professionalnye Zabolevaniia* 7 (1990): 39–42; B. A. Schwetz and R. S. Yang, "Approaches used by the U.S. National Toxicology Program in assessing the toxicity of chemical mixtures," IARC Scientific Publications 104 (Lyons, 1990): 113–20; R. S. Schoeny and E. Margosches, "Evaluating comparative potencies: Developing approaches to risk assessment of chemical mixtures," *Toxicol. & Indust. Health* 5, no. 5 (1989): 825–37; S. A. Skene et al., "Polychlorinated dibenzo-p-dioxins and polychlorinated dibenzofurans: The risks to human health, A review," *Human Toxicol.* 8, no. 3 (1989): 173–203; BEST (Board on Environmental Studies and Toxicology)/NRC Committee on Methods for the In Vivo Toxicity Testing of Complex Mixtures, *Complex mixtures: Methods for in vivo toxicity testing* (Washington, D.C.: National Academy Press, 1988).
52. It was difficult to choose insecticides for inclusion in this study among the dozens of ChE inhibiting compounds allowed to persist as residues in the food supply. Criteria for inclusion of insecticides at the time of the study included: (1) the chemicals chosen must induce the same type of biological effect, in this case plasma cholinesterase inhibition, (2) chemicals must be permitted by federal regulation to remain as residues on at least several of the foods of interest, chosen because they were normally consumed by children, (3) sufficient sample sizes of FDA residue data must exist for the chemical-food groups selected, and (4) credible estimates of the lowest observed adverse effect level for this effect must exist for each chemical. These criteria impose serious constraints, and resulted in the further study of only five insecticides among dozens allowed to remain as food residues by law. Thus most organophosphorous insecticides allowed for use on food crops were excluded from this analysis due to either poor quality toxicity or residue data.
53. Toxic equivalence was estimated by comparing the LOAEL (lowest observed adverse effect level) for chlorpyrifos to the LOAEL for each of the other pesticides (acephate, dimethoate, disulfoton, and ethion). This ratio (chlorpyrifos LOAEL/Chemical X LOAEL) is considered to be an indicator of relative potency and the ratio was then

used to adjust the laboratory-detected residue levels for each of the compounds other than chlorpyrifos. LOAEL data can be found in NAS, *Pesticides in the diets of infants and children,* tables 7–14 and 7–15, p. 302. The EPA Scientific Advisory Board has cautioned that this method is less desireable than direct biological assessment of the mixture of compounds. Similarly, the SAB was concerned that there is little evidence of correlation between levels of exposure to complex mixtures and adverse neurological effects; it has recommended additional research in this area and the use of uncertainty factors to offer margins of safety absent credible dose-response findings. See EPA SAB/SAPl, Joint Study Group on Cholinesterase, *Review of cholinesterase inhibition and its effects* (Washington, D.C., 1990).

54. USDA Nationwide Food Consumption Survey, *Food Intakes: Individuals in 48 states, year 1977–78,* report no. I-1 (Hyattsville, Md.: Consumer Nutrition Division, Human Nutrition Information Service, 1983).
55. GAO, *Nutrition monitoring: Mismanagement of nutrition survey has resulted in questionable data* (Washington, D.C., 1991), GAO/RCED-91-117.
56. USDA Nationwide Food Consumption Survey, *Continuing survey of food intakes of individuals, low-income women 19–50 years and their children 1–5 years, 1985 and 1986,* report 85 (Hyattsville, Md.: Consumer Nutrition Division, Human Nutrition Information Service, 1987), 1–5; and report 86 (Hyattsville, Md.: Consumer Nutrition Division, Human Nutrition Information Service, 1987), 1–2.
57. FDA Pesticide Program, *Computer tape of residue data* (Washington, D.C., 1989). See also FDA Pesticide Program, *Computer tape of residue data* (Washington, D.C., 1988). These data are primarily from surveillance sampling rather than compliance sampling. With only a few exceptions, the percentage of samples above the detection limit were quite small. The exceptions were all from domestic surveillance sampling, and include chlorpyrifos on oranges (34 percent positive), acephate on peas (14 percent positive), and chlorpyrifos on apples (11 percent positive). This does not mean that other samples with lower percent positive values pose little risk, because those few positive values could be quite high. Data from the two years were pooled, and residue data from compliance, surveillance, import, and domestic sampling were included. This is clearly less desirable than conducting the analysis using data collected using a single sampling design; the small sample sizes, however, precluded this refinement. Residue data were then converted to chloripyrifos equivalents by multiplying each residue data point by the toxicity equivalence ratio (Chlorpyrifos LOAEL / Chemical X LOAEL).
58. In the absence of reliable data on detected residue combinations and patterns of chemical application, the chemical distributions are assumed to be statistically independent of one another. For example, the existence of chlorpyrifos on apples is assumed to be independent from the existence of any other ChE-inhibiting compound. In reality, it is probable that these distributions are not independent, and that a pattern of substitution occurs instead. Also, some compounds are applied as mixtures. This analysis was constrained by the absence of available data on likely substitutions and combinations, and therefore it has been assumed that chemical use will be driven by a random summation across the independent chemical distributions. This approach will likely overestimate actual exposure. Until data support assumptions regarding substitution patterns, we feel that simple summation across legally allowed compounds is reasonable and prudent. Once reliable pesticide use data become available, this method could be easily modified to reflect detected combinations of specific compounds for specific foods.

59. The high proportion of zeros in FDA residue data sets could also reflect dissipation of residues following application, or FDA's decision not to report residue levels beneath certain default detection levels. Because FDA's statutory responsibility is to ensure that marketplace residues do not exceed legal tolerances, it has little incentive to precisely identify residue levels beneath tolerance levels, despite EPA's admission that tolerances were not necessarily set to protect public health.
60. A Monte Carlo summation across the residue distributions was then performed that preserves the original shape of each component distribution. This method reproduces the original chemical distributions. For example, one value will be extracted randomly from residue distribution 1, another from distribution 2, and a third from distribution 3. These three values will then be summed and the result stored. The procedure will then be repeated 5,000 times, creating 5,000 possible combinations across the three chemicals. The method was also tested using 10,000, 15,000, 20,000, and 25,000 iterations, and we discovered that negligible change in the distribution occurred after 5,000 iterations. If one were to plot the distribution of all 5,000 values extracted from the first residue distribution, the sample proportions would be exactly the same as the original distribution for chemical 1. The sampling design to estimate combinations of residue values across compounds is therefore not random but constrained to reproduce each of the component residue distributions. Whereas the extraction of values from each distribution is not random, the combination of residue values across distributions is random. This assumption is necessary due to the absence of residue data for multiple chemicals on single samples. The goal of the exposure analysis was to produce a distribution of possible person-day exposures. In this case we chose a consumption data set for two-year-olds that includes 1,831 person-day intake values for eight foods. This required the development of eight separate residue distributions for each of the compounds being analyzed. Person-day exposures are estimated using the following method: (1) intake of food 1 (F1) by person 1 (P1) on day 1 (D1) is multiplied by some randomly extracted value from the residue data set (specific to food 1), and the result is stored as an exposure value; (2) the process is repeated for n foods, still for person 1, on day 1; (3) exposure values for n foods consumed by person 1 on day 1 are summed and this sum is stored; (4) steps one through three are repeated 5,000 times using strategic simulation to extract the residue data points from each summarized food-specific residue distribution. This method reproduces the original distributions of each residue data set; (5) the 5,000 exposure values for person 1 on day 1 are stored; (6) 5,000 separate exposure values for each of 1,831 separate person-days are generated, producing 9,155,000 exposure values, all expressed in chlorpyrifos equivalents; (7) all exposure values are pooled and are plotted as a distribution; and (8) a sample proportion is estimated that falls above the reference dose (RfD) for chlorpyrifos [3 μg/kgbw/day].
61. Limit of quantitation data were provided by FDA. In this case, the average limit of quantitation—0.01 ppm—was estimated by FDA and for all chemicals and foods analyzed in this study.
62. In 1993 the NAS Committee on Pesticides in the Diets of Infants and Children proposed an additional safety factor to protect against possible heightened susceptibility of children.
63. See, for example, EPA, *Determination of routes of exposure of infants and toddlers to household pesticides: A pilot study* (Research Triangle Park: Research Triangle Institute, 1991), EPA/600/D-91/077.

Chapter 12. Fractured Law, Fractured Science

1. For each decision EPA has made, it has provided a narrative justification including the agency's interpretation of uncertain evidence of risks and benefits, published in *FR*.
2. GAO, *Pesticides on farms: Limited capability exists to monitor occupational illness and injuries,* GAO/PEMD 94-6 (Washington, D.C., 1993). See also U.S. Office of Technology Assessment, *Neurotoxic pesticides and the farmworker* (Washington, D.C., 11 Dec. 1988); G. S. Rust, "Health status of migrant farmworkers: A literature review and commentary," *Am. J. of Pub. Health* 80, no. 10 (1990): 1213–17; S. Zahm et al., "Cancer among migrant and seasonal farmworkers: An epidemiologic review and research agenda," *Am. J. of Ind. Med.* 24 (1993): 753–56.
3. D. B. Baker and R. P. Richards, "Herbicide concentration patterns in rivers draining intensively cultivated farmlands of northwestern Ohio," in D. Weigmann, ed., *Pesticides in terrestrial and aquatic environments,* Virginia Water Resources Research Center, Virginia Polytechnic Institute and State University (Blacksburg, 1989); see also E. M. Thurman et al., "Herbicides in surface waters of the midwestern U.S.: The effect of spring flush," *Envir. Sci. & Tech.* 25 (1991): 1794–96.
4. See generally chapters 5 and 6. See also NAS, *Regulating pesticides in food: The Delaney paradox* (Washington, D.C.: National Academy Press, 1987); and NAS, *Pesticides in the diets of infants and children* (Washington, D.C.: National Academy Press, 1993).
5. D. Pimentel et al., "Assessment of environmental and economic impacts of pesticide use," in D. Pimentel and H. Lehman, *The pesticide question: Environment, economics and ethics* (New York: Chapman and Hall, 1992).
6. I am not suggesting that the strategy of managing exposure evolved consciously. Instead, it appears to have evolved incrementally, in response to misunderstanding residue concentration, pathways of exposure, and pesticide toxicity.
7 See chapters 5 and 6 for a history of legislative activity.
8. R. Randle, "Safe Drinking Water Act," in J. G. Arbuckle et al., *Environmental law handbook* (Washington, D.C.: Government Institute, 1991).
9. *CFR,* vol. 40, 141.50(a). (1995)
10. *USC,* vol. 42, 1412(b)(5) (1988).
11. This level is defined as the reference dose (RfD) or the lowest "no observed effect level" divided by a safety factor, often ranging between 100 and 1,000.
12. W. H. Smith et al., "Trace organochlorine contamination of the forest floor of the White Mountain National Forest, New Hampshire," *Envir. Sci. & Tech.* 27 (1993): 2224–26.
13. *Clean Air Act, USC,* vol. 42, §7412(b) (1988).
14. Water pollution laws are governed by three statues: *Federal Water Pollution Control Act Amendments,* PL92-500, 86 Stat. 816 (1972); *The Clean Water Act of 1977,* PL95-217, 91 Stat. 1566 (1977); and *The Water Quality Act of 1987,* Pl100-4, 101 Stat. 7 (1987), enacted over presidential veto.
15. Ibid., 307, *USC,* vol. 33, §1317. See also the criteria for listing non-point source pollutants in *FR* 38 (1973): 18044. §307 includes many precautions for managing toxic and hazardous substances, including many of the most heavily used pesticides. The current list of pollutants listed in §307 contain many of the most heavily used pesticides.
16. *USC,* vol. 15, §2602(2)(B)(ii) (1988).
17. *FR* 42 (1977): 64572.
18. *FR* 51 (1986): 15096.

19. These include at least acetone, allantoin, dbutyl phthalate, dimethyl phthalate, and diethyl phthalate.
20. BNA, "Groups to sue EPA on inerts: Say some are canceled products," no. 11390506 (19 May 1994), avail. online from LegiSlate in Washington, D.C. (published by *Washington Post).*
21. *USC,* vol. 42, §6901–6982k (1988).
22. *CFR,* vol. 40,§261.33 (1995). See also 261, app. 8, for a list of hazardous constituents.
23. The Toxicity Characteristic Leaching Procedure—capable of detecting 39 hazardous compounds—is currently required to test waste streams for listed substances. This test replaced the Extraction Procedure in 1990, which was capable of identifying only 14 listed chemicals. As testing becomes more sensitive and multichemical scans expand their scope of detection potential, more facilities are likely to be added to the list of contaminated sites requiring "corrective action."
24. CERCLA (Comprehensive Environmental Response Compensation and Liability Act §107(a), *USCA* vol. 42, §7607(a).
25. Congress in 1980 established a $1.6 million trust fund to clean up abandoned hazardous waste sites if responsible parties could not be identified or were unwilling or unable to pay for site remediation. The 1986 Superfund Amendments and Reauthorization Act (SARA) provided an additional $8.5 billion for the trust fund. It was again increased by $5.1 billion in 1990 by the Omnibus Budget Reconciliation Act of 1990. See GAO, *Superfund: EPA cost estimates are not reliable or timely* (Washington, D.C., 1992), GAO/AFMD-92-40.
26. D. White et al., "Summary of hazardous waste treatment at Superfund sites," *Env't Rep.* (BNA) 18, no. 17 (21 Aug. 1987): 1122. See also "Abandoned pesticide storage facility target of $200,000 emergency cleanup by EPA," *Env't Rep.* (BNA) 23, no. 5 (29 May 1992): 434.
27. R. W. Hahn, "Reshaping environmental policy: The test case of hazardous waste," *Am. Enterprise* 2, no. 3 (May–June 1991): 79.
28. GAO, *Superfund,* 92–40. One important criticism of Superfund implementation has been the requirement to clean up contaminated sites to a pristine condition, rather than to background or adjacent land levels. Because the marginal costs of cleanup increase significantly as residues approach zero, this requirement has proven to be expensive and often unreasonable.
29. OSHA, "Field sanitation: Final rule," *FR* 52 (1987): 16050.
30. "Interagency race to regulate pesticide exposure leaves farmworkers in the dust," *Va. Envir. L. J.* 8 (1989): 293.
31. *Florida Peach Growers Assoc. v. United States Dept. of Labor,* 489 F. 2d 120, 132, 4 ELR 20170 (5th Cir. 1974).
32. *Organized Migrants in Community Action, Inc. v. Brennan,* 520 F. 2d 1161, 1163–64, 5 ELR 20681 (D.C. Cir. 1975). This suit was brought to force OSHA to adopt farmworker protection standards; the court concluded, however, that EPA's standards preempted OSHA's regulatory authority.
33. *CFR,* vol. 29, 1910.1047 (1995).
34. L. S. Gold, G. B. Garfinkel, and T. H. Slone, in C. M. Smith, D. C. Christiani, and K. T. Kelsey, eds., *Setting priorities among possible carcinogenic hazards in the workplace* (Westport, Conn.: Greenwood, 1994).
35. R. Wassertom and R. Wiles, *Field duty: U.S. farmworkers and pesticide safety* (Washington, D.C.: World Resources Institute, 1985).

36. See *CFR,* vol. 40, §170(3–5) (1995). See also "Farmworkers in jeopardy," *Ecology L.Q.* 5 (1975): 69.
37. *USCA,* vol. 42, §1801–1812. Reauthorized in 1990.
38. In hearings before the House Committee on the Environment, Energy, and Natural Resources, Subcommittee on Government Operations, EPA official Victor Kimm explained that the agency was overwhelmed by data and understaffed to manage it. See also K. Bishop, "Spill's poisonous legacy in a once-pristine town," *New York Times,* 29 Dec. 1991, sec. 1, p. 12.
39. Metam sodium in 1993 was the fourth most commonly used pesticide in the country, and roughly 50 percent of the amount applied in the field volatilizes. Although the atmospheric chemistry is not yet well understood, one of the compounds created as it volatilizes is methyl-isocyanate (MIC) and another, methyl isothiocyanate (MITC). MIC is the same compound that is believed to have killed several thousand people in Bhopal, India, following an airborne chemical release at a Union Carbide India Ltd. plant in 1984. In California, roughly 15 percent of the local population required medical care following the accident, and several miscarriages are claimed to have resulted from exposure to the herbicide. Hundreds have filed claims of adverse health effects against Southern Pacific Railway Company, and $40 million was recently awarded to government agencies, individuals who suffered loss, and for purposes of ecological restoration. See A. Bancroft, "Forty-million-dollar settlement in Dunsmuir spill," *San Francisco Chronicle,* 15 Mar. 1994, A1.
40. R. Begley, "DOT [Dept. of Transportation] weighs tighter control of chemicals," *Chemical Week,* 11 June 1991, 12.
41. Within each of these program areas, EPA has established priorities while attempting to manage "high profile" problems as they arise. Because there is often a discrepancy between public and expert perceptions of what is risky, EPA's attention is not always directed by their experts to the most threatening problems within program areas. Also, the agency has concentrated on single issues within programs such as licensing a single pesticide, cleaning up a single Superfund site, or setting a single air-emission standard (e.g., ozone or sulfur dioxide) or drinking water standard (e.g., lead). The level of anticipated benefit loss associated with any proposed regulation often determines the scale of scientific inquiry demanded.
42. Precise estimates of use data are difficult to find. This figure is most likely an underestimate, as roughly 5 billion pounds of pesticides are sold in the world each year. The weight of released pesticides may be decreasing as their potency increases. EPA has not released comprehensive use data since it was created, preferring instead to issue reports of data aggregated by major pesticide class. See OPP, *Pesticide industry sales and usage: 1989 market estimates* (Washington, D.C., July 1991); OPP, *Pesticide industry sales and usage: 1984 market estimates* (Washington, D.C., Aug. 1984); OPP, *Pesticide industry sales and usage: 1982 market estimates* (Washington, D.C., Dec. 1982).
43. See P. J. Landrigan and A. C. Todd, "Lead poisoning," *West. J. Med.* 161, no. 2 (1994): 153; H. L. Needleman, "Why should we worry about lead poisoning?" *Contemp. Ped.* (Mar. 1988): 34–56; Agency for Toxic Substances and Disease Registry, *The nature and extent of lead poisoning in children in the U.S.: A report to Congress* (Washington, D.C., 1988).
44. See J. P. Perkins, *Insects, experts, and the insecticide crisis* (New York: Plenum, 1982), 4; see also A. H. Whitaker, "A history of federal pesticide regulation in the United States to 1947," Ph.D. thesis, Emory University, 1974.

45. See, for example, D. Ehrenfeld, "Ecosystem health and ecological theories," and R. Costanza, "Toward an operational definition of ecosystem health," both in R. Costanza, B. Norton, and B. Haskell, eds., *Ecosystem health: New goals for environmental management* (Washington, D.C.: Island Press, 1992). F. H. Bormann and G. Likens's concept of resilience is one important to the definition of health, because it implies that a natural system may absorb certain types and/or intensities of stress, without losing fundamental abilities to regulate fluctuations in biological, geological, and chemical processes necessary to sustain certain patterns of life. Whether the pattern present at any time is desired by key interest groups seems to play a large role in defining health. See F. H. Bormann and G. Likens, *Patterns and processes in forested ecosystems* (New York: Springer-Verlag, 1979).
46. K. A. Vogt, J. Gordon, J. Wargo, et al., *Ecosystems: Analysis and management* (New York: Springer-Verlag, in press).
47. Ambiguity in long-term animal test results has often resulted in a demand for additional tests. A flaw in the original bioassay can delay a regulatory decision for a decade, because it normally takes five years for a carcinogenicity bioassay to be completed and undergo a peer review.
48. Cholinesterase inhibition from organophosphate and carbamate insecticides provides a good example of debate over what is "adverse." Most people rebound from slight to moderate levels of ChE depression from modest exposures to pesticides, yet some people appear to be more susceptible to chronic effects, such as delayed neuropathy. Severe ChE depression from high-level exposure quickly results in death. Thus cholinesterase depression is considered an adverse effect by some only if its level produces additional nervous system dysfunction. The threshold for dysfunction is poorly understood, may vary considerably among individuals, and may depend on the pattern of exposure. If exposure is of short duration and is modest in intensity, a rebound to normal ranges may quickly occur. If the toxic insult is frequently repeated or of high intensity, irreversible dysfunction in transmission of neurological impulses may result.
49. This effect is driven in part by the continual emergence of new toxicity evidence. Reduction of allowable contamination levels, tolerances for food, or maximum contamination levels for drinking water often lags years behind new evidence demonstrating that a pesticide is more potent than earlier believed.
50. P. A. von Hohenheim-Paracelsus, as quoted in M. O. Amdur, J. Doull, and C. Klaassen, eds., *Casarett and Doull's toxicology: The basic science of poisons,* 4th ed. (New York: McGraw Hill, 1991), 5.
51. See generally chapter 6.
52. An excellent example of this is demonstrated by the conflict over the cancer potency of Alar, a growth regulator that became famous following a *60 Minutes* television show. The television program focused on EPA's delay in regulating the compound following evidence of its carcinogenicity. See NRDC (Natural Resources Defense Council), *Intolerable risk* (Washington, D.C., 1989), and for a contrasting view, K. Smith, *Alar five years later: Science triumphs over fear* (New York: American Council on Science and Health, 1994).
53. The NAS Panel on Pesticides in the Diets of Infants and Children chose to avoid interpreting the Alar evidence. Instead it concentrated on aldicarb, a carbamate insecticide; benomyl, a fungicide; and organophosphate insecticides.
54. The NAS panel instead focused on defining appropriate methods of interpreting uncertain evidence in the expression of childhood exposures and risk.

Chapter 13. Toward Reform

1. Although it is easy to be critical of an agency with an enormous mission, few budgetary resources, and an impoverished scientific foundation, it seems more important to be constructive and to recognize the enormity of the problem that EPA inherited from USDA.
2. Tim Clark, Garry Brewer, and Janet Weiss have made me self-conscious about declaring anything to be a collective problem. The entire history of pesticide regulation might be explained as a conflict among those who perceive pesticide effects to be a problem and those who do not. Both sides appeal to science to legitimate their claims of risk, and their interpretations are infused with values, both overtly and covertly. See G. Brewer and P. de Leon, *Foundations of policy analysis* (Homewood, Ill.: Dorsey Press, 1983); J. Weiss, "The powers of problem definition: The case of government paperwork," *Policy Sciences* 22 (1989): 97–121; T. Clark, "Creating and using knowledge for species and ecosystem conservation," 36 *Pers. in Bio. & Med.* (1993): 497–525.
3. P. Slovic, "Perception of risk," *Science* (17 Apr. 1987): 280–85. See also W. Freudenburg, "Perceived risk, real risk, and the art of probabilistic risk assessment," *Science* (7 Oct. 1988): 44–49; P. C. Stern, "Learning through conflict: A realistic strategy for risk communication," *Policy Sci.* 24, no. 1 (1991): 99–119; B. Fischhoff et al., *Acceptable risk* (Cambridge: Cambridge University Press, 1991). D. Krewski et al., "Health risk perception in Canada, I: Rating hazard, sources of information, and responsibility and health protection," *Hum. & Ecolog. Risk. Assess.* 1, no. 2 (1995): 117–32; P. Slovic, B. Fischoff, and F. Lichenstein, "Why study risk perception?" *Risk Anal.* 2 (1983); 83–93.
4. NAS, *Science and judgment in risk assessment,* app. H-1: *Some definitional concerns about variability* (Washington, D.C.: National Academy Press, 1994). Three types of variability are identified: temporal, spatial, and interindividual. In addition to these types, intraindividual variability in susceptibility is important to understanding age-related risk.
5. Evidence from accidental exposures, suicide attempts, and occupational exposures provide a reasonable basis for estimating variance in susceptibility among humans.
6. Some persistent chlorinated compounds, for example, might migrate from the field where applied through a complex sequence of ecological processes: they may volatilize to the air, attach to water droplets in clouds, be deposited in soil by rain, run from the field in water to lakes, be absorbed by fish, cooked as a meal, and finally come to rest in human tissues. Understanding these movements for only one pesticide has often taken a decade or more, and full understanding of their effects en route is known for relatively few of the tens of thousands of compounds in common use.
7. This discussion of expert judgment demands more attention than I can give here. Various institutional models for achieving consensus among experts have been proposed. One of the more interesting approaches was a "science court" suggested by A. Kantrowitz in 1967. See, generally, A. Kantrowitz, "Proposal for an institution for scientific judgement," *Science* 156 (1967): 763; A. Kantrowitz, "Controlling technology democratically," *Am. Sci.* 63 (1975): 505; A. Kantrowitz, "The science court experiment: Criticisms and responses," *Bull. Atom. Sci.* 33 (1977): 44; S. Jasanoff, "Science, technology, and the limits of judicial competence," *Science* 214 (1981): 1211.
8. S. Breyer, *Breaking the vicious circle: Toward effective risk regulation* (Cambridge, Mass.: Harvard University Press, 1993). Justice Breyer's comparative analyses of risk

generally neglect differences in distributional patterns of exposure and risk that fall from different hazards. His argument rests on risk averaging techniques—criticized in this book, especially in chapter 10. His comparison also neglects the importance of distinguishing between risks generally well understood and consented to by those exposed (such as driving or flying) and those that are poorly understood by those exposed without their consent (such as pesticide risks posed by contamination of food, water, and air).

9. Although toxicity data are freely available for review, data on chemical sales and use patterns and exposure pathways have not been freely available.
10. In 1983, a NAS committee sharply criticized the variability of risk assessment standards within federal agencies in a report entitled *Risk assessment in the federal government: Managing the process* (Washington, D.C., 1983).
11. J. Shogren, Economic incentive for pesticide reduction, unpublished report to the Canadian Ministry of Agriculture, University of Iowa (1995).
12. Nuclear power also appears to fall within this category, because it produces risks of plant explosion and of leaking radioactive wastes, normally believed to be containable.
13. NAS, *Pesticides in the diets of infants and children* (Washington, D.C.: National Academy Press, 1993), 361. EPA has used three types of safety factors: ten to account for interspecies differences in vulnerability; ten to account for intraspecies differences; and ten when toxicological testing demonstrates developmental effects. This committee recommended a more cautious approach: using a tenfold safety factor when the effects of a pesticide on human development remain uncertain. In the committee's judgment, waiting for clear evidence of an adverse developmental effect is not prudent public policy. In each case the "no observed adverse effect level" would be divided by the appropriate safety factors to determine acceptable levels of exposure, which in turn are managed by setting contamination limits for various environmental media.
14. If we knew that reducing pesticide-related cancer risks from one in 10,000 to one in 1 million would cost $500 million, whereas reducing cancer risks from hazardous air pollutants by the same amount would cost only $100 million (assuming similar bounds of uncertainty), allocation of risk reduction costs would still be controversial. There would be debate over comparability of evidence across compounds, which types of adverse effects deserve public attention, which groups deserve protection, and the relative cost-effectiveness of risk-management alternatives. Yet until these types of analyses are conducted, we will not know if there are significant noncontroversial gains that might be made toward reducing risks.

Epilogue

1. My earlier call for more comprehensive cross-medium and cross compound risk analyses for pesticide management may at first seem to carry the same potential. Yet these proposals appear to be protected by my suggestion that effective risk management requires that rational analyses be simultaneously comprehensive and strategic. For a review of the strengths and limits of both the incremental and comprehensive forms of decision analysis, see H. Lasswell, "The emerging conception of the policy sciences," *Policy Sci.* 1 (1970): 3–14; H. Lasswell, "From fragmentation to configuration," *Policy Sci.* 2 (1971): 439–46; R. Brunner and W. Asher, "Science and social responsibility," *Policy Sci.* 25 (1992): 295–331; C. E. Lindblom, *The intelligence of democracy: Decision making through mutual adjustment* (New York: Free Press,

1965); C. E. Lindblom, *Politics and markets* (New York: Basic Books, 1977); C. E. Lindblom, *Inquiry and change* (New Haven: Yale University Press, 1990).

2. J. Rawls, *A theory of justice* (Cambridge, Mass.: Harvard University Press, 1971), 302–3; see also J. Rawls, "Some reasons for the maximin criterion," *Am. Econ. Rev.* 64 (May 1974): 141–46.
3. I. Shapiro, *The evolution of rights in liberal theory* (Cambridge: Cambridge University Press, 1986).
4. Such a fundamental attack on the moral foundation of those with wealth, power, and prestige was bound to inspire a heated defense of utilitarianism. J. Harsanyi described numerous hypothetical cases that made Rawls's requirement—that policies give absolute priority to the worst-off individual, regardless of the expense to others—seem unreasonable. See J. Harsanyi, "Can the maximin principle serve as a basis for morality? A critique of John Rawls's theory," *Am. Pol. Sci. Rev.* 69 (1975): 594–606.

Index

Acceptable daily intake of pesticides, 227
Acephate, 241, 242
Acetylcholine, 236, 238
Acrylonitrile, 113
Acute lymphocytic leukemia, 185
Adulterated food, 106–8, 166
Adverse effects: presumed avoidable by labeling, 71; on wildlife, 73, 90; FIFRA decision standard, 88–90; FEPCA balancing standard, 104; NOAEL, 122, 151, 247, 262; definition of, 238, 260; neurotoxicity as, 239; across species, 240; duration of, 241; continual expansion of types considered by EPA, 260; thresholds, 261; as a foundation for reform of environmental law, 285
Aedes mosquitoes, 15, 23
Aflatoxin, 156, 193, 194
Africa: and malaria, 25, 28, 45–47, 50, 54–63, 170, 186
Agricultural occupations and Ewing's sarcoma, 198
Agriculture: malaria and irrigation, 50–52; tropical deforestation, 56; continued DDT use in southern hemisphere, 137; water contamination, 147–52; crop residues, 157–68; organic farming and low input, 288; sensitive ecological regions, 288
Airplanes and pesticide application, 154
Airports: located near marshlands, 28
Alachlor, 133, 150, 151
Alar, 72, 116–17, 155, 159, 161, 258, 266, 276
Aldicarb, 149, 152
Aldrin, 88, 133, 140, 147–48, 158, 165
Amazon, 56, 58
American Cancer Society, 74
Ames, Bruce, 115, 127, 187
Analytical methods: inconsistencies in, 8, 154; residue detection, 110, 152; argument on and control of, 125–26, 275, 285; risk analysis and estimation, 199, 228, 264–66
Animal bioassay, 113, 115, 188, 190, 220, 272, 276
Anopheles: gambiae, 25, 27, 47, 50; *sacharovi,* 48, 50; *arabiensis,* 50; *culicifacies,* 50; *funestus,* 50; *nigerrimus,* 52; *subpictus,* 52; *darlingi,* 56; *labranchiae,* 62
Antagonistic effects of pesticide mixtures, 153
Antibiotics, 155, 201
Apples: processed food residue levels, 107; and Alar, 116–17, 154–55, 266; and *Les v. Reilly*, 123–24; and consumption by infants, 158–62, 212–13; and benomyl, 222–23, 230–32
Arsenic, 35, 67–69, 74, 127, 140, 179, 186
Artificial food colors, 155, 201
Atabrine, 26, 27, 31, 35, 54
Atomic weapons testing, 79
Atrazine, 133, 137, 149, 150, 151, 160

Bananas, 11, 117, 152, 158, 160, 212, 220, 222
Barium fluosilicate, 67
Beans, 81, 160, 230
Bednets and malaria control, 21
Bendiocarb, 47
Benefits of pesticides: malaria control, 15–42; farmers' financial risk management, 69–70; economic poisons, 70; FIFRA's balancing standard, 85; FEPCA's decision standard, 89; and use rights, 101–3; manufacturer's willingness to bear costs of data production as evidence, 103; FFDCA's decision standard for raw foods, 105–6; absence of consideration in Delaney clause, 106–8; need for accounting reform, 251–52, 277; understanding distribution of, 267; reasonableness of risk-

Benefits of pesticides: (*continued*)
benefit balancing standard, 282–84; utilitarianism, 293–99
Benlate. *See* Benomyl
Benomyl, 11, 98, 117, 160–62, 219–34, 268, 285
Benzo(a)pyrene, 194
BHC, 43, 44, 46, 47, 48, 137, 148, 158
Bioaccumulation, 42, 45, 71, 139–44, 149, 237, 260
Body composition, children's and adult's, 176
Boomerang effect, 163
Brazilian malaria eradication campaign, 25–26, 28, 41, 46, 51, 54, 56, 58, 61
Breast cancer, 174, 185–86
Breast milk, 11, 72, 169–70, 177, 199, 211, 213, 290
Brodeur, J., 241

Calcium arsenate, 67
Cameron, Charles, 74
Cancer risk: understanding carcinogenic potency, 188; thresholds of safety, 189; maximum tolerated dose problem, 190; models of carcinogenesis, 191–93; lessons from animal studies, 193–95; management of, 226, 247, 256, 261, 263, 266–68, 282, 285–86, 297
Captan, 158, 162
Carbamate insecticides, 149–50, 152, 160, 188, 236–38, 261
Carbendazim, 220, 221
Carbofuran, 151
Carcinogenic potency of pesticides, 111, 188, 190, 192
Carson, Rachel, 74, 78, 80–87, 169
Centers for Disease Control (CDC), 235, 236
Certified applicators, 90
Ceylon, 26, 45
Childhood cancer, 198
Childhood exposure to pesticides: driven by diet and food contamination levels, 219; tolerances not health protective, 219; danger of averaging methods, 222–25; danger of using "percent acreage treated" adjustment to estimate, 222–25; need for richer understanding of, 225; choice of residue data governs exposure and risk estimates, 228–31; childhood exposure at legal maximum residue levels, 233
Childhood susceptibility: growth rates and, 173–75; body composition and, 176; functional maturation and, 177; to pesticides as drugs, 179; to neurotoxic compounds, 180; to cholinesterase inhibiting compounds, 184–87; to cancer, 195–99
Children and pesticide toxicity, 173
China, 61, 139, 170
Chlorinated hydrocarbon pesticides: resistance of *anopheles* to, 48, 54; continued use of, 72, 167; estrogenic effects of, 74; aldrin and dieldrin, 88, 133; concentration of, 137; bonding to fats, 140, 169, 176; and contamination, 143, 147, 156–59; to control head lice, 197
Chlorinated pesticides, 44, 46, 71, 74, 137, 139, 143, 147, 169, 201, 290
Chloroquine and malaria, 54–60
Chlorpyrifos, 134, 148, 151, 158, 161, 241–45
Cholinesterase inhibition, 52, 189, 237, 240, 242, 248, 262, 274
Chromatography, 156, 157
Chrysanthemums and pyrethrum, 44
Cinchona bark and quinine, 35, 53, 54
Circle of poison, 4, 163
Clean Air Act, 254, 257
Clear Lake, California, 140
Clinton, President William, 123, 125
Colón, Panama, 21, 22
Color additives, 108, 112, 122
Communicating risk, 278
Comparative risk, 126, 128, 267, 292
Complex pesticide mixtures. *See* Mixtures
Compliance, 31, 55, 99, 123, 157, 164, 204, 228, 243
Composited food samples, 155
Constituents and trace food contaminants, 112
Continuing Survey of Food Intake by Individuals (CSFII), 203, 212
Corn, 7, 81, 88, 132–33, 149, 164, 186

Corrective action and RCRA, 255
Costs of regulation: to manufacturers, 123; of reducing risks to zero, 126; farmworker poisonings, 146; administrative, 277; clean-up of contaminated sites, 277; environmental monitoring, 277; net welfare effects, 295
Cotton and insecticide use, 7, 49, 51, 123, 133–36, 164, 281
Cottonseed oil, 123–24, 281
Council on Competitiveness, 123, 295
Crop losses, 6, 73, 133, 277, 294
Crop rotation, 288
Cryolite, 35, 67
Cumulative effects of pesticides, 263, 274
Customs records of U.S. pesticide exports, 4, 163
Cyanazine, 133, 150, 151
Cyclophosphamide, 180

D.C. Circuit Court, 87, 88, 114, 122
Daminozide, 116, 159. *See also* Alar and UDMH
DBCP, 134, 150
DCPA, 150
DDA, 168
DDD, 141, 143, 147, 168
DDE, 137, 143, 147, 157, 158, 160, 168–70, 177
DDT: malaria control before, 21–35; promise of, 35–42; post–World War II agricultural uses, 43; use of by WHO, 43–46; vector resistance to, 46–53; production in *1951,* 70; transport and fate of, 73; toxicity, 135–37; bioaccumulation of, 144; residues in estuaries, 147; residues in food, 157; foreign use of, 165; residues in human tissue, 168; residues in human breast milk, 169–71; half-life of metabolites in human body, 177; replacement by organophosphates, 236; in forest floor of New Hampshire, 254
Delaney clause, 76, 282
Delaney hearings, 71
Delaney, James, 71
Delaney paradox, 7, 10, 104, 114, 121
DES proviso, 110
Detection limits for pesticide residues, 111, 155–57, 230–31
Diazinon, 48, 158, 198
Dicamba, 150
Dieldrin, 43–51; harmful effects of, 88, 194, 276; banned, 98; residues of, 136–40, 143, 148, 157–58; detection in human food, 165, 169
Diesel fuel as a larvacide, 33
Dietary diversity, 11, 205, 207, 211
Dietary intake, average, 207, 208, 213
Dietary surveys, 7, 8, 202
Dimethoate, 161, 242
Dioxin, 134, 156, 254
Disclosure of pesticide use, 281
Discretion in risk assessment and management, 10, 87, 90, 102, 105, 107, 111, 122–27, 289
Disulfoton, 242
Drinking water: dependence of U.S. population on underground supplies, 101; contamination by herbicides, 133, 150; need for careful sampling, 147; residues in rainwater, 148; in underground aquifers, 149–51; Safe Drinking Water Act, 151; maximum contaminant levels, 151–52, 253; childhood dietary intake of, 211, 218, 247, 252–53; costs of government monitoring for pesticide residues, 277; insufficiency of single medium regulation, 286
Drug residues in food, 110, 201
Drug resistance of malaria parasites, 54–55
Drugs: resistance problem, 43–45; parasite resistance to anti-malarial compounds, 53–63; toxicity of anticancer drugs to children, 179–80; susceptibility of children to acute effects, 183; detoxification potential of children, 184
Dubois, K., 241
Dunbar, Paul, 72
Dunsmuir, California, 257
Dutch elm disease and DDT, 141

Eagles: pesticides and reproductive failure, 80, 143

EBDC, 160, 162, 178, 193
Ecological risk, 71, 81, 100, 126, 131, 236, 260, 291
Ecology of malaria, 16
Economic Research Service, 94
Ecosystem health, 260
Education, need for public, 82, 89, 105, 106, 147, 280
Eighth World Health Assembly, 45
Eisley, Lauren, 82
Emergency exemptions, 97
Empedocles, 16
Endocrine system disruption by pesticides, 5, 12, 261
Endrin, 51, 147, 165
Environmental Defense Fund v. Ruckelshaus, 88
Environmental law governing pesticides: Federal Food Drug and Cosmetic Act, 253; Federal Insecticide, Fungicide and Rodenticide Act, 253; is fractured, 253; Safe Drinking Water Act, 253; Clean Air Act, 254; Federal Water Pollution Control Act, 254; Toxic Substance Control Act, 254; Comprehensive Environmental Response, Compensation and Liability Act, 255; Resources Conservation and Recovery Act, 255; Hazardous Materials Transportation Act, 256; Occupational Safety and Health Act, 256
Environmental Working Group, 161, 162
Environmentalists, 10, 74, 92, 108, 122, 254, 258, 268, 295
EPA: as risk gatekeeper, 86–103; early lawsuits against, 87–88; birth of, 89; first use of emergency suspension power, 92; legacy of USDA, 92; reregistration delays, 92–99; data confidentiality policies, 93; emergency exemptions granted, 97–98; special local needs, 98; negotiated settlement with industry, 99; voluntary compliance by industry, 99; discretion to estimate risks and benefits, 101–3; neglect of children in tolerance setting, 104; responsibility under FFDCA, 105–8; interpretations of Delaney clause, 114–26; Alar regulation, 116; request to NAS to review Delaney clause, 117; changing permissible excess risk standards, 121; response to NAS *Delaney Paradox* report, 121; *de minimus* interpretations by, 124–26; *Les et al. v. Reilly,* 122–26; simplified food intake assumptions, 209; use of "percent acreage treated" adjustment to risk, 232; more stringent regulation of carcinogens than OSHA, 256; fractured administrative structure mirrors fractured law, 257; chemical-by-chemical and medium-by-medium regulation, 267–68; strategy of delay in pursuit of more certain evidence, 267; need for cross-program risk assessment and management, 286
Epidemics of malaria: in wars,15, 32; in Brazil and Africa, 26, 28, 35, 63; in Iran, 49; in Madagascar, 56
Epidemiology, 61, 96, 195
Ethion, 242
Ethyl parathion, 165
Evidence, quality of, 265, 270
Excess risk, 112, 119, 121, 126
Executive Order *12299,* 123
Experimentation, pesticide licensing as, 3, 38, 68, 83, 171, 180, 291
Expert Committee on Malaria, WHO, 54
Expert judgment, 274–75
Export of pesticides, 4, 163–66, 278, 284
Exposure estimates, 156, 179, 217–18, 223, 225–28, 232, 243, 263, 273. *See also* Childhood exposure to pesticides
Extrapolating human risk from animal studies, 187

Falcons: DDT and reproductive failure, 80, 142, 143
Farmworker poisonings, 146
Federal Environmental Pesticide Control Act, 89–92, 102
Federal Food Drug and Cosmetic Act: *§408* tolerances, 105–6; *§409* tolerances, 106–8, 115–16, 119; DES proviso, 110; sensitivity of method, 111–12; color additive amendment, 112–13; pesticide residues that concen-

trate and induce cancer, 116; Delaney Paradox and NAS, 120–23
Federal Insecticide, Fungicide and Rodenticide Act: passage in *1947,* 70–71; defects influence passage of Miller and Delaney amendments to FFDCA, 71; primary strategy of registration and labeling, 71; weaknesses result in *1972* passage of FEPCA, 86–89
Field trial residue data, 227–28, 231–32
Federal Water Pollution Control Act, 254
Fifth Amendment, U.S. Constitution, 93
Fleas: plague, 15; control on pets, 146
Fog, residues in, 137, 254
Fonofos, 134
Food additives, 75–77, 106–15, 122, 153, 155, 182, 201, 239
Food and Agriculture Organization (FAO), 44
Food and Drug Administration (FDA): testing health effects of DDT, 38; setting tolerances for lead and arsenic, 68–69, 78; consideration of benefits permissible in tolerance setting, 75; support for protecting tolerances of GRAS compounds, 77; opposition to passage of Delaney clause, 78; expansion of authority through Miller amendment, 106; *de minimus* interpretations of Delaney clause, 108–14; responsibility to monitor residues in food supply, 153–54; default detection limits, 156; market basket surveys, 158; differences in detections among laboratories, 160; screening imported foods, 162; GAO criticism of food monitoring program, 163–66; surveillance program residue data on benomyl, 230; surveillance data on cholinesterase inhibiting compounds, 242–43; interpreting FDA reports of "non-detectable" residue levels, 245–46; insufficiency of FDA data for exposure and risk assessment, 247
Food intake data, 201–18, 222, 242–43, 246, 272
Food safety, 85, 105, 201
Food Safety Inspection Service (FSIS), 166
Forestry practices and malaria, 56, 62, 79
Formulation of pesticides, 144, 148, 150, 278
Fractured science, fractured law entanglement, 251, 268
Fragrances as food additives, 201
Freeman, Orville, 84
Functional maturation of human organs and pesticide susceptibility, 177
Funding risk assessment and management, 288
Fungicides, 119, 132–33, 145–46, 160–62, 178, 183, 193, 220

Garden use of pesticides, 146, 196–97
Geigy Corporation, 37
General Accounting Office (GAO), 11, 164
General Agreement on Tarriffs and Trade (GATT), 167
General use pesticides, 90
Generally regarded as safe (GRAS), 76, 77, 115
Germany, 43, 142, 164, 172
Global transport and fate, 135
Gold, Lois, 127, 187, 256
Golgi, Camillo, 17
Gorgas, William, 21
Gorsuch, Anne, 98
Grape juice, 158, 232
Grapes, 68, 81, 123–24, 162, 222
Great Lakes, pesticide contamination of, 134, 136–37
Greece, malaria control in, 45, 47, 48
Groundwater contamination by pesticides, 148–51
Guadalcanal and malaria, 32, 33

Havana, Cuba, and Gorgas' sanitation efforts, 21, 23
Hayes, Wayland, and toxicity of DDT, 73
Hazardous air pollutants, pesticides as, 247, 254, 257
Hazardous Materials Transportation Act (HMTA), 256–57
HCH, 48, 50, 138, 139
Health Education and Welfare, U.S. Department of, 78
Hemoglobin, 20, 60

Heptachlor, 136, 138, 143, 147–48, 158–59, 165, 169, 183, 254
Herbicides, 3, 6, 9; threat to children, 12, 178, 183; as allowable food contaminant, 72; application of, 83, 197–198; increase in use of, 132–34, 146; detection in water, 137, 140, 149–51, 254, 260
Heredity and cancer, 186
Hippocrates, 16, 24
Hormonal effects of pesticides, 74, 110, 155, 174, 186, 201
Hueper, William, 77
Huxley, Julian, 83

Imidan, 237
Imminent hazard, 87–88, 91–92
Immune system, 55, 57, 59, 60, 63, 134
Imported foods, 4, 158, 163–68, 220
In utero, pesticide exposure, 173, 178
Incremental regulation, 14, 134, 205, 219, 257, 265, 268, 275, 291, 296
Indemnification, 92–93
India, 3, 45, 50, 138–39, 170
Individual freedom, 296
Individual rights, 291–98
Induce cancer, 7, 76, 106–8, 110, 114, 187
Inert ingredients, 3, 7, 96, 131, 152, 164, 201, 251, 255, 265
Informed consent, 281, 296
Inhalation of pesticide residues, 45, 144, 145
Initiators, of tumors, 175
Injustice, 297
Invisibility of pesticides and difficulty of monitoring and regulation, 70, 252, 278
Iran, malaria in, 49
Irrigated agriculture pesticides and malaria, 25, 51

Jackson, Richard, 235
Japan, 35, 43, 46, 51, 61, 146, 163, 164
Judgment Fund, 93
Justice, 299

Kehoe, Robert, 72
Kennedy, Donald, 113
Kennedy, Edward M., 104
Kennedy, John F., 83
Kenya, 35, 55, 56, 58
Kerosene, 24, 40, 138
Kidney filtration, 183
Korean War, 15
Krewski, Daniel, 192, 235

Labeling, 67, 71, 86–89, 102, 106, 112, 273, 279, 282, 287
Lange, William, 236
Laug, Edward, 169
Lavaran, Charles, 17
Lawns, pesticide applications to, 146, 286
Lead, 67–74, 103, 124, 127, 140, 153, 180–81
Leptophos, 241
Les et al. v. Reilly, 123–25
Lettuce, 10, 132, 162
Leukemia, 185, 193, 197–98
Liberalism, 289
Limited rationality, 289

Macedonia, 31
Malaria: discovery of, 16–19; lifecycle of, 19–21; *Plasmodia sp.,* 19; Gorgas and Panama Canal, 21–25; Brazilian epidemic, 25–29; and warfare, 29–34; quinine, cinchona, and fever tree, 36; and DDT, 35–42; eradication, 45, 48–49; control, 26, 35, 42, 44–45, 48–50, 57. *See also* Resistance of malaria parasites *and generally chapters 2 and 3*
Malathion, 47–50, 52, 148, 158, 184, 239
Manson, Patrick, 17
Margin of safety, 253, 297
Market basket residue survey, 159, 229
Mauritius, 26
Maximum contaminant level (MCL), 151
Maximum tolerated dose (MTD), 115, 180, 188–89
Meckel, Heinrich, 17
Mercury pesticides, 72, 127, 140, 143, 145, 152, 172, 179
Metabolic rate, 177, 183
Metam sodium, 101, 134, 257
Methidathion, 241
Methoxychlor, 137

Methyl parathion, 51, 184, 239, 241
Metolachlor, 133, 150, 151
Metribuzin, 150
Mexican produce, 164
Milk: DDT residues in, 72–73, 142, 144, 169–71; location of production of, 154; children's intake of, 159–60, 177–78, 207–8, 211–12
Miller amendment to FFDCA, 71, 80, 85, 106
Miller, Arthur, 75
Mining, 62
Misunderstanding pesticide use and substitution, 131
Mixtures: summary of problem, 235–36; cholinesterase inhibition of, 236–38; neurotoxic compounds, 236; war nerve gases and pesticides, 236; toxicity of, 238–39; variance in susceptibility to, 240–41; children at risk, 244–48
Models of carcinogenesis, 191
Monte Carlo computer modeling, 243
Morris, Harold, 74
Mosquitoes. See *Anopheles*
Mosquito bites, 19, 57
Mosquito larvae, 22, 35, 40, 51
Mrak Commission, 13, 105, 147, 148
Müller, Paul, 15
Multiple residues in single food samples, 159, 162, 243

Naples, Italy, typhus epidemic, 38
Natal, Brazil, malaria epidemic, 25–26, 29
National Academy of Sciences (NAS), 10, 12, 112, 115, 117–21, 201–2, 239–40, 265–66
National Cancer Institute (NCI), 115, 184, 188
National Food Consumption Survey (NFCS), 203
National Food Processors Association (NFPA), 228–29, 231–32
National Priorities List (NPL), 255
National Toxicology Program (NTP), 188, 190
Natural carcinogens, 10
Natural Resources Defense Council (NRDC), 266
Negligible risk, 10, 121, 282, 284, 297
Neurotoxicity, 95, 180–81, 261, 275
New Yorker, 78
Nicotine as pesticide, 35
Nixon, Richard M., 89
Nobel Prize, 15, 17
No observed adverse effect level (NOAEL), 122, 151, 247, 262
North American Free Trade Agreement (NAFTA), 163, 167
Nutrient intake, 7

Occupational exposure to pesticides, 113, 196–98, 256, 261, 275
Occupational, Safety, and Health Administration (OSHA), 256
Odds ratio, 179, 198
OECD, 166
Office of Pesticide Programs (OPP), 97
Office of Technology Assessment (OTA), 190
Orange juice, 158, 207–8, 222, 230, 232, 243
Orange No. 17 (food coloring), 113–14
Oranges, 7, 155, 162, 201, 222, 230
Organophosphate induced delayed neuropathy (OPIDN), 237
Organophosphate insecticides, 148, 182, 184, 188, 197, 217, 235–43, 261, 274
Ospreys: and reproductive failure, 143

Panama Canal, 21, 24, 32
Panama City, 22
Paracelsus, 172
Parasite resistance to anti-malarial drugs, 53
Parathion, 43, 51, 137, 146–48, 165, 184, 239, 241
Paris green (copper acetoarsenite), 27, 31, 33, 35, 67, 148
Pasteur, Louis, 16, 17
Peaches, 7, 81, 117, 154, 161, 162, 222, 231
Pears, 7, 117, 160, 162, 222
Pelicans, pesticide residues in, 143
Penguins, pesticide residues in, 143, 254

Pesticide law: Insecticide Act of *1910,* 67; implementation before *1947,* 67–70; FIFRA passage, 70–71; Delaney hearings and Miller amendment, 76–78; FEPCA amendments, 86–89; indemnification, 92–93; *1964* amendments, 86; *1978* amendments, 95; *1988* amendments, 97, 102
Pesticide residues: in processed foods, 106; in water, 147; in raw food, 152, 157
Pesticide substitution, 14, 120, 283
Pesticide use, 3, 131–35, 251, 253, 258–59, 281
Pesticides and cancer, 184–94; in children, 195–99
Pesticides as food additives, 114
Pesticides as neurotoxins, 181
Pesticides in the Diets of Infants and Children, 12, 157, 218, 235, 240
Phorate, 134
Pineapples, 220, 222
Plasmodia sp.: falciparum, 19, 54–61; *malariae,* 19; *ovale,* 19; *vivax,* 19, 59–60
Plums, 123, 222, 231
Poisoning, 68, 80–81, 146, 152, 167, 169, 178, 183, 236–40
Pollution prevention, 288
Potatoes, 11, 160, 207
Pregnancy, 178, 181–82, 197–98, 290
Private residue testing lab, 161, 230, 232
Processed foods, 5; lack of study of, 9–10, 155, 230–32; residue levels in, 107–9, 114–23, 243; estimation of intake, 204, 227, 230–32; international distribution of, 281
Promoters of tumors, 175, 176
Property rights, 4, 93, 125–26, 278, 291
Prophylactic anti-malarial drug use, 28, 35, 55, 57–58, 63
Prostate cancer, 185, 186
Public Citizen v. Young, 122
Public Health Service (PHS), 38, 68, 73
Punjab, India, 26
Pyrethrum, 27, 35, 38

Quantitative risk assessment, 111–12, 114
Quinine, 26–27, 30–31, 35, 53, 58–59

Rain, residues in, 135, 138
Rawls, John, 293, 296
Reagan, Ronald, 96, 98, 101, 121, 123, 295
Recipes of complex foods, 110, 204
Red No. *19,* 113, 114
Reference dose (acceptable daily intake), 227, 247
Reforms needed: quality of evidence, 270; expressing uncertainty, 271; understanding distribution of risk, 271; richer exposure estimates, 272; expert judgment, 274; strategic attention, 274; study of cumulative effects, 274; burden of proof, 275–76; accounting for costs and benefits, 277; risk communication, 278–9; education, 280; international pesticide use disclosure, 281; Delaney clause, 282; risk-benefit balancing standard, 282–83; negligible risk standard, 284–85; precaution in standard setting, 285; cross-media law, 286; avoiding high-risk technologies, 286–87; strategic risk reduction, 287; funding risk assessment and management, 288; pollution prevention, 288
Registration, 84, 99, 102, 283
Reilly, William, 101, 104, 121
Relative risk, 179, 266, 270, 275–76, 279, 284
Repellents, 21, 28, 31, 33, 35
Reregistration, 9, 92, 94–100, 102–3
Residues: concentration during processing, 9, 108, 116, 123, 135, 153, 223, 230; knowledge of pesticide use necessary for effective environmental monitoring, 131–35; global transport and fate, 135–39; bioaccumulation of, 139–44; sources of human exposure to, 144–46; in water, 147–52; in domestically produced foods, 152, 157; sampling and detection problems, 153–57; in imported foods, 162–68; in human tissue, 168; in human breast milk, 169; residue pathways and fate, 259
Resistance of malaria parasites, 53–58
Resource Conservation and Recovery Act, 93, 255
Ribicoff, Abraham, 82, 86

Rice, 49, 51, 52, 134, 138, 151
Rights, 4, 86–87, 93, 101–2, 125–26, 289–98
Risk averaging, 232, 282, 297
Risk ceilings, 285, 286
Risk distribution, 263, 294, 298
Risks and rights, 101
Ross, Ronald, 18
Rotenone, 35
Rural development and malaria, 62, 70

Saccharin, 112
Safe Drinking Water Act (SDWA), 151, 253
Safety net, 293, 297
Salivary ducts of *Anopheles* mosquitoes, 18, 20
Sample size, 144, 159, 161, 188, 207, 217, 223, 228–32, 242
Sampling designs: seeking control of, 14, 126, 144, 252; modifications to FDA's types of, 153, 252; different types of, 157–59, 228–30, 243; Ames and Gold's, 187; effect on outcome of study, 247; defects of, 264, 272
Science of pesticides: is specialized and fractured, 258; pesticide use data, 259; residue pathways and fate, 259; ecological risk, 260; defining adverse effects, 261–62; relative potency, 262–63; exposure variance, 263; risk distribution, 263; mixtures, 263–64; sources of uncertainty in risk estimates, 264; quality of evidence, 265, relative risk, 266; regulatory delay as effect of uncertainty, 267
Screens: for insects, 21, 23, 33, 74; residue testing, 154, 157, 160, 161, 164
Seals, residues in tissues, 139, 143
Secretary of the Interior: pesticide management responsibility, 91
Selinunte, Sicily, 16
Sensitivity of method, 111–12
Shraden, Gerhard, 236
Silent Spring: atomic weapons testing and milk contamination, 79; Senate hearings in response, *1963,* 79–80; Monsanto response, *Desolate year,* 80; USDA response, 81, 84; President Kennedy's Science Advisory Committee response, 83; Ribicoff hearings, 86
Simazine, 133, 150
60 Minutes, 10, 266
Soil, 136–41, 148–50, 170, 251–54
Solomon Islands, 32
Sources of human exposure, 144
Southeast Asia, 50, 58, 61
Soybean oil, 205, 212
Soybeans, 132, 133, 164
Spear, L., 241
Special Local Needs, 97
SPf66, 60
State regulation, 71
Strategic planning, 275
Strategic risk reduction, 287
Stratification, 230
Substitution of pesticides, 14, 67, 120, 134, 212, 279, 283, 287
Summary exposure and risk statistics, dangers of, 154, 164, 208, 213, 226, 234, 271
Superfund, 124, 255, 295
Surveillance, Epidemiology and End Results (SEER, NCI program), 184–85
Surveillance residue sampling by FDA, 133, 157, 164, 228, 230, 243
Susceptible groups, 263, 282
Swimming pools and pesticides, 131, 252
Synergism of pesticide mixtures, 182, 239, 242

Thorburn, Patrick, 18
Thresholds for adverse effects, 244, 247, 262
Tobacco, 149, 185
Todhunter, John, 98
Tolerances: overview, 7–9; *§408,* 105–6; *§409,* 106–8, 115–16, 119; Delaney Paradox and NAS, 120–23; reforms needed in setting, 253, 285, 294
Toxaphene, 43, 51, 134, 165
Toxic Substance Control Act (TSCA), 254
Toxicity Equivalence Factor (TEF), 241
Toxicity tests, 13, 38, 95, 290
Train, Russell, 88
Trifluralin, 150

Trivial risk, 9; with color additives, 114, 122; with pesticides, 235, 272, 282–83, 284, 287; distinguishing types of risks, 263
Typhus, 13, 15, 37–38, 45, 72, 267

UDMH, 116, 155
Uncertainty, expression of, 271
Uncertainty, sources in risk estimates, 264
Union Carbide India, Ltd., 3
Union Carbide U.S., 3
United Nations, 43, 44, 281
Unreasonable adverse effects, 89–91, 105
Unreasonable risk, 89, 292
Usual intake, 204
Utilitarianism, 293, 295, 299

Vaccines, 59
Variance in risk distribution, 271
Variance in susceptibility, 240
Vietnam War, 86, 134
Vitamin/Mineral Supplement Intake Survey, 203
Volatility, 47, 138
Vulnerability of children and infants, 211

Waxes, 140, 154, 155
Weeds, 6, 13, 267
Weisner, Jerome, 83
Wenatchee, Washington, 68
Western grebes, 141
Wheat, 123, 124, 132, 164, 201, 204, 206, 207
Woodwell, George, 140
World Health Organization (WHO), 15, 44–51, 54, 58, 60–61, 63, 236
World War I, 15, 30, 37, 54
World War II: introduction of new pesticides in, 13; combat deaths in, 15; and control of malaria, 29, 31, 35; and chloroquine synthesis, 54; relation of pesticides to nerve gas, 236

Yellow fever, 13, 21, 22, 23, 24, 41

Zeidler, Othmar, 37
Zero risk, 107, 124
Zineb, 43